Magmas. Rocks and Planetary Development

Magmas, Rocks and Planetary Development

A survey of magma/igneous rock systems

ERIC A. K. MIDDLEMOST
Department of Geology and Geophysics, University of Sydney

LONGMAN

Addison Wesley Longman Limited
Edinburgh Gate, Harlow
Essex CM20 2JE
England

and Associated Companies throughout the world.

First published 1997

ISBN 0 582 23089 6

British Library Cataloguing-in-Publication Data
A catalogue record for this book is
available from the British Library.

Library of Congress Cataloging-in-Publication Data
A catalog entry for this title is
available from the Library of Congress.

Set by 16 in 9/11pt Times
Produced by Longman Singapore Publishers (Pte) Ltd.
Printed in Singapore

Contents

Preface

In the following pages the reader will find a smorgasbord of information about igneous rocks, magmas and volcanic eruptions. The idea of this book has been to bring together petrological data from different times, different tectonic settings, and even different planets, and attempt to devise a coherent global interpretation of the many interlinked magma/ igneous rock systems. Each separate magmatic system can be visualised as containing long sequences of regularities that arise from the fundamental laws of physics and physical chemistry. Periodically these systems change and become more complex when this order is disrupted by seemingly chance events that reset the pattern of magmatic evolution.

It is only when we compare the Earth with other planetary bodies that we realise how extraordinary our 'pale blue dot' is. It contains exceptionally long and narrow neovolcanic zones, active subduction systems, moving lithospheric plates, oceans of liquid water, a reactive oxygen-rich atmosphere, singular geochemical anomalies called orebodies, and surface layers that teem with life. Like Venus it remains an active planet and much of the information on past events that was once encoded in its rocks has been obliterated. Fortunately the Moon has been dormant for a long time, and its rocks contain an excellent store of data suitable for unravelling the early history of the Earth–Moon system.

Chapter 2 contains a long section on the naming of igneous rocks and their magmas. This topic is considered important because much of the current confusion in igneous petrology stems from the inept, or incorrect, naming of magmas and igneous rocks. At this stage in the development of igneous petrology, an ideal classification would have to be comprehensive to account for all the igneous rocks in the Solar System, yet controlled so that it contained fewer rock names than those in current use.

The book systematically examines the composition, origin and evolution of the common igneous rocks. It also contains several dissertations on rare rocks because there is no direct correlation between the current surface abundance of igneous rocks and their importance in developing an understanding of the global magma/igneous rock system. For example, there are many rare alkaline rocks that contain a wealth of exceptional information about magmatic materials and processes in the upper mantle. It is also thought provoking to consider how the global magma/igneous rock system works on planets and planetoids where some physical and chemical conditions are considerably different from those of Earth.

After considering the origin and evolution of the normal igneous rocks the book concludes with a short chapter that considers the age and origin of the chemical elements, and the role of magma in planetary evolution. Each planet and planetoid currently contains information about different aspects of the evolution of the global magma/igneous rock system. By combining these data with current speculations on the evolution of the Sun one can compose a tentative synopsis of the magmatic and volcanic history of the Earth for the next 5 billion years. Readers are challenged to think about this scenario and try to imagine the full majesty of the global magma/igneous rock system evolving through time.

Eric A. K. Middlemost

Acknowledgements

We are grateful to the following for permission to reproduce copyright material:

Figure 1.3 from *Volcanism in Hawaii*, Vol. 1, United States Geological Survey Professional Paper 1350, 839 pp. US Department of Industry, Geological Survey (Decker *et al.*, 1987); Fig. 1.4 from a sketch map of Campí Flegreì caldera compex west of Napoli in southern Italy. Scale 1:15,000 Consiglio Nazionale Delle Richerche, Progetto finalizzeato Geodinamaica Roma, Italy (After Rosi *et al.*, 1986); Fig. 1.9 from *The Solid Earth: An Introduction to Global Geophysics*, p. 6, Cambridge University Press (Fowler, 1990); Fig. 1.11 from *Volcanism Associated with Extension at Consuming Plate Margins*, pp. 3–28, The Geological Society Publishing House, London, Special Publication No. 81, 293 pp. (Hamilton, 1994); Fig. 1.12 from sketch map of the *Tharsis Dome on Mars Geological Map of Mars*, United States Geological Survey, Miscellaneous Investigations Series, Map I-1083, US Department of Industry, Geological Survey (adapted from Scott and Carr, 1978); Fig. 2.1 from 'Nomenclature of pyroxenes,' *Mineralogical Magazine*, 52, pp. 535–550, Mineralogical Society (Morimoto *et al.*, 1988); Fig. 2.6 from 'IUGS Subcommission on the Systematics of Igneous Rocks: classification and nomenclature of volcanic rocks,' *Neues Jahrbuch für Mineralogie*, Stuttgart, 143, pp. 1–14 (Streckaisen, 1978); Fig. 2.7 from 'A chemical classification of volcanic rocks . . . ,' *Journal of Petrology*, 27, pp. 745–750 (adapted from Le Bas *et al.*, 1986) and from 'A classification of Igneous rocks . . . ,' *Journal of Petrology*, 17 (4), pp. 589–637 (Le Maitre *et al.*, 1976); Fig. 2.9 from 'The frequency distribution of igneous rocks,' *Mineralogical Magazine*, 19, pp. 303–313, Mineralogical Society (Richardson and Sneesby, 1922); Fig. 2.17 from 'Naming materials in the magma/igneous rock system,' *Earth Science Reviews*, 37, pp. 215–24, Trans A: 271, 1972, The Royal Society (Middlemost, 1994); Fig. 2.19 from 'Meteorites' p. 241–250 (Wood, 1990) in *The New Solar System*, 3rd Edition, Cambridge University Press (Beatty and Chaikin, Editors); Fig. 2.21 reprinted from 'Trace element variation during fractional crystallisation . . . ,' *Geochimica et Cosmachimica Acta*, 6, pp. 90–99 (Neumann *et al.*, 1954) with kind permission from Elsevier Science Ltd., The Boulevard, Langford Lane, Kidlington OX5 1GB, UK; Fig. 2.22 reprinted from 'Tectonic setting of basic volcanic rocks using trace element analyses,' *Earth and Planetary Science Letters*, 19, pp. 290–300 with kind permission of Elsevier Science,

— NL, Sara Burgerhartstraat 25, 1055 KV Amsterdam, The Netherlands (Pearce and Cann, 1973); Fig. 2.23 from 'Hawaiian alkaline volcanism' (Clague, 1987) in *Alkaline Igneous Rocks*, (Fitton and Upton, Editors) Geological Society of London Special Publication No. 30, 568 pp; Fig. 3.1 from *A Continent Revealed: the European Geotraverse*, p. 212, Cambridge University Press (Blundell *et al.,* Editors 1992); Fig. 3.2 from 22.1 *Understanding the Earth*, The Open University (Clifford, 1971); Fig 3.3 from 'Eruptive environment and geochemistry of Archaean ultramafic, mafic and felsic volcanic rocks in Eastern Yilgarn Craton,' *IAVCEI Excursion Guide*, Canberra, Record 1993/62, Australian Geological Survey Organisation 40 pp. (Morris *et al.,* 1993); Fig. 3.4 from *Neodymium Isotope Geochemistry: an Introduction*, p. 100 © Springer Verlag GMbH and Co. KG, Berlin, 187, pp. (De Paolo, 1988); Fig. 3.5 reprinted from 'The Yellowstone hotspot,' *Journal of Volcanology and Geothermal Research*, 61, pp. 121–187 (Fig. on p. 176) with kind permission from Elsevier Science — NL, Sara Burgerhartstraat 25, 1055 KV Amsterdam, The Netherlands (Smith & Braile, 1994); Fig. 4.9 from 'Volcanological study of the great Tambora eruption of 1815,' *Geology*, 12 pp. 659–663, Geological Society of America, (Rampino *et al.,* 1984); Figs 4.14, 4.16, 4.17 and Plates 4.1 and 6.3 from Universities Space Research Association, Houston, Texas — all in the public domain; Fig. 5.2 from *Volcanism in Hawaii*, United States Geological Survey, Professional Paper 1350, 1667 pp. US Department of Industry (Decker *et al.,* 1987); Fig. 5.3 from 'Long Valley caldera and Mon-Inyo Craters volcanic chain, eastern California,' *New Mexico Bureau of Mines and Mineral Resources Memoir*, 47, pp. 227–254, New Mexico Bureau of Mines and Mineral Resources (Bailey *et al.,* 1989); Fig. 5.4 from 'The anatomy of a batholith,' *Journal of the Geological Society*, London, 135, pp. 157–182 (Pitcher, 1978); Fig. 6.1 from *The Evolution of Igneous Rocks: Fifteenth Anniversary Perspectives*, © 1979, reprinted/reproduced by permission of Princeton University Press (Yoder, Editor, 1979); Fig. 6.6 from *The Regional Geological Setting of the Bushveld Complex*, Economic Geology Research Unit, University of the Witwatersrand, Johannesburg, South Africa, 18 pp. (Hunter, 1975); Figs 7.2 and 7.3 from 'Origins of basalt magmas: An experimental study of the natural and synthetic rock systems,' *Journal of Petrology*, 3 (3), pp. 342–532, Oxford University Press (Yoder and Tilley, 1962); Figs 7.4 and 11.9 from *A Classification of Igneous Rocks and Glossary of Terms: Recommendations of the International Union of Geological Sciences Subcommission on the Systematics of Igneous Rocks*, Blackwell Science Ltd, Oxford (Le Maitre *et al.*, Editors, 1989); Fig. 7.6 from *Generation of Basaltic Magma*, National Academy of Sciences, © 1976 by the US National Academy of Sciences, Courtesy of the National Academy Press, Washington, DC (Yoder, 1976); Fig. 8.1 from 'Geochemistry of Upper Cretaceous Volcanic Rocks from the Pontic Chain, Northern Turkey,' *Bulletin Volcanologique*, 39(4), pp. 1–13 © Springer Verlag GMbH & Co. KG (Peccerillo and Taylor, 1975); Fig. 8.7 from 'Structure and eruptive history of the Tarawera Volcanic Complex, New Zealand,' *Journal of Geology and Geophysics*, 13(4) pp. 879–902, The Royal Society of New Zealand (Cole, 1970); Fig. 10.1 from 'Recent high temperature research on silicates and its significance in igneous geology,' *American Journal of Science*, 33, pp. 1–21 (Bowen, 1937) and also in 'The alkali feldspar join . . . ,' *Journal of Geology*, 58, pp. 512–517, University of

Chicago Press (Schairer, 1950); Fig. 10.2 from sketch map of Eastern Australian Volcanic Province, in *The Warrumbungle Volcano: a geological quide to the Warrumbungle National Park*, Australian Geological Survey Organisation (AGSO), (Duggan and Knutson); Fig. 2.21 reprinted from 'A mildly depleted upper mantle beneath Southeast Norway: evidence from basalts in the Permo-Carboniferous Oslo Rift,' *Tectonophysics*, 178, pp. 89–107 with kind permission from Elsevier Science — NL, Sara Burgerhartstraat 25, 1055 KV Amsterdam, The Netherlands (Neumann *et al.,* 1954); Fig. 11.1 from 'Tertiary–Quaternary extension-related alkaline magmatism in western and central Europe,' *Journal of Petrology*, 32, pp. 811–849, Oxford University Press (Wilson *et al.,* 1991); Fig. 11.6 from 'Petrogenesis of potassium rich lavas from the Roccamonifina Volcano, Roman Region, Italy,' *Journal of Petrology*, 13 pp. 425–456, Oxford University Press (Appleton, 1972); Fig. 11.10 from *Kimberlites: Mineralogy, Geochemistry and Petrology*, Plenum Publishing Corporation (Mitchell, 1986).

Whilst every effort has been made to trace the owners of copyright material, in a few cases this has proved impossible and we take this opportunity to offer our apologies to any copyright holders whose rights we may have unwittingly infringed.

CHAPTER 1

Historical account of magma/ igneous rock system

The Earth is itself essentially a composite igneous rock...

Daly, 1933, p. vii.

1.1 Molten rock

The awesome power and beauty of volcanic eruptions epitomise the global magma/igneous rock system. As one observes the spectacular ejection of incandescent material, and feels the heat and harmonic tremor, one senses the presence of hot, pulsating molten material gushing up from within the Earth. During a volcanic eruption one may discern the turbulent interface between magma and the materials, such as lava, pyroclasts and gases, that are explosively released as magma thrusts towards the surface. Magmas and magmatic rocks belong to a complex, dynamical system that operates within planets (see Fig. 1.1). Although this system contains the potential to evolve in a bewildering variety of ways, a particular magma usually settles into an ordered pattern of evolution, and the number of common magmas is remarkably small.

Valuable insights into the operation of the magma/ igneous rock system may be gleaned from investigating the eruption cycle of an accessible volcano that erupts frequently. Kilauea on Hawaii is just such a volcano (see Fig. 1.2). Research by seismologists and volcanologists has established how the magma system in this volcano operates from source to vent (Tilling and Dvorak, 1993, p. 125). Magma is generated at a depth of approximately 80 km. It migrates up a vertical conduit until it reaches the level where its density almost equals that of the surrounding rocks. This happens at a depth of about 7 km beneath the summit caldera. It is here that a magma reservoir begins to amass. The reservoir beneath Kilauea has three main outlets. Two are near-horizontal and follow the present trends of the southwest and east rift zones, while the third outlet discharges upwards into the summit caldera. When considered on a regional scale, Kilauea Volcano is perceived as a young, active member of the 6000 km long Hawaiian-Emperor chain of volcanic islands and seamounts (see Fig. 1.3). It is likely that this dogleg-shaped volcanic chain developed because of the passage of the Pacific Plate over a stationary magma-generating anomaly.

As concepts seem to have moved ahead of explanations and definitions, we will now scrutinise some common terms and ideas encountered in the study of magmas and magmatic rocks. Rocks are naturally produced bodies of solid, inanimate matter. They usually reveal some degree of internal homogeneity. Essentially they are the solid materials that make up the planets, natural satellites and other similar cosmic bodies. The Greek word for rock is *petra* and it occurs as a component in many terms, such as petrography, petrogenesis and petrology. Petrography is the systematic classification and description of rocks, while petrogenesis is mainly to do with the evaluation of ideas concerning the origin and evolution of rocks. Petrology is the science of rocks, and it subsumes both petrography and petrogenesis. The term igneous is derived from the Latin word *ignis* that means fire.

Magma is derived from the Greek word *massein* that means 'to knead'. This implies that magmas are likely to have a thick, dough-like consistency. Magma is thus hot, mobile, rock-forming material generated by natural processes within a planet. Some magmas are wholly liquid, but most contain suspended crystals, rock fragments, and dissolved, or exsolved, gases. In the opening description of a volcanic eruption it was implied that magmas only exist below the surface. This is true, but in petrogenetic discussions magmas are deemed to have a tangible existence. They have a definite chemical

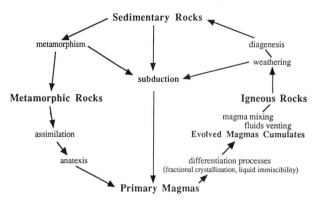

Fig. 1.1 Flow chart illustrating the operation of the magma–igneous rock system on Earth.

Fig. 1.2 Basaltic lava from the East Rift of Kilauea shield volcano entering the sea to the east of Kamoamoa Campground on the big island of Hawaii in July 1989.

composition that is calculated as equal to the sum of the chemical components in the rocks they produce, plus the volatile components discharged during eruption and solidification. Magmatic rocks are produced when magma is extruded through a volcanic vent, or intruded and cooled beneath the surface.

Most magmas contain more than 43 per cent SiO_2. There are some rare magmatic rocks, such as the natrocarbonatites that have recently extruded from Oldoinyo Lengai Volcano in Tanzania, which usually contain less than 0.1 per cent SiO_2, but high abundances of CO_2, Na_2O, CaO and K_2O (Keller and

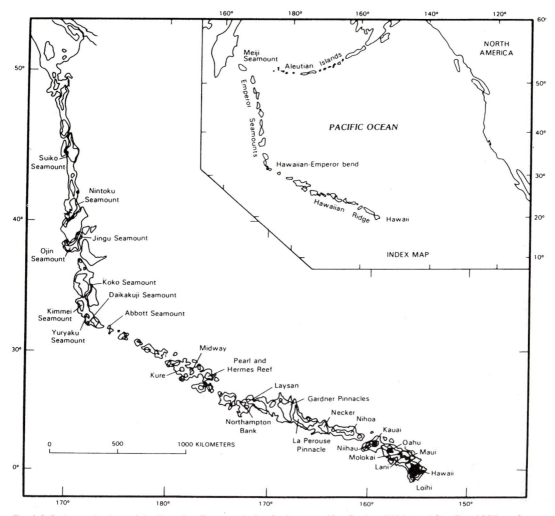

Fig. 1.3 Bathymetric chart of the Hawaiian-Emperor chain of volcanoes. After Decker, Wright and Stauffer, 1987, p. 6.

Hoefs, 1995, p. 117; see Table 12.1). Natrocarbonatite lavas are exceptional because they erupt at temperatures as low as 580°C thus providing a lower temperature limit for normal igneous processes. The term igneous rock has a wide meaning as it traditionally includes both magmatic rocks and a variety of 'igneous-looking' rocks produced by deuteric, metasomatic or metamorphic processes (Le Maitre *et al.*, 1989, p. 3). Two Greek words, *pyr* meaning fire and *klastos* meaning broken, explain the etymology of the term pyroclast. Pyroclasts are usually crystal, glass or rock fragments produced by direct volcanic action. Rocks essentially composed of assemblages of unconsolidated, or consolidated, pyroclasts are called pyroclastic deposits.

Over the years large volumes of dark-coloured lava has been extruded from Kilauea Volcano on Hawaii. Most of this lava is basalt (see Table 1.1). Basalts are the most abundant group of volcanic rocks found on the surface of the Earth and the other terrestrial planets (Basaltic Volcanism Study Project, 1981, p. xxvii). If one was to collect a sample of basalt from an active lava flow, it would be natural to call it a volcanic magmatic rock. When given a sample of an older basalt from an unknown source, one would only be able to identify it as a lava or basalt if one had prior knowledge of the usual appearance of these materials. If given a crystalline rock, one might be able to identify the minerals present and use these data to name the rock. A lava that

Table 1.1 Basaltic lavas from Kilauea Volcano, Hawaii. After Basaltic Volcanism Study Project, 1981, pp. 166–167.

Number	HAW-2	HAW-3	HAW-21	HAW-27	Mean
SiO_2	50.51	50.22	50.09	50.94	50.44
TiO_2	2.63	2.54	2.79	2.57	2.63
Al_2O_3	13.45	13.31	13.37	13.53	13.42
Fe_2O_3	1.78	1.39	1.79	2.48	1.86
FeO	9.59	9.68	9.52	8.74	9.38
MnO	0.17	0.18	0.18	0.17	0.18
MgO	7.41	7.28	7.26	7.16	7.28
CaO	11.18	11.03	11.6	11.39	11.3
Na_2O	2.28	2.31	2.31	2.29	2.3
K_2O	0.49	0.47	0.56	0.48	0.5
P_2O_5	0.28	0.27	0.29	0.26	0.28
Total	99.77	98.68	99.76	100.01	99.57
CIPW norm					
Quarz	1.37	1.53	0.32	1.19	1.1
Orthoclase	2.9	2.78	3.31	2.84	2.96
Albite	19.29	19.55	19.55	19.38	19.44
Anorthite	25.02	24.56	24.21	24.51	24.58
Diopside	23.42	23.24	23.92	23	23.4
Hypersthene	20.09	19.63	18.88	19.2	19.45
Magnetite	2	1.95	1.95	1.92	1.96
Ilmenite	5	4.82	5.17	4.73	4.93
Apatite	0.65	0.63	0.7	0.6	0.65

contained only glass would be more difficult to identify and one probably would have to perform a chemical element analysis before one could accurately name the rock.

Once a sample, such as a basalt from an unknown source, is named, one can use the whole corpus of geological knowledge to hypothesise about its origin. In this example, geological knowledge would attest (1) that materials of this composition are extruded from many present-day volcanoes, and (2) that the phases usually found in basalt crystallise, or congeal, within the 1190°C to 980°C range of temperatures. Before the development of experimental petrology, geologists often found it difficult to decide whether a particular rock had congealed from a magma. Experimental petrology is now an essential part of igneous petrology because field and petrographic observations by themselves do not usually provide definitive proof of the magmatic nature of a particular rock. Wyllie (1983, p. 13) has suggested that experimental petrology provides limits for the conditions of melting and crystallisation of mineral assemblages and also for the exchange of elements among minerals, liquids and vapours. He summarised his views by asserting that the results of experimental petrology can help distinguish between likely and dubious petrogenetic processes.

The mutual relations among the component phases in an igneous rock decide its texture. Many intrusive rocks have textures that evolved over the long time they took to cool. Such textures are often less related to the rock's magmatic origin than to its subsequent recrystallisation as it cooled. The evolution of complex textures in intrusive rocks will be evaluated when considering the evolution of the granitic rocks (see Section 9.6).

1.2 A snippet of history

Historical records from many cultures contain descriptions of volcanoes and volcanic eruptions. For example, the writings of the Greek geographer Strabo in about AD 0007 contain powerful descriptions of the volcanoes of southern Italy and Sicily. These writings also described a large seismic sea wave, or tsunami, activated by a volcanic eruption on the island of Ischia, southwest of Naples. According to Greek legend their god of fire, Hephaistos, worked at his forge beneath the volcano Etna in Sicily, hammering out splendid weapons for his fellow gods. The Romans adopted this legend, changed the name of the fire-god to Vulcan, and found it desirable to move his workshop to a location nearer Rome. The new site was beneath Hiera Volcano in the Aeolian Islands between Sicily and Naples. Hiera is now called Vulcano. In the twelfth century the word volcano acquired its present meaning.

Greek and Egyptian literature contains many oblique references to the huge, explosive eruption of Santorini (Thera) Volcano in the Cyclades archipelago to the north of Crete (Vitaliano, 1976, pp. 179–271; Francis, 1993, pp. 63–65; White and Humphreys, 1994, pp. 184–185). This cataclysmic eruption took place at sometime between 1625 and 1630 BC. It ejected close to 30 km^3 of pyroclastic material, excavated an enormous caldera, buried the Minoan city of Akrotiri, and triggered seismic sea waves that decimated coastal areas bordering the eastern Mediterranean. In Egypt, 800 km downwind, the materials released by the eruption disrupted agriculture and caused widespread famine. The effects of this huge eruption on global weather patterns are recorded in the official dispatches of King Chieh of the Xia dynasty in China (Anon, 1990), in tree rings from the western USA and northern Ireland, and in the Greenland ice-cores which show a enormous increase in acidity at about this time.

In AD 0079 during the reign of Emperor Titus, the citizens of Neapolis (now Naples), and adjoining areas of southern Italy, were startled by the destructive eruption of the long-dormant volcano Vesuvius (Somma-Vesuvio). The eruption continued for about 30 h, and buried the towns of Herculanium, Pompeii and Stabiae (Lirer *et al.*, 1993, p. 133). Somma-Vesuvio developed on the Campanian plain to the east of Naples. West of the city lies the Campi Flegrei (also called Campi Phlegrae or Phlegraean Fields) volcanic complex (see Fig. 1.4). In the past this large caldera, which is about 13 km in diameter, has been the source of enormous explosive eruptions that produced huge pyroclastic deposits, such as the Campanian Tuff (80 km^3 of dense rock equivalent).

Before the AD 0079 eruption of Somma-Vesuvio, local earthquakes caused widespread damage in the Campanian region. One of these premonitory earthquakes interrupted a theatrical performance given by Emperor Nero. An important reason why the AD 0079 eruption of Somma-Vesuvio has elicited so much subsequent attention is that a lucid report on the event has been preserved. The dispatch was written by Pliny the Younger. He starts by describing the tall eruption column he saw from Misenum, across the Bay of Naples. The shape of the column is compared to a Mediterranean pine tree. Pliny's report has

the spontaneity of a direct radio broadcast; and because of its appealing immediacy and objectivity, volcanologists call all cataclysmic explosive eruptions of similar type, and magnitude, plinian eruptions (Newhall and Self, 1982, pp. 1231–1238).

During the sixteenth and seventeenth centuries the great ocean voyages of discovery produced much new information on the abundance and distribution of volcanoes. In Europe the scientific community was forced to admit that volcanoes were more abundant than they had previously suspected. In the early part of the eighteenth century, excavations commenced at Pompeii and Herculanium and visitors flocked to see the vast array of Roman buildings and artifacts on display. Most visitors were also aware of the awesome presence of Somma-Vesuvio because during the eighteenth and nineteenth centuries it erupted frequently. During the excavations a fresco was discovered. It depicted the pre-AD 0079 Somma-Vesuvio as a high, steep-sided, single cone covered by vines and trees (Krafft, 1991, p. 39). A similar image of the pre-AD 0079 volcano is found in the reports of the Gladiatorial War against Rome (BC 0073–0071) because, during the early stages of this war, Sparticus and his followers used the summit of the volcano as a natural fortress.

In 1767 Sir William Hamilton started producing his comprehensive reports on the volcanoes of the Neapolitan area. His trailblazing work included a chronology of the main eruptions of Somma-Vesuvio (Hamilton, 1776). He obtained the data for his chronology from a church register that recorded the dates on which the relics of the patron saint of Naples were carried through the streets in procession. Such processions only took place when Somma-Vesuvio erupted. Hoffer (1982) has written a popular account of Somma-Vesuvio. It includes details of the many prominent people who visited the volcano when on the 'grand tour' circuit. For example, in 1828 the young Babbage (mathematician and designer of the analytical engine) and the young Lyell (geologist) were among the visitors. As one might have expected, both scientists made perceptive observations about the eruptive activity of Somma-Vesuvio. Babbage was the more adventurous and climbed into the crater to examine an erupting vent from close-by.

The Vulcanists

In the mid-eighteenth century, while the excavations of Pompeii and Herculanium were at their height,

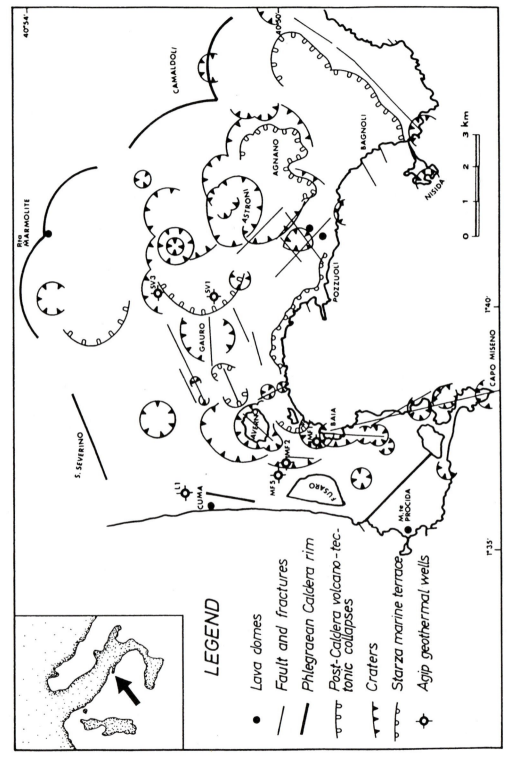

Fig. 1.4 Sketch map of the principal volcanic and structural features of the Campi Flegrei caldera complex, west of Napoli in southern Italy. After Rosi *et al.* 1986.

Guettard was travelling around France with a royal warrant to examine and collect rocks, minerals and fossils. While in central France, near the town of Moulins, he was surprised to find posts made of a hard, black rock that looked like lava. Enquiries revealed that the rock came from the quarries of Volvic. The name excited his curiosity. It is reported that he exclaimed '*Volcani vicus!*' (seat of a volcano), and then set off to find the source of the black rock. Near Volvic he saw a cone-shaped hill, with a breached summit crater containing lava. Although he had not previously seen an active volcano, he was familiar with illustrations of volcanoes, such as Somma-Vesuvio, Etna and Teide in the Canary Islands. He immediately concluded that he had discovered the remains of an old volcano. Further exploration revealed a chain of volcanic cones and domes. This Chaine des Puys (see Fig. 1.5) includes the large and steep-sided Puy de Dome (Guettard, 1752). After making these momentous discoveries, Guettard still believed that columnar basalts were different from lavas and that the former had precipitated from an ocean. Later Desmarest (1771) made a more detailed examination of the volcanic rocks of the Auvergne, and succeeded in tracing the columnar basalts back to their original vents. He also suggested that the spectacular columnar basalts from the Giant's Causeway in county Antrim, Northern Ireland, were of volcanic origin (see Fig. 1.6). Desmarest and his supporters were dubbed the Vulcanists.

The Neptunists

During the eighteenth century there was a rapid increase in the rate at which new information was accumulating in the Earth sciences. Unfortunately much of what was written at this time was distorted by erroneous opinions, often generated by careless observations, and a widespread ambition by many scholars to reconcile real data with mythological events, such as a worldwide, catastrophic flood, or an awkwardly short geologic time scale. There was clearly a need for a critical evaluation of the database, and the introduction of a new way of interpreting how minerals and rocks formed. In the last quarter of the eighteenth century, two people of very different temperament and training tried to tackle this task. They were Werner of Freiberg in Saxony and Hutton of Edinburgh in Scotland.

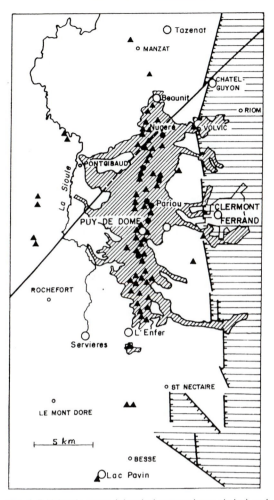

Fig. 1.5 Volcanic cones (triangles), maars (open circles) and volcanic domes (closed circles) of the Chaine des Puys in the Auvergne, France. After Autran and Dercourt, 1980, p. 143.

Werner studied both at the Mining Academy in Freiberg (1769–1771) and at the University of Leipzig. While at the latter institution he studied history, languages, law and philosophy. He maintained his childhood passion for mineralogy, and during his stay in Leipzig he published a paper that proposed a new systematic method for describing minerals. On his return to Freiberg in 1775 he was appointed lecturer in metallurgy and mining. He was a practical man who was heir to a great store of practical knowledge accumulated over many generations by the mining community in the Erzgebirge. His particular interests were in the fields of mineralogy, ore deposits and mining geology. As a mineralogist, Werner was

Fig. 1.6 Columnar jointed Tertiary tholeiitic basalts of the Giant's Causeway in county Antrim, Northern Ireland.

pre-eminent. His mineralogical research was later continued by Mohs, his successor at Freiberg (cf. the Mohs scale of hardness). Werner was also a charismatic lecturer. He inspired many students to go out into the field and produce methodical observations. In the second paragraph of his *Kurze Klassifikation* (1789), he remarks on the need for clear definitions, and suitable classifications, for all the materials of the mineral kingdom.

Two of his most illustrious students were Alexander von Humboldt and Leopold von Buch. Humboldt was a wealthy polymath who in his early career was greatly influenced by Werner's idea of 'universal rock formations' that extended around the globe and occurred in a fixed sequence (Botting, 1973). Many of Werner's petrological statements were derived from the earlier work of the Swede, Bergman. According to the Bergman–Werner explanation of how the rocks formed, the initial surface of the Earth was enveloped in a deep and calm ocean. The first rocks were crystalline precipitates and they adhered to, and took their shape from, the uneven primeval floor of the ocean. This oceanic floor was part of the unknown nucleus of the Earth. Granite

formed first, and, as the waters receded, gneiss, mica schist, and clay slate emerged, one after the other. Later the 'floetz gebirge' or fossil-bearing stratified rocks formed in the shrinking ocean. Werner regarded the alluvial and volcanic rocks as products of local conditions. He considered all basalts to be of aqueous origin (Werner, 1789, p. 86 in the Ospovat translation, 1971). Because Werner had advocated that all crustal rocks had either precipitated out of water, or were deposited in water, he and his supporters were dubbed the Neptunists. In the period between the publication of *Kurze Klassifikation* (1789) and the publication of Lyell's *Principles of Geology* (1830), the Neptunist theory was the dominant petrogenetic paradigm.

The Plutonists

In marked contrast to Werner, Hutton envisaged the whole Earth as operating as a single system that endlessly recycled materials. His hypothesis implied that heat and gravity drove the system. To appreciate the intellectual climate in which Hutton lived, one must appreciate that he was a contemporary of Newton, and that David Hume (the empiricist philosopher) and Adam Smith (the political economist) were close friends. Hutton was first to conceive the magma/igneous rock system. Before embarking on these speculations he studied law, chemistry, medicine and agriculture in Edinburgh, Paris and Leyden in the Netherlands. In 1768 he moved to Edinburgh, a city dominated by the remnants of the Carboniferous Arthur's Seat Volcano. On arrival he immediately became enmeshed in the intellectual life of the city. Hutton regarded himself as a philosopher, and aspired to use philosophy to unite the many disparate facts then known about the Earth. His first action was to become well acquainted with all the available publications on rocks and their origin. He then evaluated these data and as a consequence he developed what he regarded as important insights into the way the Earth works. Like Newton he seems to have regarded his theory as a set of working assumptions to be accepted hypothetically for as long as their consequences provided additional insights into unexplained phenomena. In his *System of the Earth*, and his more complete *Theory of the Earth*, Hutton (1785, 1788, 1795) developed the idea that all geological phenomena can be explained by processes currently active. At the surface, rocks are weathered and eroded, while, at depth, 'subterranean heat' produces melting and generates new rocks. Some rocks congeal from molten materials, while others are changed by their contact with hot materials.

In 1785 Hutton presented his ideas to the Royal Society of Edinburgh. His ideas were not well received. Some in his audience were troubled by his sweeping philosophical approach, whereas others were irritated by his unorthodox opinions. Kirwan (1794), Murray (1802) and Jameson (1808) were among those who tried to ridicule Hutton's *Theory of the Earth*. Kirwan labelled Hutton's ideas the 'plutonic theory', and Hutton and his supporters were dubbed the plutonists. Many of Hutton's contemporaries were critical of his proposal that granites were intrusive and crystallised from a hot, molten material. Hutton later conceded that when he first proposed his theory, he was undecided whether granite was a stratified rock that been transformed by fusion, or a molten material that had moved up from deep within the Earth and invaded the overlying strata.

In the 3 years after 1785 he travelled widely in search of evidence to support his conjectures. He particularly wished to examine the contacts between granites and the rocks that surrounded them. During these travels he discovered intrusive contacts at Glen Tilt, Galloway and Arran. Unfortunately Hutton's ideas only became widely known after his death when his friend John Playfair (1802) wrote a straightforward account of his geological speculations. Later Lyell promoted and refined Hutton's ideas, particularly his uniformitarian paradigm, and his speculations on the age, emplacement and origin of granite. Lyell insisted that not all granites were primordial, and that field evidence showed that the various granites had formed during different periods in Earth history. Geikie also championed Hutton's ideas; and just over a hundred years after Hutton's death, Geikie was responsible for persuading the Geological Society of London to publish the then unpublished third volume of Hutton's *Theory of the Earth* (1899).

The last decade of the eighteenth century was a productive time in the solid Earth sciences. For the first time Earth scientists acquired quantitative chemical analyses of silicate minerals and rocks (Loewinson-Lessing, 1954, p. 33). In 1798 Cavendish determined the density of the Earth. This was also when Hall, a close friend of Hutton, began his important experiments on the melting of rocks and cooling them under controlled conditions. Hall's experiments (1798, 1800) were initially prompted by an accident at the Leith glass-works, near Edinburgh, where a batch of molten bottle-glass happened to cool slowly, producing an opaque, partly crystalline rock-like material (Geikie, 1905, p. 319).

Hall's initial experiments revealed that the rate of cooling determined whether clear glass or a crystalline material formed. He then repeated these experiments using holocrystalline basaltic rocks (Hall, 1805). Later he experimented with granites, and a variety of volcanic rocks from active volcanoes in southern Italy (Eyles, 1961). These experiments showed that the common igneous rocks all melted at temperatures above 850°C. In other experiments he constructed a box that enabled him to control the amount of vertical and horizontal compression he could exert on horizontal layers of clay. This enabled him to simulate a variety of fold structures. He completed at least 500 experiments. Some of his most innovative experiments used a sealed pressure-vessel made from the barrel of a gun, and a novel pyrometer. Experiments on rocks and rock-forming minerals that simulate the pressures and temperatures anticipated within the Earth are now acknowledged as essential for understanding magmatic processes.

The nineteenth century

Systematic petrography flourished during the nineteenth century (cf. Loewinson-Lessing, 1954; Robson, 1986; Yoder, 1993). New systematic classifications were introduced by Haüy (1811), Brongniart (1813), von Leonhard (1823), Abich (1841) and Beaumont (1847). In 1825 Scrope published a rambling dissertation on volcanoes. This work is important because it was the first to consider the question of primary magmas, and the way they evolved into other magmas. Petrography was revolutionised by Nicol, Williamson, Sorby, Oschatz and Zirkel, who introduced and developed a technique for preparing thin sections of rocks and examining them under a polarising microscope (Loewinson-Lessing, 1954, pp. 3–5). Their innovative research greatly expanded the database of igneous petrography, and it stimulated a new quantitative approach to petrography. Rosenbusch, Fouque, Michel-Levy, Teall, Brögger and Iddings were foremost among this new band of petrographers.

Le Maitre *et al.* (1989, pp. 38–39) have examined the distribution of new igneous rock-names with time. They found that only 37 igneous rock-names were used before 1800, 523 were introduced between 1800 and 1899, and 974 new names found their way into the literature between 1900 and 1988. Johannsen, the author of the four-volume work entitled *A Descriptive Petrography of the Igneous Rocks*, coined the greatest number of these new igneous rock-names. His volume on the feldspathoidal and ultramafic rocks (1938) contains 42 new rock-names.

Petrological highlights of the second half of the nineteenth century included the following: (1) Airy (1855) postulated that mountains should have roots of lighter material, and he introduced the concept that later became known as isostasy; (2) Hall (1876) was first to describe basalt from a mid-oceanic ridge; (3) Dutton (1880) recognised that magma is usually generated in one of three ways – from a local increase in heat, from a local release in pressure, or from a lowering of the melting point of a rock by the addition of water; (4) Rosenbusch (1882) announced a set of empirical rules for interpreting the order of crystallisation in igneous rocks; (5) Judd (1886) introduced the idea of petrographic provinces; (6) Iddings (1890) proposed the particularly convenient silica-versus-other-oxides variation diagram; and (7) Becker (1897) evaluated the role of fractional crystallisation in convecting magma reservoirs. Not all petrologists supported the idea of petrographic provinces. Washington and Iddings preferred to emphasise areas dominated by a particular magma, instead of extensive areas containing similar rocks. Washington (1906) introduced the term *comagmatic region*, whereas many other petrologists of this period preferred the terms *suite* or *rock association*. In 1893 the maturity of petrology was acknowledged by the University of Chicago, when it appointed Iddings to occupy the first-ever chair in petrology (Loewinson-Lessing, 1954, p. 1).

As we have already remarked the sixteenth and seventeenth centuries are notable for a series of great ocean voyages of discovery. The observations made during these voyages enabled cartographers to exchange fact for fable, and the scientific community acquired a new global perspective. During the nineteenth century the voyages of vessels, such as the *H.M.S. Beagle* (1831–1836) and the *H.M.S. Challenger* (1872–1876), brought back a magnificent store of new biological, geographical, geological and oceanographic information. Darwin was on board the *H.M.S. Beagle*, and in his book (1844, pp. 117–124) describing his geological observations, he addressed the conundrum of the diversity of igneous rocks. He suggested that the composition of an igneous liquid could be changed by removing the minerals that crystallised early. This important idea ultimately led to an understanding of the processes of fractional crystallisation and magmatic differentiation. These are the processes that enable different types of magmas to be derived from a single parental magma. One of the most important details uncovered by the Challenger Expedition was that the ocean floor was not flat, but irregular, with a complex morphology.

The twentieth century

During the twentieth century the number of Earth scientists has increased dramatically, and their collective exertions have generated a huge database. Fortunately, as this database has grown, computers have changed from being people employed to make complex calculations, to being powerful electronic machines, guided by innovative programs, that can sort and manipulate the ever-increasing flood of new information. In the early part of the present century great changes took place in all branches of the Earth sciences. In 1902 Cross, Iddings, Pirsson and Washington (CIPW) introduced their quantitative system for classifying and naming igneous rocks. Their system was based on the use of a set of standard, or normative, minerals. These ideal minerals were derived from whole rock chemical analyses by using a set of standard calculations formulated by the system's originators (CIPW). The normative compositions of igneous rocks are still frequently calculated and used, but the CIPW classification of rocks has failed the test of time. In the year that saw the introduction of CIPW norm, Doelter (1902) published the first quantitative estimate of the relative viscosities of molten natural rocks. Ideas on the origin of the Earth took another sizeable step forward in 1905 when Chamberlin published his paper on the fundamental problems of geology, and proposed that the Earth and other planets developed from the accretion of planetesimals. During the same year, Day *et al.* (1905) published their pioneering phase diagram illustrating solid solution in the plagioclase feldspars.

Between 1906 and 1913 there was a revolution in geophysics when Oldham (1906), Mohorovičić (1909) and Gutenberg (1913) revealed that the Earth was layered, and contained a core, mantle and crust. The inner core was only found by Lehmann in 1936. In 1908 a comparable change appeared in geochemistry with the publication of the first edition of Clarke's unique compendium *The Data of Geochemistry*. Ideas on the emplacement of igneous rocks were irrevocably changed in 1909 when Clough *et al.* described ring dykes, and the process of cauldron-subsidence, from Glencoe in Scotland. Fermor (1913) also radically changed petrological thinking when he suggested that basaltic rocks can be transformed into eclogites at high pressures. In 1915 another critical petrological idea was put in place when Harper proposed that basaltic magmas were likely to be generated by the partial melting of subcrustal ultramafic rocks.

Norman L. Bowen

The foremost contributor to the progress of igneous petrology during the first half of the twentieth century was Bowen. In 1910 he became the first predoctoral fellow in the newly established Geophysical Laboratory (Gee-whizz-lab) in Washington. His initial research was for his Ph.D. program at the Massachusetts Institute of Technology. It encompassed an investigation (1912) of the binary system of $Na_2Al_2Si_2O_8$ (nepheline, carnegieite) and

$CaAl_2Si_2O_8$ (anorthite). After graduation he joined the staff of the Geophysical Laboratory, and according to Yoder (1992, p. 53) his prestige arose from 'his clarity of presentation of simple physicochemical concepts and their application to complex geological field problems'. In 1915 he published two trailblazing papers. The first (1915a) described a series of experiments on the sinking of olivine in silicate melts. In the second (1915b) he suggested that basalt is the only common type of magma, and that most, if not all, magmatic rocks can be derived from it by fractional crystallisation. He also argued that volatile components play an important role in the evolution of alkaline rocks.

These reports were followed by a surge of innovative resnearch that examined a wide range of petrological problems. Topics included: the problem of the origin of anorthosites (1917), crystallisation differentiation (1919), diffusion in silicate melts (1921), the reaction principle in petrogenesis (1922a), the alnöites of Quebec (1922b), the behaviour of inclusions in igneous magmas (1922c), the genesis of melilite (1923), the alkaline rocks and carbonatites of the Fen area, Norway (1924, 1926), the origin of ultrabasic and related rocks (1927), and geothermometry (1928a). A remarkable coincidence occurred in 1922 when Bowen and Goldschmidt (the renowned geochemist) independently described the reaction principle in igneous petrogenesis. Both identified two different types of reaction series; a continuous reaction series for the plagioclase feldspars, and a discontinuous reaction series for the common ferromagnesian minerals. To furnish concrete illustrations of their ideas, both devised flow-sheets that depicted an ideal reaction series for cooling basaltic magmas (see Figs. 1.7 and 1.8).

In 1928 Bowen published his thought-provoking book, *The Evolution of the Igneous Rocks*, in which he showed that the principles of physical chemistry are readily applied to mineralogy and igneous petrology (1928b). The following topics were surveyed in this book: (1) the problem of the diversity of igneous rocks; (2) liquid immiscibility in silicate magmas; (3) fractional crystallisation; (4) crystallisation in silicate systems; (5) the reaction principle; (6) fractional crystallisation of basaltic magma; (7) liquid lines of descent and variation diagrams; (8) glassy rocks; (9) rocks whose composition is determined by crystal sorting; (10) assimilation; (11) the formation of magmatic liquids very rich in potash feldspar; (12) the alkaline rocks; (13) lamprophyres and related rocks; (14) the fractional resorption of complex minerals

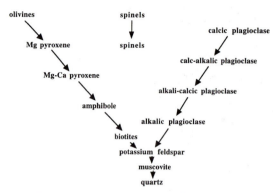

Fig. 1.7 Bowen's reaction series. After Bowen, 1922a, and 1928b, p. 60.

and the formation of strongly femic alkaline rocks; (15) further effects of fractional resorption; (16) the importance of volatile constituents; (17) petrogenesis and the physics of the Earth, and finally (16) the classification of igneous rocks. This book set the agenda in igneous petrology for the next 50 years. Even now it remains essential reading for all who seriously wish to evaluate the origin and evolution of igneous rocks.

Some 50 years after the publication of Bowen's book, 18 leading petrologists produced a book (Yoder, 1979) with chapters paralleling those in the original publication. This new work revealed that Bowen's approach to igneous petrogenesis has stood the test of time for 50 years. Bowen remained an active research scientist until his formal retirement in 1952. Many of his post-1928 publications were written in collaboration with colleagues. Even after his death, some of the research he had started, continued. This culminated in the publication of *Origin of granite in the light of experimental studies in the*

system *NaAlSi₃O₈–KAlSi₃O₈–SiO₂–H₂O* (Tuttle and Bowen, 1958). It showed that the alkali feldspars form a complete solid solutions series above 660°C, but unmix when held at lower temperatures. This work also revealed that rhyolitic magmas may originate from the fractional crystallisation of more basic liquids, or from the partial melting of appropriate sedimentary or metamorphic rocks. Other highlights of Bowen's research led to his powerful idea of petrogeny's residua system (1937, p. 11), and his cogitations on magmas (1947, 1950), alkaline rocks (1938, 1945) and the origin of granite (1948). In his 1945 paper, on phase equilibria bearing on the origin and differentiation of alkaline rocks, he (p. 88) introduced the idea of the 'plagioclase effect'. This effect arises because the plagioclase that crystallises from a magma always contains more calcium and aluminium (i.e. contains more of the anorthite component) than its host magma; thus the fractional crystallisation of plagioclase depletes a residual magma in calcium and aluminium and enriches it in sodium and silica.

Time and the new petrology

In the twentieth century, Earth scientists have also come to recognise that time is the very essence of their discipline. Many geological processes require vast periods of time to produce results that are inexplicable on the shorter time scale of laboratory experiments, or individual human experience. At the beginning of the century most informed, non-Hindu people believed that the total span of Earth history was between 20 and 100 million years. This all changed when Rutherford and Soddy (1902),

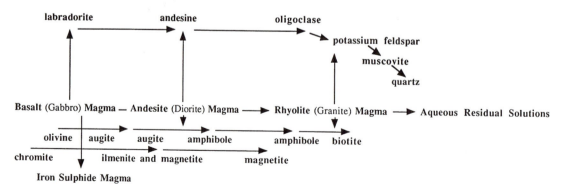

Fig. 1.8 Evolution of magmas through fractional crystallisation. After Goldschmidt, 1922, p. 6.

Boltwood (1907) and Holmes (1913) introduced radiometric dating. During the past 50 years there have been substantial improvements in the equipment and techniques used in dating minerals and rocks. This has resulted in the development of a standard geologic time scale that begins at 4.56 giga-years (billion years) at the base of the Hadean Era (Harland *et al.*, 1982, p. 10: see Table 1.2). With the stretching out of the geologic time scale, ideas that had once seemed whimsical became feasible. For example, the movements of continents at fingernail growth speeds were now deemed able to close oceans within the time span allocated to them by regional stratigraphy.

During the second half of the twentieth century there has been a great deal of exceptional research and scholarship in the field of igneous petrology. The following are some of the publications that have changed petrological perceptions: (1) Tilley's (1950) astute evaluation of magmatic differentiation; (2) Turner and Verhoogen's (1951) pioneering book that changed the teaching of igneous and metamorphic petrology; (3) Goldschmidt's (1954) innovative book on geochemistry (edited by Muir and published posthumously); (4) Pecora's (1956) original and stimulating evaluation of carbonatites; (5) Read's comprehensive and witty series of papers on granites and granitisation that was republished as a book in 1957; (6) Kuno's (1959) novel paper that showed that the composition of primary magmas varied with depth in an east–west section across central Japan; (7) Buddington's (1959) excellent review paper on the emplacement of granite; (8) King and Sutherland's (1960) inspiring papers on the African alkaline rocks; (9) Hess's (1960) comprehensive, quantitative study of the Stillwater Layered Complex of Montana (note, this research had previously been presented as a presidential address to the Mineralogical Society of America); (10) Yoder and Tilley's (1962) trailblazing paper on the origin of basaltic magmas; (11) a most valuable book by Wyllie (editor, 1967) on the nature and origin of various types of ultramafic and ultrabasic rocks; (12) Wager and Brown's (1968) thought-provoking work on the layered igneous rocks and the many processes that operate in magma chambers; (13) the Viljoen brothers' (1969) paper that showed that ultramafic extrusive rocks exist; (14) Sørensen's (editor, 1974) comprehensive book on alkaline rocks that sparked a new generation of petrologists to examine this enigmatic group of rocks, and (15) Yoder's (1976) erudite book on the generation of basaltic magmas. Later in

Table 1.2: Standard geologic time scale.

Giga-years	Eon	Era	
0.0		Cenozoic	
0.065			
		Mesozoic	
0.245	Phanerozoic		
		Paleozoic	
0.57			
		Sinian	
0.8			
1.0			
		Riphean	
	Proterozoic		
1.65			
		Animikean	
2.0			
2.2			
		Huronian	
2.45			
		Randian	
2.8			
3.0			
	Archean	Swazian	
3.5			
		Isuan	
3.8			
4.0			
		Hadean	
	Priscoan		
4.56		— Origin of Earth —	
		Pre-Hadean	
5.0			

1981 the many geochemists, planetologists and petrologists of the Basaltic Volcanism Study Project published their voluminous (1286 pp.) and extremely informative book *Basaltic Volcanism on the Terrestrial Planets*. The appearance of this publication seems to have closed a phase in the history of petrology. Since 1981 there has been a flood of innovative petrological research. This research has been boosted by improvements in technology, particularly electronics, and by phenomenal advances in the use of computers in petrology. An excellent example of this is the program MELTS that performs interactive thermodynamic modelling in magmatic systems (Ghiorso *et al.*, 1994). In the post-1981 period there has also been a impressive increase in isotope studies and this would seem an opportune time to consider this topic in more detail.

Isotopes

The isotopes of the same element have similar external electron configurations and similar chemical properties, but different nucleon numbers or mass numbers. Isotope geochemistry explores the abundance of radioactive and stable isotopes in planetary materials. Initially these studies were directed towards determining the age of the Earth and meteorites, establishing the age relationships between different rocks, and setting up a chronometric scale for the Earth and the Moon. A chronometric scale consists of units of equal duration; therefore the scale exists by virtue of the definition of a unit of duration, such as the second, or the year. The International System of Units (SI) uses the second as its base unit of time. A second is the duration of $9\,192\,631\,770$ periods of the radiation corresponding to the transition between the two hyperfine levels of the ground state of the caesium-133 atom. The year (general symbol $=$ a) is equal to 31.56 megaseconds ($1a = 31.56$ Ms). Ages are normally given in years before the present (BP). To avoid constantly changing the datum, the present is taken to be 1950 which is approximately the date when modern isotopic dating techniques were introduced.

The abundance ratios of radioactive and stable isotopes are of value in solving a wide range of geochemical and petrological problems. They are especially useful in identifying the sources of magmas, determining the extent to which minerals or rocks have reacted with natural waters, determining the

equilibrium temperatures of some fluids, minerals, or rocks, and also as indicators of the provenance of ore-forming fluids. As the sensitivity and precision of the analytical methods has improved, so the field of application has expanded.

In the study of stable isotopes one determines the abundance ratios of isotopes of the same element, and attempts to explain how natural processes cause isotopic fractionation. The physical properties of isotopes of the same element are essentially the same, except for those properties determined by the mass. Such physical differences are significant in the isotopes of light elements, such as hydrogen, helium, carbon, nitrogen and oxygen. At present these stable isotopes, plus sulphur, are the only ones commonly used in the study of igneous rocks. For example, oxygen has three isotopes, 16, 17 and 18, and oxygen-16 is the most abundant (99.76 per cent). In a typical report on oxygen isotopic composition the results would be transmitted in terms of a parameter called the delta value. Delta ^{18}O is the difference between the $^{18}O/^{16}O$ ratio of the sample and a ratio for standard mean ocean water (SMOW) expressed in parts per thousand. Fresh rocks derived from the mantle typically have delta values that range from $+5.5$ parts per thousand to $+6.0$ parts per thousand, whereas crustal rocks are likely to have larger delta values. Oxygen isotopes thus provide a method for detecting crustal assimilation, or contamination, in primary magmas.

The principal radioactive isotopes used in igneous petrogenesis are potassium-40, rubidium-87, samarium-147, rhenium-187, thorium-232, uranium-235 and uranium-238. Fractionation processes operating within the Earth generate variations in the concentrations of radioactive (isotopes that undergo spontaneous nuclear decay) and radiogenic (stable isotopes produced by radioactive decay) isotopes. When rocks that contain uranium and lead are mobilised and redeposited, isotopic ratios are usually able to assist in identifying the source, or sources, of the younger material, provided that the genetic processes were relatively simple. Radiogenic isotopes are also able to provide information about the chemical evolution of the upper mantle. This is because the isotopic ratios of the heavier elements are essentially unchanged by normal magmatic differentiation processes.

Rubidium has an ionic charge and size that makes it an incompatible element during the fractional crystallisation of basaltic magmas. It tends to concentrate in residual liquids, and over time it has concentrated in the upper continental crust of the Earth. This means that the generation of radiogenic strontium-87 is greater in the upper continental crust than in the upper mantle, and the upper continental crust is likely to have a relatively high strontium 87/86 ratio. This simple model may prove valuable in establishing the source of rhyolitic magmas, because most magmas of this composition are usually claimed to be either the product of the differentiation of a mantle-derived magma or the product of anatexis. Anatectic magmas tend to have initial strontium 87/86 ratios of 0.704 or greater, whereas magmas derived from the mantle are likely to have initial ratios of less than 0.704. This simple model assumes that (1) the mantle is isotopically homogeneous; (2) silicic magmas are derived either from the mantle or the crust, through simple, essentially single-stage processes; and (3) isotopic ratios are not affected by post-crystallisation processes. Experimental studies have shown that rubidium is often quite volatile during hydrothermal alteration, or mantle metasomatism. The rubidium/strontium ratio of metasomatised mantle xenoliths ranges from below 0.01 for depleted samples to 1.4 for highly enriched samples. Mantle rubidium/strontium ratios are so variable that it was only after the oceanic mantle strontium/neodymium ratio was established that a bulk Earth reservoir value of 0.7045 for the strontium 87/86 ratio, and 0.0827 for the rubidium-87/strontium-86 ratio, were generally accepted (Shirey, 1991, p. 113).

Wasserburg (1987, p.129) states that rock samples are like a library of past experiments, and the 'art' of the isotope geochemists is 'to select those materials which have a persistent memory and which together can bring testimony to the natural experiment of interest'. He has also likened the isotope geochemist to a detective on the top floor of a tall building interrogating the passengers as they walk off a lift or elevator. The detective's task is to discover the position of each floor and who normally resides on them.

The samarium–neodymium isotope system is based on the alpha decay of samarium-147 to neodymium-143. This system is critical in tracing the enrichment/depletion history of the light lanthanoids during magmatic differentiation. It is also important in studying the geochronology of Archean rocks. This is because samarium and neodymium are resistant to metamorphic remobilisation on the whole-rock scale, and samarium-147 has a remarkably long half-life (about 106.3 giga-years). Chronometric nuclide pairs, such as samarium-147/neodymium-143, have also enabled geochemists to establish the timing of major events in the evolution of the Earth. Such isotopes are able to probe the interior of a planet because they can detect the existence of domains with different chemical compositions, and under optimal conditions they can trace the evolution of these domains back through time (see DePaolo, 1988).

The rhenium–osmium system is of particular value in determining the age of mafic and ultramafic rocks because it usually remains closed under greenschist grade metamorphism. For example, osmium 187/186 ratios have been used to show that the Stillwater Layered Complex of Montana developed from several distinct sources. The ultramafic magmas were derived from a mantle source with low initial osmium 187/186 ratios, whereas the anorthosites were shown to contain assimilated radiogenic upper crustal material. Areas of deep mantle upwelling, such as Hawaii and Iceland, also appear to have distinctive osmium 187/186 signatures.

Continents adrift

Continental drift, and the more comprehensive plate tectonics paradigm, were both conceived in the twentieth century. The former developed from the latter as more information became available about the geology and geophysics of the oceanic crust. Both postulates have been invaluable in providing a framework for bringing the various branches of the Earth sciences together. The paradigm of plate tectonics has explained many inadequately understood geologic phenomena, and it has also helped in research planning because it has been used to predict the probable geology of poorly explored areas. The first concrete contributions to the debate on continental drift were made by Pickering (1907), Taylor (1910), Baker (1912) and Wegener (1912).

In 1915, during a long convalescence from war wounds, Wegener wrote the first edition of his classic work on continental drift, entitled *Die Entstehung der Kontinente und Ozeane* (the origin of continents and oceans). With the clear perspective afforded by hindsight it is instructive to note that Wegener postulated that the North Atlantic Ocean rifted apart at an average rate, not of centimetres, but of tens of metres per year. At first, few solid Earth scientists accepted the idea of continents moving, but gradually more paleoclimatic, paleontological, and paleomagnetic data that supported the theory were amassed. In 1926

Daly published his book *Our Mobile Earth*, in which he evaluated the various ideas on migrating continents and subcrustal convection currents. Later in the 1920s Wadati (1928) searched for a pattern in the seismic foci database then available. He discovered that many of these foci clustered on a plane that appeared to begin in the ocean-floor trenches and then dip away beneath the adjoining continental margins. The full importance of this observation was not generally appreciated until Benioff (1949) described the same seismic phenomenon and proposed that this active seismic zone (Wadati–Benioff Zone) represented a large-scale fault system.

Du Toit (1928, 1937) and Holmes (1928, 1931, 1933, 1944) were the most enthusiastic supporters of continental drift. Du Toit had a vast knowledge of the geology of southern Africa, and his book, *Our Wandering Continents*, contains a skilful reconstruction of the geology of the continents that had once formed the supercontinent of Gondwana. Holmes made many original proposals. In 1928 he suggested that as the continents move they displace the heavier oceanic rocks by a stoping process. In 1931 and 1933 he discussed subcrustal convection currents driven by radiogenic heat, and suggested that such currents were likely to move continental blocks. In his book *Principles of Physical Geology* (1944, pp. 505–509) he summarised his ideas on global tectonics. He proposed that subcrustal convection currents moved the continents, and that the oceanic crust trapped by this movement was displaced downwards. In the diagram illustrating his ideas (p. 506) he shows basaltic oceanic crust being dragged down beneath a continental block, where it is transformed into eclogite. When Holmes presented this global tectonic model he was unaware that the mid-oceanic ridge system was continuous.

In 1939 Griggs discussed the possibility of a single convection cell covering the whole Pacific basin with the currents sinking beneath what he called the circum-Pacific mountains. Later, after the Wadati–Benioff Zone had been rediscovered by Benioff (1949), Earth scientists began to realise that most of the dissipation of mechanical energy in the crust and upper mantle occurred in a few narrow seismic belts, and much of the remaining energy was expended in activating large-scale vertical movements. In 1951 Amstutz, who was working in the Alps, recognised that in this area huge plates of rock had converged, with one plunging beneath the other. He coined the term subduction to describe the process. At this time

the idea of continental drift was attracting fresh interest as new paleomagnetic data was showing that the continents had moved large distances during the Phanerozoic Eon. Much of this paleomagnetic data was collected by two geophysicists, Irving (1956) and Runcorn (1956), who were flippantly called the 'paleomagicians'.

In the present century, instruments, such as precision echo sounders for measuring water depth, marine magnetometers, and the deep-sea drilling equipment, have changed marine geology. This new phase in exploration began during World War Two. In 1943 Jagger, the founder of the Hawaiian Volcano Observatory, proposed that Earth science and engineering societies should use their influence to promote the idea of drilling a thousand core-producing holes in the deep ocean floor (Bascom, 1961, p. 43). This idea remained fallow until 1953 when Ewing of the Lamont Geological Observatory attempted to raise funds for a similar, but less ambitious, scheme. Whilst Jagger was promoting his radical proposal, Hess, in command of the *U.S.S. Cape Johnson*, was criss-crossing an area in the mid-Pacific on a peculiar zig-zag track, collecting data from a new precision deep-sea echo sounder. During this episode in his unconventional naval service, Hess discovered the tablemounts of the central Pacific. He suggested that they were originally volcanic islands that formed near the crest of an old mid-oceanic ridge. Once the volcanoes became inactive their surfaces were flattened by wave action. Eventually they became deeply submerged beneath the ocean.

In 1956 Estabrook proposed a remarkably ambitious project of drilling through the Earth's crust to examine the composition, properties and physical condition of the upper mantle. In his paper to the journal *Science* he even suggested that the mantle rock might contain diamonds. During the following year, at the International Union of Geodesy and Geophysics meeting in Toronto, the idea of launching a spectacular new Earth sciences project was discussed. The meeting adopted a resolution highlighting the importance of attempting to obtain actual samples of the upper mantle. In mid-1958 the US National Science Foundation set up a committee to examine the feasibility of drilling through the Mohorovicic Discontinuity and into the mantle. In the April 1959 edition of *Scientific American* the project was dubbed the 'Mohole'. Hess was one of the originators of this project, but at this time he was actively carrying out research in a number of diverse fields.

Spreading and disappearing oceans

In 1959, and again in 1962, Hess presented his cogitations on the evolution of the Earth and in particular the oceanic basins. He labelled his musings 'geopoetry'. In these speculations he perceived the Earth as a single system that has evolved through time. His specialist knowledge of the processes within large igneous magma chambers led him to propose that immediately after the Earth had formed, the planet was disrupted by a single-cell convective overturn on a global scale. This primordial process generated the core, mantle and a single large continent, the latter emerging at the apex of the rising limbs of the global convective cell. He also speculated that this process may have caused the asymmetrical surface configuration of the planet, including the precursors to the present oceanic basins.

Hess (1962, p. 608) then went on to suggest that the mid-oceanic ridges were likely to represent 'the traces of the rising limbs of convective cells, while the circum-Pacific belt of deformation and volcanism represents descending limbs'. He also insisted that the continents did not plough through the solid oceanic crust, but float passively on a current of mantle materials. Hess also predicted that if his speculations on the evolution of the oceanic basins were correct then most submarine sediments were likely to be Mesozoic or younger in age. His 1962 paper concludes with the idea that the ocean basins are impermanent features, whereas the continents are essentially permanent, though they occasionally collide, get torn apart or deform, particularly along their margins.

Dietz (1961) proposed a broadly similar dynamic model for the evolution of the lithosphere. He claimed that the mid-oceanic ridges marked the sites where magma welled up resulting in sea-floor spreading. Sea-floor trenches were regarded as zones of convergence where oceanic lithospheric was consumed. Further ideas that refined the ocean-floor spreading concept were introduced by Vine and Matthews (1963). They set out to explain why bands of normal and reversely magnetised rock occur on the mid-oceanic ridges. They suggested that as each new segment of ocean floor erupted on the mid-oceanic ridges it acquired the magnetic polarity characteristic of the Earth's field at that time. Each reversal in the planet's magnetic field produced a new band of rock with a different polarity; thus each parallel band of normal or reversely magnetised rock represents a unit of time. Together these units show how the oceans grew through time. If Vine and Matthews were correct, one would expect (1) magnetic anomaly patterns to be parallel with the mid-oceanic ridges, and (2) the anomaly pattern found on one side of a ridge to be repeated in mirror image on the other side of the ridge. In 1966 Pitman and Heirtzler recorded these critical features in the anomaly pattern for the Pacific–Antarctic Ridge.

Wilson (1965) was the first to show that there were clear-cut geometrical constraints on the manner in which rigid blocks can move relative to one another over the surface of a spherical body of fixed size. He (p. 343) proposed that young mountain belts and island arcs, mid-oceanic ridges and major faults with horizontal displacements, were all connected. These features formed a continuous network of mobile belts that girdled the Earth and divided its surface into large rigid plates. To enable all these plates to interact through time, he suggested that the various components that make up the mobile belts can change from one type into another. For example, a strike-slip fault may end by transforming into a mid-oceanic ridge. Such transformations are common on the mid-oceanic ridges where the horizontal shear motion of a strike-slip fault ends abruptly and is changed into an expanding tensional motion across the ridge. In 1967 Sykes used seismic data to find the movements taking place on faults cutting the mid-oceanic ridges. This research confirmed that Wilson's ideas on transform faults were essentially correct.

The Mohole project reached its apogee in 1961 when the drilling ship *Cuss 1*, operating off Guadalupe Island in an area where the ocean was 3570 m deep, drilled five holes into the ocean floor. This drilling expedition was a huge success. It was widely reported in the popular press. *Life* magazine had the distinguished author John Steinbeck on board the drilling ship to describe the event in vivid detail. After this impressive achievement the project lost many of its supporters. Costs escalated, contractors became embroiled in political controversy, and some institutions that had formerly given it their whole-hearted support formed a new consortium called the Joint Oceanographic Institutions Deep Earth Sampling (JOIDES) syndicate.

Within a relatively short time the new group had secured funds to build a large (10 500 ton) drilling ship designed to sample the upper layers of the deep-ocean floor. The voyages of the *Glomar Challenger* began in August 1968 when it set out to drill in the Gulf of Mexico and the Atlantic Ocean. This drilling program added the dimension of time to our understanding of how the oceans evolved. Data from the second leg provided evidence in support of ocean-floor spreading, but conclusive evidence was gathered during the third leg (sites 14–22: Maxwell *et al.*, 1970). In this traverse across the South Atlantic, to the east of Rio de Janeiro, it was discovered that the age of the sediments directly overlying what was called the basaltic basement became progressively younger as the axis of the mid-Atlantic ridge was approached.

Extraterrestrial petrology

At much the same time that some petrologists were grappling with the problems of sea-floor spreading, others were examining a torrent of novel data about the Sun and its retinue of planets and satellites (cf. Beatty and Chaikin, 1990). This new era of exploration began in 1959 with the outstanding *Luna Three* voyage to the far side of the Moon. Ten years later, the *Apollo 11 Mission* landed on the Moon and 58 samples, with a combined weighed of 21.6 kg, were collected. Nineteen-sixty-nine was also the year the *Voyager 1* and *2* spacecraft had their spectacularly successful encounters with Jupiter and its satellites. During the next decade these spacecraft continued to produce a succession of bizarre and beautiful images of the saturnian and neptunian systems. There has also been a succession of missions to the terrestrial planets.

In 1994 the remarkably efficient *Clementine Mission* used a new generation of lightweight technologies to digitally image (ultraviolet to infrared) the whole lunar surface under constant geometry and lighting conditions. Its battery of instruments, including a super-fast computer, were able to produce images that depict the slight differences inherent in the various rocks on the lunar surface (Bruning, 1994; Talcott, 1994). Processing the huge volume of data gathered over the past 25 years has yielded a series of outstanding images that have profoundly and irrevocably changed our perception of the Earth and its place in the Solar System. Our idea of the normal has been successfully challenged. Petrologists, volcanologists and geophysicists are being forced to ask new and searching questions, because on every planet some parameters that governed planetary processes differ from those currently active on Earth, whereas others are almost the same, or were the same in the past.

We are now beginning to acknowledge that the planets and their satellites may hold the clues that will enable us to predict the future evolution of the Earth. Igneous petrology is a pivotal science in the quest to describe and explain how the Earth functions. Some would regard the history of igneous petrology as a saga that traces the development and use of new tools for investigating rocks, but one must always remember that the essential skills needed by successful petrologists have always been, and remain, deductive reasoning, three-dimensional perception, imagination and inventiveness (Yoder, 1992, p. 45).

1.3 Physical structure and composition of the Earth

When discussing Kilauea Volcano in Hawaii it was implied that seismologists had found that the magma responsible for the current cycle of volcanic activity originated about 80 km below the summit caldera. Seismologists are geophysicists who study the passage of elastic waves through rocks. At present, seismology provides the most robust method for studying the internal layering within the planets. By the 1950s seismology had established that the physical character of the Earth's crust beneath the oceans and continents was fundamentally different. The significance of these differences was more fully appreciated when it was discovered that most of the oceanic crust had been produced during the last 200 million years, whereas the continents contain some rocks that are older than 3.5 giga-years (billion years). Time is the very essence of the solid-Earth sciences. It is what sets the solid-Earth sciences apart from chemistry and most of physics. These other disciplines are mainly concerned with phenomena that can be studied directly (in real time) in the laboratory.

The elastic, or seismic, waves that travel through the body of a planet are called body waves. Their propagation is similar to that of light as they comply with the laws of refraction (Snell's Law). There are two types of body waves. The P-waves, or primary waves, are dilatational waves that involve the compression and rarefaction of the rocks they pass through. Secondary waves, or S-waves, are rotational waves that involve the shearing and rotation of the rocks they pass through. Because S-waves do not produce a change in volume they are sometimes called equivoluminal waves.

Table 1.3: Chemical composition of the Earth's crust. After Taylor, 1992, pp. 200–202.

	Upper continental crust (per cent)	Lower continental crust (per cent)	Bulk continental crust (per cent)	Ocean crust (per cent)		Upper continental crust (ppm/ppb)	Lower continental crust (ppm/ppb)	Bulk continental crust (ppm/ppb)	Ocean crust (ppm/ppb)
SiO_2	66	54.4	57.3	49.5	Ag	50 ppb	90 ppb	80 ppb	26 ppb
TiO_2	0.5	1	0.9	1.5	Cd	98 ppb	98 ppb	98 ppb	130 ppb
Al_2O_3	15.2	16.1	15.9	16	In	50 ppb	50 ppb	50 ppb	72 ppb
FeO (total)	4.5	10.6	9.1	10.5	Sn	5.5	1.5	2.5	1.4
MgO	2.2	6.3	5.3	7.7	Sb	0.2	0.2	0.2	17
CaO	4.2	8.5	7.4	11.3	Cs	3.7	0.1	1	30
Na_2O	3.9	2.8	3.1	2.8	Ba	550	150	250	25
K_2O	3.4	0.34	1.1	0.15	La	30	11	16	3.7
Total	99.9	100.04	100.1	99.45	Ce	64	23	33	11.5
	ppm/ppb	ppm/ppb	ppm/ppb	ppm/ppb	Pr	7.1	2.8	3.9	1.8
Li	20	11	13	10	Nd	26	12.7	16	10
Be	3	1	1.5	0.5	Sm	4.5	3.17	3.5	3.3
B	15	8.3	10	4	Eu	0.88	1.17	1.1	1.3
Sc	11	36	30	38	Gd	3.8	3.13	3.3	4.6
V	60	285	230	250	Tb	0.64	0.59	0.6	0.87
Cr	35	235	185	270	Dy	3.5	3.6	3.7	5.7
Mn	600	1670	1400	1000	Ho	0.8	0.77	0.78	1.3
Co	10	35	29	47	Er	2.3	2.2	2.2	3.7
Ni	20	135	105	135	Tm	0.33	0.32	0.32	0.54
Cu	25	90	75	86	Yb	2.2	2.2	2.2	5.1
Zn	71	83	80	85	Lu	0.32	0.29	0.3	0.56
Ga	17	18	18	17	Hf	5.8	2.1	3	2.5
Ge	1.6	1.6	1.6	1.5	Ta	2.2	0.6	1	0.3
As	1.5	0.8	1	1	W	2	0.7	1	0.5
Se	0.05	0.05	0.05	160 ppb	Re	0.5 ppb	0.5 ppb	0.5 ppb	0.9 ppb
Rb	112	5.3	32	2.2	Ir	0.02 ppb	0.13 ppb	0.1 ppb	0.02 ppb
Sr	350	230	260	130	Au	1.8 ppb	3.4 ppb	3 ppb	0.23ppb
Y	22	19	20	32	Tl	750 ppb	230 ppb	360 ppb	12 ppb
Zr	190	70	100	80	Pb	20	4	8	0.8
Nb	25	6	11	2.2	Bi	127 ppb	38 ppb	60 ppb	7 ppb
Mo	1.5	0.8	1	1	Th	10.7	1.06	3.5	0.22
Pd	0.5 ppb	1 ppb	1 ppb	<0.22 ppb	U	2.8	0.28	0.91	0.1

The Earth's crust

The crust makes up less than 1 per cent of the volume and less than 0.4 per cent of the mass of the Earth. Its physical structure and chemical composition are extremely variable. The continental crust has a mean thickness of 35 km and a mean P-wave velocity of 6.5 km/s. A thinner, layered crust has developed beneath the oceans. Layer one is at the top, and it typically contains unconsolidated sediments with a mean thickness of about 0.45 km and a P-wave velocity of about 2 km/s (Raitt, 1963). The upper unit in layer two is usually an irregular-shaped body of altered pillow lavas, volcanic rubble and intermingled sediments. It passes downwards into a more consoli-

dated unit comprised of basaltic lavas. Layer two is usually about 1.7 km thick, with P-wave velocities of about 4 km/s in the upper unit, but the velocity increases in the lavas to about 5.1 km/s. Layer three is about 4.9 km thick with a mean P-wave velocity of 6.7 km/s. It mainly contains microgabbroic and gabbroic rocks (i.e. medium- and coarse-grained rocks of basaltic composition). A more complete picture of the petrology of the oceanic lithosphere can be gleaned from the study of ophiolites, or slabs of oceanic lithosphere thrust towards the surface by colliding lithospheric plates.

Table 1.3 provides estimates of the chemical composition of the upper continental crust, lower continental crust, total continental crust and oceanic crust (Taylor, 1992, pp. 200–202). A more recent estimate

Table 1.4: Chemical composition of the whole continental crust. After Wedepohl, 1994, p. 960.

	per cent		ppm		ppm		ppm
SiO_2	61.5	Li	17.5	Ga	15.5	Eu	1.3
TiO_2	0.68	B	9.3	Rb	76	Gd	4.1
Al_2O_3	15.1	F	526	Sr	334	Tb	0.65
Fe_2O_3	6.3	S	725	Y	24	Ho	0.78
MnO	0.1	Sc	16	Zr	201	Yb	2
MgO	3.7	V	101	Nb	18.5	Lu	0.36
CaO	5.5	Cr	132	Ba	576	Hf	4.9
Na_2O	3.2	Co	25	La	25	Ta	1.1
K_2O	2.4	Ni	59	Ce	60	Pb	14.8
P_2O_5	0.18	Cu	26	Nd	27	Th	8.5
Total	98.66	Zn	66	Sm	5.3	U	1.7

of the whole continental crust is given in Table 1.4. The main difference between the data in the two tables is that in the first table it is assumed that the lower crust comprises 75 per cent of the total crust, whereas in the second table Wedepohl (1994) used a ratio of 5:3:2 for the upper crust, lower felsic crust and lower mafic crust, respectively. Wedepohl derived his model of the continental crust from data obtained by the European Geotraverse (Blundell *et al.*, 1992).

It is convenient to divide the crust of the Earth into six major regions:

1 continental shields,
2 continental margin arcs and island arcs,
3 continental rifts and paleorifts,
4 active mid-oceanic ridges and oceanic crust,
5 aseismic ridges, and
6 oceanic islands and seamounts.

The continental shield areas occupy about 20 per cent of the surface of the Earth. They are usually tectonically stable, and often contain old nuclei of Archean age. A typical continental shield area contains a 0.5 km thick layer of sedimentary rocks, overlying a much thicker layer of granites, granodiorites and tonalites (see Chapter 9). Granulite facies metamorphic rocks often underlie this granitic layer. All the continents also contain vast bodies of continental flood basalts (see Section 7.6), such as the Columbia River flood basalts of North America. Since the Archean the continents have gradually grown out from their Archean nuclei with the new rocks being similar to those presently forming in continental margin arcs and island arcs. The latter are seismically and volcanically active zones of plate convergence where oceanic lithosphere is being subducted. Such areas contain a wide range of volcanic and plutonic rocks, such as basaltic andesites, andesites, dacites, diorites, granodiorites and granites (see Chapter 8).

Continental rifts and paleorifts are typically linear zones where the entire thickness of the lithosphere has been ruptured under extension. They are usually long, narrow troughs bounded by normal faults. Such structures are also called grabens. Modern examples of these structures are the Rhine Graben of Germany and the huge East African Rift System. Continental rifts usually contain a remarkable selection of alkaline igneous rocks (see Chapter 11). Most of the trachytes and phonolites found on Earth crop out in the East Africa Rift System.

Active mid-oceanic ridges and oceanic crust cover over half the surface of the Earth. The mid-oceanic ridge system is a continuous submarine mountain range. This unique and remarkable feature of the Earth is between 1 and 3 km in elevation, about 1500 km wide and over 84 000 km long (see Fig. 1.9). According to the sea-floor spreading hypothesis the oceanic crust is growing as the result of convective upwelling of magma into the feeder dykes and magma chambers of the mid-oceanic ridge system (see Section 7.5).

Aseismic ridges are linear elevated volcanic chains that rise between 2 and 4 km above the surrounding ocean floor. They usually range between 250 and 400 km in width, and are between 700 and 5000 km in length. These ridges comprise about 25 per cent of the area of the ocean floor and appear to represent chains of volcanic islands and seamounts that have subsided in elevation since they became inactive. Examples of this topographic feature include the Iceland-Faeroe Ridge, Walvis Ridge, and the Ninety-East Ridge in the Indian Ocean. Oceanic islands and seamounts occur in all the oceans and contain a remarkable range of igneous rocks (see Section 7.9).

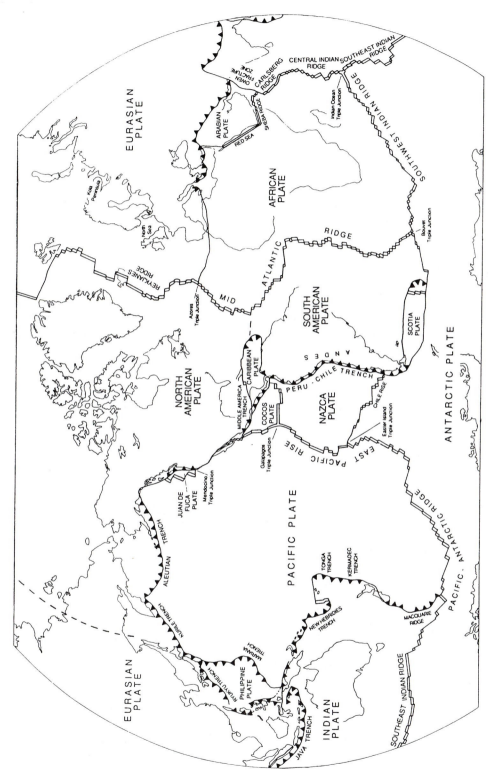

Fig. 1.9 Sketch map of the global mid-oceanic ridge system, oceanic trenches, tectonic plates and transform faults. After Fowler, 1990, p.6, and others.

The lithosphere

The crust and the 10 to 50 km thick uppermost layer of the mantle form a single, relatively strong unit called the lithosphere. In continental regions it is usually a complex, layered unit that varies in thickness, composition and physical properties. When considering the movement of magma through the lithosphere one must remember (1) that the lithosphere is composed of at least two major layers with very different densities, and (2) when the common mantle–derived magmas arrive at the mantle–crust interface their densities are likely to exceed the densities of the average continental crust. If such magmas are able to delaminate the join between the crust and mantle, they are likely to produce a magma chamber that underplates the crust.

Beneath the lithosphere is the structurally weak asthenosphere on which the lithospheric plates slide. The interface between these layers is usually both a mechanical and a thermal boundary. It is generally assumed that the low-velocity zone (a geophysical concept) and the asthenosphere (a tectonic concept) are essentially the same. The former is generally some 100 km thick, and seismic waves travel within this zone at velocities (approximately 7.7 km/s) that are approximately 6 per cent slower than one would expect in normal uppermost mantle. In different tectonic settings one finds variations in the physical properties, thickness and depth of this zone. It is usually well developed in active tectonic areas, but poorly developed in old cratonic areas. As the low-velocity zone strongly attenuates seismic waves, it is usually interpreted as a layer where partial melting is likely to occur.

The Earth's mantle

The mantle makes up 84 per cent of the volume and 67 per cent of the mass of the Earth. It extends from the base of the crust down to a depth of 2900 km (see Fig. 1.10). The name 'mantle' was introduced by Wiechert in 1897, but it was only in this century that its boundaries were clearly delineated. Its upper boundary is the Mohorovičić Discontinuity, or Moho. This discontinuity separates rocks in which the P-wave velocity is between 6 and 7 km/s for the lower crustal rocks, from the underlying mantle rocks in which the P-wave velocity is normally 8.1 km/s. At 400 km the first of the deep mantle dis-

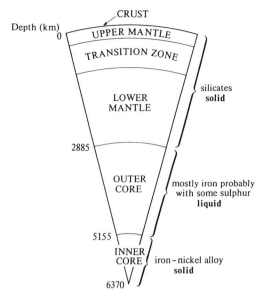

Fig. 1.10 Sketch illustrating the internal structure of the Earth. After Brown and Mussett, 1981, p. 2.

continuities occurs. It is followed by an even stronger discontinuity at a depth of 670 km. Both of these discontinuities are likely to represent phase changes in a mantle of broadly constant composition. This model suggested that as depths increase, pressure increases; this causes minerals to convert to denser crystal structures. If the intra-mantle discontinuities do not represent substantial changes in chemical composition, then it is doubtful whether they represent barriers to convection. The upper mantle is usually defined as a shell of the Earth, extending down to the second intra-mantle discontinuity at a depth of 670 km.

In the lowermost mantle there is a layer with variable physical properties. It ranges in thickness between 5 and 300 km and it is called the D'' (D double prime) layer. Ultra-high-pressure (greater than 100 gigapascals) laboratory simulations have established that this layer probably represents a zone of reaction and mixing between the silicate rock of the lower mantle and an immiscible oxide liquid that evolves in the outer core (Ito *et al.*, 1995). The pascal (Pa) is the derived SI unit of pressure and it is equal to 1 newton per square metre. Convection is likely to occur in the liquid outer core, and as the hot liquid rises and impinges on the lowermost mantle it will probably invoke reaction between the phases in the abutting layers (Knittle and Jeanloz, 1991, p. 1438). Thermal perturbations in the D'' layer are probably responsible for generating mantle plumes, and modulating the convection system that originates in the deep mantle.

Many geochemical and geophysical observations impose constraints on our models of the chemical composition of the mantle. These models will be examined in Section 3.4. All that is required to be noted at this stage is that the normal upper mantle is a magnesium-rich rock that contains essential amounts of olivine, orthopyroxene and, usually, clinopyroxene (see Table 1.3).

The Earth's core

The core makes up 15 per cent of the volume and 32 per cent of the mass of the Earth. Its mean density is $11\,000\,kg/m^3$, and it increases to over $13\,000\,kg/m^3$ in the inner solid core. Geophysical data, experimental investigations of pertinent phases at high pressures, and chemical data from iron meteorites, have all contributed to our understanding of the composition of the core. Mason and Moore (1982, p. 51) suggested that it had the following composition: 86.3 per cent iron, 7.4 per cent nickel, 0.4 per cent cobalt and 5.9 per cent sulphur. Other geochemists have proposed that the more massive outer core contains between 10 and 15 per cent of one or more light element, such as silicon or sulphur. There is no consensus about the amount of potassium stored in the outer core. It may contain up to one tenth of a per cent of potassium. If it does, the radioactive decay of potassium-40 is likely to provide at least part of the heat required to maintain the convection currents within the outer core. All the terrestrial planets are likely to possess iron-rich cores.

1.4 The magma/igneous rock system

The terrestrial planets are formed by the accretion of small planetesimals and planetoids (see Chapter 13). Taylor (1992, p. 25) has estimated that over 50 per cent of the present mass of the Earth accreted from massive planetoids that had already gone through a melting stage which produced a separation of silicate, sulphide and metal phases. If this model is correct it means that there may have been little high-pressure equilibration between the core and the mantle of the Earth. Core formation in the terrestrial planets probably started when the particular planet was large enough for melting to occur.

The Earth–Moon system probably formed as the result of a collision between the Earth, which had already developed a core, and a planetoid of about 0.15 Earth masses. During the collision much of the Moon and the Earth melted. On the surface of the Moon a magma/lava ocean crystallised circa 4.44 giga-years ago. This is the minimum age for this catastrophic heating event. Since the Hadean Era (about 3.8 giga-years) the terrestrial planets have been cooling and crystallising. Mass transport in the mantle usually occurs as the result of mantle convection, and this process is also the key to understanding heat transport. Silicate rocks are poor conductors of heat and their flow (rheology) is dependent on temperature. This means that once heat starts to build up in the interior of a planet, such as Earth, the heated rocks will almost inevitably become pliable enough to convect, and this establishes a temperature-regulating mechanism.

Heat extraction from the interior of a planet is a fundamental planetary process, and on a planetary scale the actual composition of the heat-bearing fluid is incidental. These fluids, or magmas, are the quintessence of igneous petrology. The Earth can be likened to a complex, self-regulating heat pump. At its simplest, fluids extract heat from the outer fluid core and lower mantle, traverse most of the mantle, and later discharge both mass and heat into the lithosphere, hydrosphere or atmosphere. Igneous petrology is primarily concerned with the composition, evolution and movement of heat-bearing fluids, as they pass through the upper mantle and crust.

Several indirect methods are used to trace convective systems within the Earth. It is often postulated that there are three major interrelated convective systems. One operates within the liquid outer core; the second is in the deep mantle, below 670 km; whilst the third is a more complex system operating within the upper mantle where it interacts with the rigid lithospheric plates. On Venus the crust is hotter and less rigid. This has influenced its system of heat transference through the crust and also the surface distribution volcano-magmatic landforms. The idea of simple, symmetrical deep-mantle convection is supported by maps and sections determined by seismic tomography. This technique shows the elastic properties of slices through the mantle (Gardini *et al.*, 1987). Zones of slower than average seismic velocities are usually interpreted as being hot, whereas those with faster velocities are interpreted as zones of colder material. In such tomographs the Hawaiian Islands are depicted as broad, high-temperature foci, whereas a cross section of the subduction zone in northern Japan and China shows a cold slab of oceanic lithosphere overlain by a hotter zone where partial melting is probably taking place.

1.5 Magmas and igneous rocks on Venus, Earth and Mars

On Earth the cycle of plate growth, suboceanic cooling, subduction and reheating accounts for most of the heat lost by the mantle (Davies, 1992: see Fig. 1.11). In addition the presence of lithospheric plates usually regulates the fine pattern of convective flow in the upper mantle. Magmatic plumes are likely to transport less than 10 per cent of the heat through the mantle and into the crust. The magmas generated within the Earth are usually products of decompression melting, heating from below, or the influx of fluids from subducted rocks.

Venus is without rigid plates, oceans, surface water, or any process that enables volatile-rich materials to be subducted. The surface temperature on Venus is 475°C, and its atmospheric pressure is 92 times higher than Earth. These harsh conditions result in the development of a variety of singular landforms that seem to depict a merging of the processes normally associated with volcanism and the intrusion of magma in the near-surface environment. Products of volcano-magmatic activity dominate the Venusian landscape (Head *et al.*, 1992). Its numerous volcanic landforms are distributed globally. They are not confined to narrow, linear belts as on Earth. On Venus most heat loss is a result of the upward flow and partial eruption of volcano-magmatic plumes. One might visualise the process of losing heat from the Venusian mantle as being akin to slowly boiling porridge.

Mars also has many distinct volcanic landforms. They range in size from several hundred kilometres across down to the resolution limits imposed by the instruments that collected the data. Most of the young volcanism is concentrated in the Tharsis and the Elysium volcanic provinces. Both provinces developed on huge topographic bulges. The Tharsis Bulge is over 7 km high and more than 7500 km long (see Fig. 1.12). It occupies nearly a quarter of the surface area of the planet. This dome is probably underlain by a low-density mantle residuum generated by partial melting, and the removal of large quantities of iron-rich magma. Four huge volcanoes, including the 24 km high Olympus Mons, are directly related to this huge dome. Sleep (1994) has proposed that a system of plate tectonics was active on Mars during its early development, circa 3.5 giga-years ago. Spreading is considered to have taken place within an 8000 km long equatorial belt. It produced the younger crustal layer that makes up the northern lowlands. This crustal spreading event was probably responsible for most of the cooling of this small planet. As the huge Tharsis volcanoes lie astride a postulated subduction zone, their origin and the origin of the Tharsis Dome may be related to ongoing magma generation at a defunct subduction site. This would have occurred after crustal spreading had ceased, and the surface of the planet had become completely rigid.

An appraisal of the operation of the magma/igneous rock systems on Venus and Mars offers pointers to the performance of the magma/igneous rock system on Earth. On Venus the magma/igneous rock system is dominated by widespread volcanic activity caused by the upward movement of plumes of magma. This Venusian pattern of magmatic activity may simulate the volcano-magmatic conditions that prevailed on Earth during the Hadean Era. Mars is portrayed as a post-plate-tectonics planet, and our grasp of its magma/igneous rock system should enable us to anticipate how the Earth is likely to develop when sea-floor spreading ceases.

References

Abich, H.W., 1841. Geologische Beobachtungen über die vulkanischen Erscheinungen und Bildungen in Unter- und Mittel-Italien, braunschweig. Friedrich Vieweg und Sohn, Brunswick, 134 pp.

Airy, G.B., 1855. On the computation of the effect of mountain masses as disturbing the apparent astronomical latitude of stations in geodetic surveys. *Philosophical Transactions of the Royal Society, London*, 154, pp. 101–104.

Amstutz, A., 1951. Sur l'evolution des structure alpines. *Archives Sci.*, 4, pp. 323–329.

Anon., 1990. Ancient Chinese records help date Mediterranean eruption. *Geotimes*, March, p. 14.

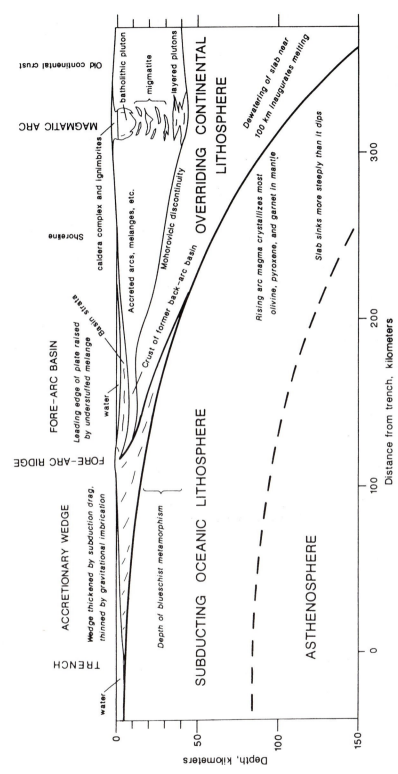

Fig. 1.11 Sketch section through a subduction system, such as the one currently active beneath the island of Sumatra in Indonesia. After Hamilton, 1994, p. 6.

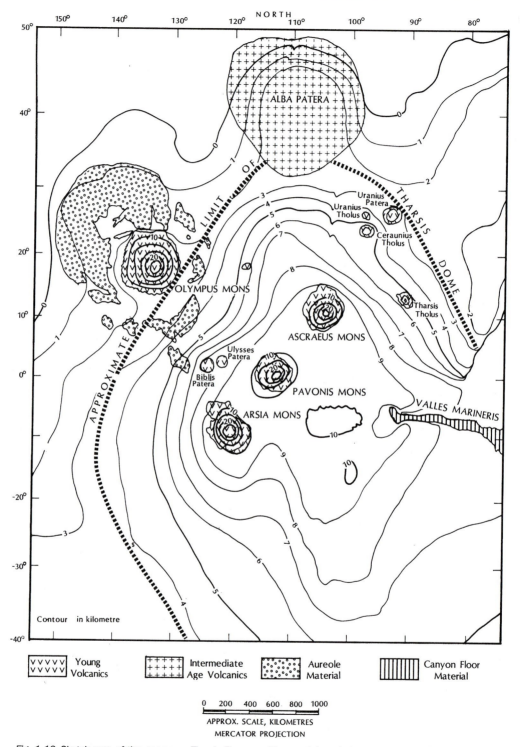

Fig. 1.12 Sketch map of the enormous Tharsis Dome on Mars and the relationship between it and the various large volcanoes in this region. Adapted from the geologic map of Mars by Scott and Carr, 1978.

Autran, A. and Descourt, J. (editors), 1980. Evolutions geologiques de la France. *Memoire du Bureau de Recherches Geologiques et Minieres*, 107, 355 pp.

Baker, H.B., 1912. The origin of continental forms, 2. Annual Report of the Michigan Academy of Science for 1911–1912, pp. 116–141.

Basaltic Volcanism Study Project (BVSP), 1981. *Basaltic Volcanism on the Terrestrial Planets.* Pergamon Press, Inc., New York, 1286 pp.

Bascom, W., 1961. A *Hole in the Bottom of the Sea: The story of the Mohole Project.* Weidenfeld and Nicolson, London, 352 pp.

Beatty, J.K. and Chaikin, A., 1990. *The New Solar System.* Third Edition. Cambridge University Press, Cambridge, 326 pp.

Beaumont, E. de, 1847. Note surles emanations volcaniques et metalliferes. *Bulletin Société Geologie France*, 4, pp. 1249–1333.

Becker, G.F., 1897. Fractional crystallisation of rocks. *American Journal of Science*, Series 4, 4, pp. 257–261.

Benioff, H., 1949. Seismic evidence for the fault origin of ocean deeps. *Bulletin of the Geological Society of America*, 60, pp. 1837–1856.

Blundell, D.J., Freeman, R. and Mueller, S. (editors), 1992. *A Continent Revealed: The European Geotraverse.* Cambridge University Press, Cambridge, 275 pp.

Boltwood, B.B., 1907. On the ultimate disintegration products of the radioactive elements. *American Journal of Science*, 4(23), pp. 77–88.

Botting, D., 1973. *Humboldt and the Cosmos.* Sphere Books Ltd, London, 295 pp.

Bowen, N.L., 1912. The binary system $Na_2Al_2Si_2O_8$ (nepheline, carnegieite) – $CaAl_2Si_2O_8$ (anorthite). *American Journal of Science*, 33, pp. 551–573.

Bowen, N.L., 1915a. Crystallisation differentiation in silicate liquids. *American Journal of Science*, Series 4, 39, pp. 175–191.

Bowen, N.L., 1915b. The late stages of the evolution of the igneous rocks. *Journal of Geology*, 23, Supplement, pp. 1–89.

Bowen, N.L., 1917. The problem of anorthosites. *Journal of Geology*, 25, pp. 209–243.

Bowen, N.L., 1919. Crystallisation-differentiation in igneous magmas. *Journal of Geology*, 27, pp. 393–430.

Bowen, N.L., 1921. Diffusion in silicate melts. *Journal of Geology*, 29, pp. 295–317.

Bowen, N.L., 1922a. The reaction principle in petrogenesis. *Journal of Geology*, 30, pp. 177–198.

Bowen, N.L., 1922b. Genetic features of alnöite from Isle Cadieux, Quebec. *American Journal of Science*, 3, pp. 1–34.

Bowen, N.L., 1922c. The behaviour of inclusions in igneous magmas. *Journal of Geology*, 30, pp. 513–570.

Bowen, N.L., 1923. The genesis of melilite. *Journal of the Washington Academy of Sciences*, 13, pp. 1–4.

Bowen, N.L., 1924. The Fen area in Telemark, Norway. *American Journal of Science*, 8, pp. 1–11.

Bowen, N.L., 1926. The carbonate rocks of the Fen area in Norway. *American Journal of Science*, 12, pp. 499–502.

Bowen, N.L., 1927. The origin of ultrabasic and related rocks. *American Journal of Science*, 14, pp. 89–108.

Bowen, N.L., 1928a. Geologic thermometry. In Fairbanks, E.E. (editor) *The Laboratory Investigation of Ores.* McGraw-Hill, New York, pp. 172–199.

Bowen, N.L., 1928b. *The Evolution of the Igneous Rocks.* Princeton University Press, Princeton, New Jersey, 334 pp.

Bowen, N.L., 1937. Recent high-temperature research on silicates and its significance in igneous geology. *American Journal of Science*, 33, pp. 1–21.

Bowen, N.L., 1938. Lavas of the African Rift Valleys and their tectonic setting. *American Journal of Science*, 35-A, pp. 19–33.

Bowen, N.L., 1945. Phase equilibria bearing on the origin and differentiation of alkaline rocks. *American Journal of Science*, 243-A, pp. 75–89.

Bowen, N.L., 1947. Magmas. *Bulletin of the Geological Society of America*, 58, pp. 263–280.

Bowen, N.L., 1948. The granite problem and the method of multiple prejudices. *The Geological Society of America, Memoir*, 28, pp. 79–90.

Bowen, N.L., 1950. The making of a magmatist. *The American Mineralogist*, 35, pp. 651–658.

Brongniart, A., 1813. Essai d'une classification mineralogique des roches melangees. *Journal des Mines*, 34(199), pp. 5–48.

Brown, G.C. and Mussett, A.E., 1981. *The Inaccessible Earth.* George Allen and Unwin, London, 235 pp.

Bruning, D., 1994. Clementine maps the Moon. *Astronomy*, 22(7), pp. 36–39.

Buddington, A.F., 1959. Granite emplacement with special reference to North America. *Bulletin of the Geological Society of America*, 70, pp. 671–748.

Chamberlin, T.C., 1905. *Fundamental Problems of Geology.* Carnegie Institution of Washington Year Book, 3 (1904), pp. 195–254.

Clarke, F.W., 1908. *The Data of Geochemistry*. 1st Edition Bulletin No. 330; 2nd Edition, 1911, No. 491; 3rd Edition, 1916, No. 616; 4th Edition, 1920, No. 695; 5th Edition, 1924, No. 770, 841 pp. United States Geological Survey, Washington, D.C.

Clough, C.T., Maufe H.B. and Bailey, E.B., 1909. The cauldron-subsidence of Glen Coe. *Quarterly Journal of the Geological Society, London*, 65, pp. 611–678.

Cross, C.W., Iddings, J.P., Pirsson, L.V. and Washington, H.S., 1902. A quantitative chemico-mineralogical classification and nomenclature of igneous rocks. *Journal of Geology*, 10, pp. 555–690.

Daly, R.A., 1926. *Our Mobile Earth*. Charles Scribner, New York, 342 pp.

Daly, R.A., 1933. *Igneous Rocks and the Depths of the Earth*. McGraw-Hill Book Company, Inc., New York, 598 pp.

Darwin, C., 1844. Geological observations on the volcanic islands visited during the voyage of *H.M.S. Beagle*, together with some brief notices on the geology of Australia and the Cape of Good Hope; being the second part of the geology of the voyage of the Beagle under the command of Capt. Fitzroy, R.N., during the years 1832 to 1836. Smith, Elder, London, 175 pp.

Davies, G.F. 1992. Plates and plumes: dynamos of the Earth's mantle. *Science*, 257: pp. 493–494.

Day, A.L., Allen, E.T. and Iddings, J.P., 1905. *The Isomorphism and Thermal Properties of the Feldspars*. Carnegie Institution of Washington, Publication 31, 95 pp.

Decker, R.W., Wright, T.L. and Stauffer, P.H., 1987. *Volcanism in Hawaii*, Volume 1. United States Survey Professional Paper 1350, 839 pp.

DePaolo, D.J., 1988. *Neodymium Isotope Geochemistry*. Springer-Verlag, Berlin, 187 pp.

Desmarest, N., 1771. Sur l'origine et la nature du basalte a grandes colonnes polygones, determinees par l'histoire naturelle de cette pierre, observee en Auvergne. *Memoire Royal Academie Sciences, Paris*, 87, pp. 705–775.

Dietz, R.S., 1961. Continental and ocean basin evolution by spreading of the sea floor. *Nature*, 190, pp. 854–857.

Doelter, C., 1902. *Die chemische Zusammensetzung und die che Petrographische Mitteilung*, 21, pp. 191–225.

Du Toit, A.L., 1928. Some reflections upon a geological comparison of South Africa with South America. Anniversary address by the president for 1927, *Transactions and Proceedings of the Geological Society of South Africa*, 31, pp. xix–xxxviii.

Du Toit, A.L., 1937. *Our Wandering Continents: An hypothesis of continental drifting*. Oliver and Boyd, Edinburgh, 366 pp.

Dutton, C.E., 1880. *Geology of the High Plateaus of Utah*. U.S. Geological and Geographical Survey, Washington, D.C., 307 pp.

Estabrook, F.B., 1956. Geophysical research shaft, *Science*, 124, p. 686.

Eyles, V.A., 1961. Sir James Hall, Bt, 1761–1832. *Endeavour*, 20, pp. 210–216.

Fermor, L.L., 1913. Preliminary note on garnet as a geological barometer and on an infra-plutonic zone in the Earth's crust. *Geological Survey of India*, 43(1), pp. 41–47.

Fowler, C.M.R., 1990. *The Solid Earth: An Introduction to Global Geophysics*. Cambridge University Press, Cambridge, 472 pp.

Francis, P., 1993. *Volcanoes: A Planetary Perspective*. Clarendon Press, Oxford, 443 pp.

Gardini, D., Xiang-Dong, L. and Woodhouse, J.H., 1987. Three-dimensional structure of the Earth from splitting in free oscillation spectra. *Nature*, 325, pp. 405–411.

Geikie, A., 1905. *The Founders of Geology*. Second Edition. Macmillan and Company, London, 486 pp.

Ghiorso, M.S., Hirschmann, M. and Sack, R.O., 1994. MELTS: Software for thermodynamic modelling of magmatic systems. *EOS*, 74, p. 338.

Goldschmidt, V.M., 1922. Stammestypen der Eruptivgesteine. *(Norske) Videnskaps-Akademi* i Oslo, Matematisk-naturvidenskapelig Klasse, Skrifter, 10, 6 pp.

Goldschmidt, V.M., 1954. *Geochemistry*. Muir, A. (editor). Oxford University Press, Oxford, 730 pp.

Griggs, D., 1939. A theory of mountain building. *American Journal of Science*, 237, pp. 611–650.

Guettard, J.E., 1752. Surquelques montagnes de France qui ont ete des volcans. *Memoire Royal Academie Science, Paris*, pp. 1–59.

Gutenberg, B., 1913. Üeber die konstitution des Erdinnern, erschlossen aus Erdbebenbeobachtungen. *Physikalishe Zeit schrift*, 14, pp. 1217–1218.

Hall, J., 1798. Curious circumstances upon which the vitreous or the stony character of whinstone and lava respectively depend. *Nicholson's Journal*, 2, pp. 285–288.

Hall, J., 1800. Experiments on whinstone and lava. *Nicholson's Journal*, 4, pp. 8–18, 56–65.

Hall, J., 1805. Experiments on whinstone and lava. *Transactions of the Royal Society of Edinburgh*, 5, pp. 43–75.

Hall, M., 1876. Note upon a portion of basalt from Mid-Atlantic. *Mineralogical Magazine*, 1, pp. 1–3.

Hamilton, W., 1767. Two letters containing an account of the late eruption of Mount Vesuvius. *Philosophical Transactions of the Royal Society (London)*, 57, pp. 192–200.

Hamilton, W., 1776. Campi Phlegraei; observation on the volcanoes of the Two Sicilies, as they have been communicated to the Royal Society of London, 2 Volumes, Naples [s.n.], 54 leaves of plates, one folded leaf of plates, and map.

Hamilton, W., 1994. Subduction systems and magmatism. In Smellie, J.L. (editor), *Volcanism Associated with Extension at Consuming Plate Margins*, pp. 3–28, The Geological Society, London, Special Publication Number 81, 293 pp.

Harland, W.B., Armstrong, R.L., Cox, A.V., Craig, L.E., Smith, A.G. and Smith, D.G., 1982. *A Geologic Time Scale 1989*. Cambridge University Press, Cambridge, 263 pp.

Harper, L.F., 1915. Geology and mineral resources of the Southern Coalfield, Part 1, The South Coast portion. *Geological Survey of New South Wales, Memoir*, 7, 410 pp.

Haüy, R.-J., 1811. A mineralogical classification of rocks, in an unpublished letter to von Leonhard.

Head, J.W., Crumpler, L.S., Aubele, J.C., Guest, J.E. and Saunders, R.S., 1992. Venus volcanism: classification of volcanic features and structures, associations, and global distribution from Magellan data. *Journal of Geophysical Research*, 97 (E8), pp. 13153–13197.

Hess, H.H., 1959. Nature of the great ocean ridges. In *International Oceanographic Congress, Abstracts*, American Association for the Advancement of Science, Washington, D.C., pp. 33–34.

Hess, H.H., 1960. Stillwater Igneous Complex, Montana: A quantitative mineralogical study. *The Geological Society of America, Memoir*, 80, 230 pp.

Hess, H.H., 1962. History of ocean basins. In Engel, A.E.J., James, H.L. and Leonard, B.F. (editors), *Petrological Studies: A volume in honour of A.F. Buddington*, Geological Society of America, New York, pp. 599–620.

Hoffer, W., 1982. *Volcano: The Search for Vesuvius*. Summit Books, New York, 189 pp.

Holmes, A., 1913. *The Age of the Earth*. Harper and Brothers, London.

Holmes, A., 1928. Radioactivity and continental drift. *Geological Magazine*, 65, pp. 236–238.

Holmes, A., 1931. Radioactivity and Earth movements., *Transactions of the Geological Society of Glasgow*, 18, pp. 559–606.

Holmes, A., 1933. The thermal history of the Earth. *Journal of the Washington Academy of Sciences*, 23, pp. 169–195.

Holmes, A., 1944. *Principles of Physical Geology*, Thomas Nelson and Sons Ltd, London, 532 pp.

Hutton, J., 1785. Abstract of a dissertation concerning the System of the Earth, its duration, and stability – Hutton's address to the Royal Society of Edinburgh on his Theory of the Earth. Edinburgh, 32 pp.

Hutton, J., 1788. Theory of the Earth; or an investigation of the laws observable in the composition, dissolution, and restoration of land upon the globe. *Transactions of the Royal Society of Edinburgh*, 1, pp. 209–304.

Hutton, J., 1795. *Theory of the Earth with Proofs and Illustrations*, Volumes 1 and 2. Cadell, Davies and Creech, London and Edinburgh.

Hutton, J., 1899. *Theory of the Earth with Proofs and Illustrations*. Volume 3. Geikie, A. (editor) Geological Society of London, Burlington House.

Iddings, J.P., 1890. The mineral composition and geological occurrence of certain igneous rocks in Yellowstone National Park. *Bulletin of the Philosophical Society of Washington*, 11, pp. 191–220.

Irving, E., 1956. Paleomagnetic and palaeoclimatological aspects of polar wandering. *Pure and Applied Geophysics*, 33, pp. 23–41.

Ito, E., Morooka, K., Ujike, O. and Katsura, T., 1995. Reaction between molten iron and silicate melts at high pressure; implications for the chemical evolution of Earth's core. *Journal of Geophysical Research*, 100, B4, pp. 5901–5910.

Jameson, R., 1808. *System of Mineralogy, Volume 3, Treatise on Geognosy*. Constable, Edinburgh.

Johannsen, A., 1938. *A Descriptive Petrography of the Igneous Rocks*, Volume 4. The University of Chicago Press, Chicago, 511 pp.

Judd, J.W., 1886. On the gabbros, dolerites and basalts of Tertiary age in Scotland and Ireland. *Quarterly Journal of the Geological Society, London*, 42, pp. 49–97.

Keller, J. and Hoefs, J., 1995. Stable isotopic characteristics of recent natrocarbonatites from Oldoinyo Lengai. In Bell, K. and Keller, J. (editors), *Carbonatite Volcanism: Oldoinyo Lengai and the petrogenesis of natrocarbonatites*, pp. 113–123. Springer-Verlag, Berlin, 210 pp.

King, B.C. and Sutherland, D.S., 1960. Alkaline rocks of eastern and southern Africa. *Science Progress*, 48, pp. 298–321, 504–524, 709–720.

Kirwan, R., 1794. Examination of the supposed igneous origin of stony substances. *Transactions of the Royal Irish Academy*, 5, pp. 51–81.

Knittle, E. and Jeanloz, R., 1991. Earth's core–mantle boundary: results of experiments at high pressures and temperatures. *Science*, 251, pp. 1438–1443.

Krafft, M., 1991. *Volcanoes: Fire from the Earth*. Thames and Hudson, London, 207 pp.

Kuno, H., 1959. Origin of the Cenozoic petrographic provinces of Japan and surrounding areas. *Bulletin Volcanologique*, Series 2, 20, pp. 37–76.

Lehmann, I., 1936. *Bureau Central Seismologyique International*, Series A, Travaux Scientifiques, 14, pp. 3–31.

Le Maitre, R.W., Bateman, P., Dudek, A., Keller, J., Lameyre, J., Le Bas, M.J., Sabine, P.A., Schmidt, R., Sørensen, H., Streckeisen, A., Woolley, A.R. and Zanettin, B., 1989. *A Classification of Igneous Rocks and Glossary of Terms: Recommendations of the International Union of Geological Sciences Subcommission on the systematics of igneous rocks*. Blackwell Scientific Publications, Oxford, 193 pp.

Lirer, L., Munno, R., Petrosino, P. and Vinci, A., 1993. Tephrostratigraphy of the A.D. 79 pyroclastic deposits in perivolcanic areas of Mt. Vesuvio (Italy). *Journal of Volcanology and Geothermal Research*, 58, pp. 133–149.

Loewinson-Lessing, F.Y., 1954. *A Historical Survey of Petrology*. (Translated from the Russian by Tomkeieff, S.I.). Oliver and Boyd, Edinburgh, 112 pp.

Lyell, C., 1830–1883. *Principles of Geology*; being an attempt to explain the former changes of the Earth's surface by reference to causes now in operation, 3 Volumes. Murray, London.

Mason, B. and Moore, C.B., 1982. *Principles of Geochemistry*. Fourth Edition. John Wiley and Sons, New York, 344 pp.

Maxwell, A.E., von Herzen, R.P., Hsü, K.J., Andrews, J.E., Saito, T., Percival, S.F., Milow, E.D. and Boyce, R.E., 1970. Deep sea drilling in the South Atlantic. *Science*, 168, pp. 1047–1059.

Mohorovicic, A., 1909. Das Beben vom 8/10/1909. *Jahrd des meteorologischen Observ in Zagreb für das Jahr 1909*, 9(4), pp. 1–63.

Murray, J., 1802. *A Comparative View of the Huttonian and Wernerian System of Geology*. Ross and Black, Edinburgh.

Newhall, C.G. and Self, S., 1982. The Volcanic Explosivity Index (VEI): An estimate of explosive magnitude for historical volcanism. *Journal of Geophysical Research (Oceans and Atmosphere)*, 87, pp. 1231–1238.

Oldham, R.D., 1906. The constitution of the Earth as revealed by earthquakes. *Quarterly Journal of the Geological Society, London*, 62, pp. 456–475.

Ospovat, A. M., 1971. *Kurze Klassifikation* (Short Classification and Description of the Various Rocks). A facsimile and English translation of the original text by A.G. Werner, 1789. Hafner Publishing Company, New York, 194 pp.

Pecora, W.T., 1956. Carbonatites: A review. *Bulletin of the Geological Society of America*, 67, pp. 1537–1555.

Pickering, W.H., 1907. The place of origin of the Moon: the volcanic problem. *Journal of Geology*, 15, pp. 23–38.

Pitman, W.C. and Heirtzler, J.R., 1966. Magnetic anomalies over the Pacific–Antarctic Ridge. *Science*, 154, pp. 1164–1171.

Playfair, J., 1802. *Illustrations of the Huttonian Theory of the Earth*. Cadwell and Davies, London, and William Creech, Edinburgh, 528 pp.

Raitt, R.W., 1963. The crustal rocks. In Hill, M.E. (editor), *The Sea*, Volume 3, pp. 85–102, Interscience.

Read, H.H., 1957. *The Granite Controversy*. Thomas Murby and Company, London, 430 pp.

Robson, D.A., 1986. *Pioneers of Geology*. The Natural History Society of Northumbria, Special Publication, The Hancock Museum, Newcastle Upon Tyne, UK, 73 pp.

Rosenbusch, H., 1882. Ueber das Wesen der körnigen und porpyrischen Struktur bei Massengesteinen. *Neues jahrbuch*, 2, pp. 1–17.

Rosi, M., Sbrana, A. and Zan, I., 1986. *Carte Geologica e Gravimetrica dei Campi Flegrei, Scale 1:15 000*, Consiglio Nazionale Della Ricerche, Progetto Finalizzato Geodinamica, Roma, Italy.

Runcorn, S.K., 1956. Paleomagnetic comparisons between Europe and North America. *Proceedings of the Geological Association of Canada*, 8, pp. 77–85.

Rutherford, E. and Soddy, F., 1902. The cause and nature of radioactivity. *The Philosophical Magazine*, 6(4), pp. 569–585.

Scott, D.H. and Carr, M.H., 1978. *Geological Map of Mars*, Miscellaneous Investigations Series, Map I-1083. United States Geological Survey, Washington D.C.

Scrope, G.P., 1825. *Consideration on Volcanoes*; the probable causes of their phenomena, the laws which determine their march, the disposition of their products, and their connection with the present state and past history of the globe; leading to the establishment of a new theory of the earth. W. Phillips, London, 470 pp.

Shirey, S.B., 1991. The Rb–Sr, Sm–Nd and Re–Os isotope systems: A summary and comparison of their applications to cosmochronology and geochronology of igneous rocks. In Heaman, L. and Ludden, J.N. (editors), *Applications of Radiogenic Isotope Systems to Problems in Geology*, pp. 103–162. Mineralogical Association of Canada, Toronto, Short Course Handbook, Volume 19, 498 pp.

Sleep, N.H., 1994. Martian plate tectonics. *Journal of Geophysical Research*, 99 (E3), pp. 5639–5655.

Sørensen, H. (editor), 1974. *The Alkaline Rocks*. John Wiley and Sons, London, 622 pp.

Sykes, L.R., 1967. Mechanism of earthquakes and nature of faulting on the mid-oceanic ridges. *Journal of Geophysical Research*, 72, pp. 2131–2153.

Talcott, R., 1994. The Moon comes into focus. *Astronomy*, 22(9), pp. 42–47.

Taylor, F.B., 1910. Bearing of the Tertiary mountain belt on the origin of the Earth's plan. *Bulletin of the Geological Society of America*, 21, pp. 179–226.

Taylor, S.R., 1992. *Solar System Evolution: A New Perspective*. Cambridge University Press, Cambridge, 307 pp.

Tilley, C.E., 1950. Some aspects of magmatic evolution. *Quarterly Journal of the Geological Society, London*, 106, pp. 37–61.

Tilling, R.I. and Dvorak, J.J., 1993. Anatomy of a basaltic volcano. *Nature*, 363, pp. 125–133.

Turner, F.J. and Verhoogen, J., 1951. *Igneous and Metamorphic Petrology*. McGraw-Hill Book Company, Inc., New York, 602 pp.

Tuttle, O.F. and Bowen, N.L., 1958. Origin of granite in the light of experimental studies in the system $NaAlSi_3O_8$–$KAlSi_3O_8$–SiO_2–H_2O. *The Geological Society of America, Memoir*, 74, 153 pp.

Viljoen M.J. and Viljoen, R.P., 1969. Evidence for the existence of a mobile extrusive peridotite magma from the Komati Formation of the Onverwacht Group. *The Geological Society of South Africa, Special Publication*, 2, pp. 87–112.

Vine, F.J. and Matthews, D.H., 1963. Magnetic anomalies over oceanic ridges. *Nature*, 199, pp. 947–949.

Vitaliano, D.B., 1976. *Legends of the Earth: Their Geologic Origins*. The Citadel Press, Secaucus, New Jersey, 305 pp.

von Leonhard, K.C., 1823–1824. *Charakteristik der Felsarten* 2 Volumes. J. Engelmann, Heidelberg.

Wadati, K., 1928. Shallow and deep earthquakes. *Geophysics Magazine, Tokyo*, 1, pp. 102–202.

Wager, L.R. and Brown, G.M., 1968. *Layered Igneous Rocks*. Oliver and Boyd, Edinburgh, 588 pp.

Washington, H.S., 1906. *The Roman Comagmatic Region*. Carnegie Institution of Washington, Publication Number 57, 140 pp.

Wasserburg, G.J., 1987. Isotopic abundances: inferences on solar system and planetary evolution. *Earth and Planetary Science Letters*, 86, pp. 129–173.

Wedepohl, K.H., 1994. The composition of the continental crust. *Mineralogical Magazine*, 58A, pp. 959–960.

Wegener, A.L., 1912. Die entstehung der kontinente. *Petermanns geographische Mitteilungen*, 58, pp. 185–195.

Wegener, A.L., 1915. *Die Entstehung der Kontinente und Ozeane*. F. Vieweg, Braunschweig, 144 pp.

Werner, A.G., 1789. '*Kurze Klassifikation*'; a facsimile of the original text, and translation into English by Ospovat, A.M. 1971. *Short Classification and Description of the Various Rocks*. Hafner Publishing Company, New York, 194 pp.

White, R.S. and Humphreys, C.J., 1994. Famines and cataclysmic volcanism. *Geology Today*, September–October 1994, pp. 181–185.

Wiechert, E., 1897. *Ueber die Massenverheilungim Innern der Erde*. Naturforschende Gesellschaft Wissenschaften, Göttingen, pp. 221–243.

Wilson, J.T., 1965. A new class of faults and their bearing on continental drift. *Nature*, 207(4995), pp. 343–347.

Wyllie, P.J. (editor), 1967. *Ultramafic and Related Rocks*. John Wiley and Sons, New York, 464 pp.

Wyllie, P.J., 1983. Experimental studies of biotite- and muscovite-granites and some crustal magmatic sources. In Atherton, M.P. and Gribble, C.D. (editors), *Migmatites, melting and metamorphism*, pp. 12–26. Shiva Publishing Limited, Nantwich, UK, 326 pp.

Yoder, H.S., 1976. *Generation of Basaltic Magmas*. National Academy of Sciences, Washington, D.C., 265 pp.

Yoder, H.S. (editor), 1979. *The Evolution of Igneous Rocks: Fiftieth Anniversary Perspectives*. Princeton University Press, Princeton, New Jersey, 588 pp.

Yoder, H.S., 1992. Norman L. Bowen (1887–1956), MIT class of 1912, first predoctoral fellow of the Geophysical Laboratory. *Earth Sciences History*, 11(1), pp. 45–55.

Yoder, H.S., 1993. Timetable of petrology. *Journal of Geological Education*, 41, pp. 447–487.

Yoder, H.S. and Tilley, C.E., 1962. Origin of basaltic magmas: An experimental study of natural and synthetic rock systems. *Journal of Petrology*, 3(3), pp. 342–532.

Zirkel, F., 1866. *Lehrbuch der Petrographie*, 2 Volumes. Marcus, Bonn, 635 pp.

CHAPTER 2

Classifying materials in the magma systems

Whether we regard it as a duty or a pleasure, our task is to search for order and sometimes to invent it when we cannot find it...

Chayes, 1949, p. 239.

2.1 Naming rocks and magmas

Rocks are essential because they provide the raw materials used in constructing the buildings, transport systems and artifacts of an industrialised society. They are also important because they contain information about the major events that happened in our neighbourhood in the cosmos during the last six billion years. Rock classifications are needed to describe and aid in the discovery of new materials. They also spur on petrologists in their quest for information about the continuing evolution of our planet and the Solar System.

Before one can evaluate the magma/igneous rock system of Earth, or set out to find new information about the early Solar System, one has to formulate a consistent method for naming magmas and magmatic rocks. If such a classification is to be effective, it must be comprehensive, because rocks or magmas that are rare in one physical environment may be abundant and important in another. For example, magmatic materials that are seldom found on the surface of the Earth are considered to hold the key to understanding magmatic processes in the deep mantle. Rock classifications also must be comprehensive if they are to name magmas and rocks from extraterrestrial sources.

Classification is implicit in language. When one uses a common noun, or a name for a group (such as a troop of monkeys, or an exaltation of larks) one acknowledges that one is dealing with a class of objects. In science classification helps promote the search for order in a particular system, or in the whole cosmos. An ideal classification of magmas and magmatic rocks would group these materials in an order that directs attention to petrogenetic relationships between individual rocks and magmas, and also between larger comagmatic groups.

Igneous petrography lacks formal rules to regulate the naming process (cf. Tomkeieff *et al.*, 1983; Mitchell, 1985; Larouziere, 1989; Le Maitre *et al.*, 1989). In the first century AD Pliny the Elder used the rock names basalt, obsidian, syenite and tephrite. Basalt is also a term of great antiquity but its derivation is uncertain. It possibly comes from the ancient Egyptian word for a hard, dark rock. Obsidian was named after the Roman consul, Obsidius, who collected some natural volcanic glass from Ethiopia. Syenite was named after the town of Syene (Asswan) in Egypt. Tephrite is derived from the Greek word *tephra* meaning ash. The term basanite is probably derived from the same Egyptian word as basalt, but the direct antecedent of this term is the Greek word *basanos* meaning touchstone. A touchstone was usually a dark rock used to test the purity of alloys of gold or silver. The streak left by a sample on a touchstone was compared with that of an alloy of known composition.

Diorite comes from the Greek word *diorizein* meaning to distinguish. This is because the essential minerals in this rock are usually easy to distinguish. Phonolite is derived from two Greek words, *phone* meaning sound and *lithos* meaning rock. They were supposed to make a ringing sound when struck with a hammer. Rhyolite is derived from the Greek word *rheo* meaning to flow. Many rhyolites display a fluidal or flow texture. Trachyte is derived from the Greek word *trachys* meaning rough to the touch. The name of the common plutonic rock granite is at least 400 years old. It probably comes from the Latin word *granum* meaning a grain.

Many common rocks are named after the places where they were originally described. Examples include andesite (the Andes mountains), benmoreite (Ben More on the island of Mull, Scotland), dacite (Dacia – the old Roman name for Transylvania, Rumania), gabbro (Gabbro – a village in Tuscany, Italy), hawaiite (Hawaiian islands), komatiite (Komati River near Barberton, South Africa), lherzolite (Lake Lherz, Pyrenees, France), monzonite (Monte Monzoni in the Tyrolean Alps, Italy), mugearite (Mugeary on the island of Skye, Scotland), tholeiite (Tholei, Saarland, Germany) and tonalite (Alpe de Tonale in the Tyrolean Alps, Italy). Compound or compressed names like basaltic andesite, basaltic trachyandesite, phonotephrite, picrobasalt, tephriphonolite, trachyandesite, trachybasalt and trachydacite, are usually self-explanatory (see Figs 2.6 and 2.7).

It is difficult to reform igneous nomenclature because the names of many common rocks, such as basalt, gabbro, granite and syenite, have been used and abused for such a long time. Many of these names are held in such reverence that they dominate most petrographic classifications. Early this century Shand (1917) introduced an innovative system of petrography. It dispensed with the traditional names and replaced them with a series of symbols that accurately depicted the main characteristics of the rocks. This system is both powerful and coherent, but it has never been popular, and it is now seldom used. Igneous petrography is thus in many ways a hostage to the way in which it developed historically.

The International Union of Geological Sciences Subcommission on the Systematics of Igneous Rocks has spent over 20 years trying to develop a rational igneous rock nomenclature. Their cogitations and recommendations provide a valuable resource, yet their system of classification still requires some fine tuning. In the foreword to the Subcommission's book (Le Maitre et al., 1989) the original chairperson of the Subcommission recounts the highlights of 20 years of consultation, negotiation and debate. He records that these deliberations initially focused on developing an internationally acceptable classification of the plutonic rocks because these rocks were considered easier to classify than the volcanic rocks (Streckeisen in Le Maitre et al., 1989, p. vii). By following this procedure, the discussions and recommendations of the Subcommission developed a predisposition towards giving priority to the use of modal data. One of their formal recommendations (p. 3) is that 'the primary classification of igneous rocks should be based on their mineral content or mode'. The Subcommission's modal classification is now widely used in the study of holocrystalline rocks (see Fig. 2.2).

A compelling reason for using modal classifications is that the common igneous rocks contain a limited number of phases (minerals + glass or glasses). Goldschmidt's mineralogical phase rule explains why this is so. This rule was first systematised in Goldschmidt's doctoral thesis of 1911 (Mason, 1992, pp. 14–17). It states that in a rock that crystallises under equilibrium conditions the number of phases (minerals) will not exceed the number of independent chemical components. Igneous rocks normally contain eight components, or major elements (O, Si, Al, Fe, Mg, Ca, Na and K), and eight, or fewer, phases. An important reason why igneous rocks usually contain fewer than eight minerals is that many common rock-forming minerals are isomorphous mixtures. Their compositions vary between two or more end-members. It is interesting to record that the various rocks found in carbonatite–nephelinite complexes frequently evolve from a series of complex magmas that contained large numbers of different chemical components. As one might foresee the rare rocks that form from these magmas often contain a bewildering number of different minerals (Heinrich, 1966, pp. 157–160; see Chapter 12).

2.2   Minerals and isomorphism

Before continuing our appraisal of the merits of using modal data in the systematic classification of rocks, we will briefly review the characteristics of the common rock-forming minerals (cf. Deer et al., 1992). According to Wedepohl (1969, p. 248) the average mineral composition of the upper continental crust of the Earth is plagioclase 41 per cent, potassium feldspar 21 per cent, quartz 21 per cent, amphibole 6 per cent, biotite 4 per cent, pyroxene 4 per cent, iron-titanium oxides 2 per cent, olivine 0.6 per cent and fluorapatite 0.5 per cent. From data on the average chemical composition of the oceanic crust (Taylor, 1992, p. 202) it is estimated that this layer is likely to have the following average mineral composition: plagioclase 55 per cent, pyroxene 27 per cent, olivine 15 per cent and iron-titanium oxides 3 per cent. It is thus evident that the most abundant rock-forming minerals in the accessible upper crust of the Earth are the feldspars, particularly the plagioclase feldspars, the pyroxenes, quartz, olivine, the amphiboles,

Table 2.1 The common rock-forming minerals. After Clark, 1993, pp. 789–838.

Sulphides			
Galena	PbS		
Pyrite	FeS_2		
Oxides			
Quartz	SiO_2		
Ilmenite	$FeTiO_3$		
Chromite	$Fe^{2+}Cr_2O_4$		
Magnetite	$Fe^{2+}Fe^{3+}{}_2O_4$		
Hematite	Fe_2O_3		
Carbonates			
Calcite	$CaCO_3$		
Dolomite	$CaMg(CO_3)_2$		

Silicates not containing aluminium

Silicates of magnesium
Forsterite Mg_2SiO_4 *(Olivine)*
Enstatite $Mg_2Si_2O_6$ *(Pyroxene)*
Silicates of calcium with magnesium
Diopside $MgCaSi_2O_6$ *(Pyroxene)*
Silicates of zirconium
Zircon $ZrSiO_4$
Silicates of iron
Fayalite $Fe^{2+}{}_2SiO_4$ *(Olivine)*
Ferrosilite $Fe^{2+}{}_2Si_2O_6$ *(Pyroxene)*
Silicates of iron and alkali metals
Aegirine $NeFe^{3+}Si_2O_6$ *(Pyroxene)*
Riebeckite $Na_2(Fe^{2+},Mn,Mg)_5Si_8O_{22}(OH)_2$
    (Amphibole)
Silicates of iron and magnesium
Olivine $(Mg,Fe)SiO_4$
Silicates of iron and calcium
Hedenbergite $CaFe^{2+}Si_2O_6$ *(Pyroxene)*

Silicates of iron and alkalis
Arfvedsonite $Na_3Fe^{2+}{}_4Fe^{3+}Si_8O_{22}(OH)_2$ *(Amphibole)*

Silicates containing aluminium and other metals
Aluminosilicates of sodium
Albite $NaAlSi_3O_8$ *(Feldspar)*
Aluminosilicates of potassium:
Kalsilite $KAlSiO_4$ *(Feldspathoid)*
Leucite $KAlSi_2O_6$ *(Feldspathoid)*
Microcline $KAlSi_3O_8$ *(Feldspar)*
Orthoclase $KAlSi_3O_8$ *(Feldspar)*
Sanidine $KAlSi_3O_8$ *(Feldspar)*
Muscovite $KAl_2(Si_3Al)O_{10}(OH)_2$ *(Mica)*
Aluminosilicates of sodium and potassium
Nepheline $(Na,K)AlSiO_4$ *(Feldspathoid)*
Anorthoclase $(Na,K)AlSi_3O_8$ *(Feldspar)*
Aluminosilicates of magnesium and alkalis
Phlogopite $KMg_3AlSi_3O_{10}(F,OH)_2$ *(Mica)*
Aluminosilicates of calcium
Anorthite $CaAl_2Si_2O_8$ *(Feldspar)*
Tschermakite $Ca_2(Mg,Fe^{2+})_3Al_2(Si_6Al_{12})O_{22}(OH)_2$
    (Amphibole)
Edenite $NaCa_2(Mg,Fe^{2+})_5Si_7AlO_{22}(OH)_2$
    (Amphibole)
Pargasite $NaCa_2(Mg,Fe^{2+})_4Al(Si_6Al_2)O_{22}(OH)_2$
    (Amphibole)
Aluminosilicates of iron, magnesium and alkalis
Biotite $K(Mg,Fe^{2+})_3(Al,Fe^{3+})Si_3O_{10}(OH,F)_2$ *(Mica)*
Aluminosilicates of iron, calcium and magnesium
Augite $(Ca,Mg,Fe)_2Si_2O_6$ (Pyroxene)
Hornblende $Ca_2(Mg,Fe^{2+})_4AlSi_7AlO_{22}(OH)_2$ *(Amphibole)*
Aluminosilicates of iron, magnesium, calcium and alkalis
Aegirine-augite $(Na,Ca)(Fe^{3+},Fe^{2+},Mg,Al)Si_2O_6$ *(Pyroxene)*

the micas and the iron-titanium oxides (see Table 2.1).

Minerals are naturally occurring inorganic solids that contain atoms arranged in an orderly repetitive manner. Groups of minerals with analogous formulas, and ions of similar sizes, often have closely related crystal structures and are said to be isomorphous. Isomorphism is important in the classification of the amphiboles, feldspars, garnets, olivines, pyroxenes and spinels. These minerals contain a few distinct structural sites. Each site is suitable for the entry of ions of the appropriate size and charge. This plainly means that ions of more than one element can enter a particular structural site.

In the olivine group there is continuous chemical and physical variation from forsterite, Mg_2SiO_4, to fayalite, Fe_2SiO_4. An olivine from a common basalt may have the formula $Mg_{1.7}Fe_{0.3}SiO_4$. This implies that Mg^{2+} (ionic radius $= 0.066$ nm) and Fe^{2+} (ionic radius $= 0.074$ nm) that have broadly similar ionic radii and the same charge can substitute for each other. The plagioclase feldspars belong to a more complex isomorphous series in which there is continuous substitution of (Na^+ plus Si^{4+}) for (Ca^{2+} plus Al^{3+}) as one proceeds from the high-temperature anorthite, $CaAl_2Si_2O_8$, to lower temperature albite, $NaAlSi_3O_8$. Bowen (1913) was first to publish a phase equilibrium diagram of the plagioclase feldspars. This diagram shows how these minerals respond under anhydrous conditions. When water is added the shape of the liquidus and solidus curves remains much the same, but they are both displaced to lower temperatures. If the system albite–anothite–quartz–water is investigated the positions of the curves are again changed in response to the change in the bulk chemical composition of the system.

It is convenient to discuss the composition of an isomorphous series in terms of one of its end-members. The end-members of the plagioclase feldspars series are albite and anorthite, and the

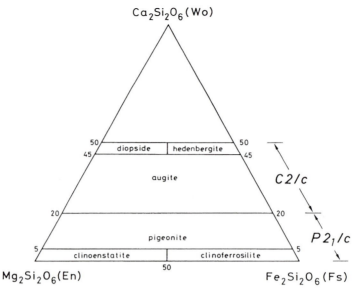

Fig. 2.1 Compositional ranges of the common (Ca-Mg-Fe) clinopyroxenes and orthopyroxenes. After Morimoto et al., 1988, p. 542.

series is subdivided in the following way: albite An_{00-10}, oligoclase An_{10-30}, andesine An_{30-50}, labradorite An_{50-70}, bytownite An_{70-90} and anorthite An_{90-100}. Minerals in the olivine group are usually named by referring to the magnesium-rich end-member. A typical olivine from an alkali basalt might be described as Fo_{72} (forsterite seventy-two). The common pyroxenes are usually described in terms of the end-members Wo (wollastonite, a phase that is not structurally related to the pyroxene group), En (enstatite) and Fs (ferrosilite). A typical augite from a Hawaiian tholeiitic basalt might be described as $Wo_{40.3}En_{49.3}Fs_{10.4}$. Fig. 2.1 shows that the common pyroxenes belong to two main groups. They are the orthopyroxenes with orthorhombic symmetry and the clinopyroxenes with monoclinic symmetry. The orthopyroxenes form a simple isomorphous series that ranges in composition between magnesium-rich enstatite and iron-rich ferrosilite. Fig. 2.1 and Table 2.1 both reveal that the clinopyroxenes have a more complex chemical composition. Augite is the clinopyroxene typically found in basaltic rocks.

2.3 Modal compositions

To understand the implications of the decision by the Subcommission on the Systematics of Igneous Rocks to use modal data in their primary classification, one has to recognise that there are many difficulties inherent in carrying out petrographic modal analyses. These problems relate to extremes in grain size, the complexity of some rock textures, the presence of megacrysts, heteromorphism, and the unwillingness of many research scientists to carry out this tedious task. Chayes (1956) has shown that it is not feasible to accurately measure the modal composition of rocks of all grain sizes and textures. It is self evident that modal data cannot be collected from volcanic rocks with glassy, or cryptocrystalline, mesostases. Problems also arise in determining the modes of some coarse-grained rocks, particularly those with complex textures. Chayes (1956, p. 1) observed that the number of reliable modes obtained by measurements made on polished slabs is 'almost vanishingly small'. Many granitic rocks contain large, and often complexly zoned, alkali feldspar megacrysts, such as the famous horses-teeth feldspars from the Flamanville granite in France. In his book on the nature and origin of granites Pitcher (1993, pp. 60–77) devotes a whole chapter to the evolution of granitic textures, and shows how many common textural features seen in granites evolved through both magmatic and metasomatic stages. In modern research the trend is away from the quantitative modal analysis of coarse-grained rocks because this process requires 'tedious slabbing, messy staining, and mindless point-counting' (Clarke, 1992, p. 6). Contemporary

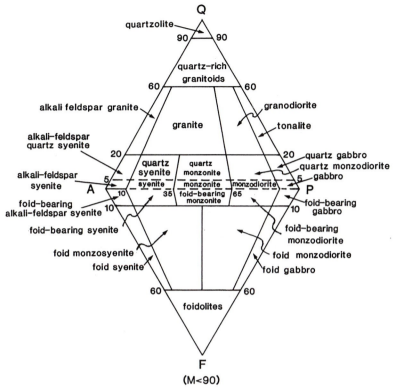

Fig. 2.2 Modal classification and nomenclature of plutonic rocks. Adapted from Streckeisen, 1976, Figure 1a.

research papers on igneous rocks are more likely to contain major element data than accurately determined modal data.

Heteromorphism

It is also difficult to use the same modal classification for volcanic and plutonic rocks because some minerals that crystallise in one class of rocks do not crystallise in the other. This is heteromorphism. It occurs when two magmas of similar chemical composition crystallise to produce rocks with different essential phases. Rittmann (1973, p. 72) has studied its influence on the modal classifications of igneous rocks. He showed that heteromorphism was likely to occur in magmatic rocks that had similar chemical compositions, but experienced different cooling histories. For example, if most of the sodium and potassium in a high-level granite is allocated to alkali feldspar (perthite), such a rock may have the modal composition of an alkali feldspar granite (see Fig. 2.2). In a

rock of similar composition emplaced at greater depths most of the sodium may join with calcium to form plagioclase and the greater part of the potassium may concentrate in biotite, thus producing a rock with the modal composition of a granodiorite. Another common and excellent example of heteromorphism is the pair andesite–diorite. The high temperature, quickly cooled, andesites typically contain pyroxenes and glass; whereas the plutonic diorites invariably contain amphibole or mica. The diorites usually contain visible quartz but in most andesites the excess of silica is concealed in the groundmass.

Heteromorphism is essentially what sets the charnockitic rock series apart from normal plutonic rocks. This series of enstatite-bearing rocks is named after the rock used as a tombstone for Job Charnock in St John's Churchyard, Calcutta, India. In the lamprophyric rocks (see Chapter 11) heteromorphism confounds all simple classifications. Rittmann (1973) and Streckeisen and Le Maitre (1979) have all tried to resolve the problem of heteromorphism in the modal classification of igneous rocks. It is likely that as long as modal classifications are used hetero-morphism will remain a formidable problem.

Occult minerals

Another problem that needs to be unravelled con-cerns igneous rocks that contain normative minerals (ideal minerals calculated from the bulk chemical composition of the rock) that differ from their modal phases. The modal minerals that appear to be missing are called occult minerals. Quartz is usually an occult mineral in andesites that contain a glassy matrix; while many monzogabbros, monzo-diorites, monzonites and syenites contain occult feld-spathoids, typically nepheline. In volcanic rocks the chemical components that go to form the occult minerals are usually present in the groundmass, whereas the occult feldspathoids are often concealed in the chemical composition of the mafic minerals. It is particularly revealing to calculate the norms of some mafic minerals that are not used in the standard CIPW normative calculations (see Section 1.2). Phlo-gopite contains 52.9 per cent normative olivine, 27.3 per cent normative leucite and 19.8 per cent norma-tive kalsilite (see Table 2.1). A typical biotite (Deer *et al.*, 1992, p. 285, No. 6) contains 35.3 per cent nor-mative leucite, 5.9 per cent normative kalsilite and 1.9 per cent normative nepheline.

2.4   Textures

Textures describe the general physical appearance or character of a rock. They are about the shape and size of the various phases found in a rock, and they also show how the various phases come together to form a rock. The study of textures is important because they usually provide valuable information about the physical conditions that existed when an igneous rock congealed (cf. MacKenzie *et al.*, 1982; McPhie *et al.*, 1993). In the past textures have been used in some classifications of igneous rocks. The common or abundant minerals are usually called the essential minerals because they are essential in classification. Other minerals that are usually present in minor amounts are called the accessory minerals. Fluorapatite $\{Ca_5(PO_4)_3F\}$, titanite ($CaTiSiO_5$), and zircon ($ZrSiO_4$) are typical accessory minerals.

A particular igneous texture is usually related to the rate and order in which the essential minerals crystallise from a magma. This in turn depends on the initial temperature, pressure, rate of cooling, composition, volatile content and viscosity of the magma. To describe the texture of an igneous rock one has to consider: (1) crystallinity; (2) granularity; (3) shape of crystals; and (4) the mutual relationships between the various phases. Crystallinity is the pro-portion of crystalline to non-crystalline material in a rock. Plutonic rocks are normally composed wholly of crystalline phases and are thus often called holocrystalline (Greek, *holos* = complete) rocks. Granularity is the term used to describe the size of the mineral grains found in a rock. Many lavas contain minute mineral grains and such mineral aggregates are called microcrystalline or crypto-crystalline. In contrast to these rocks some pegma-tites contain minerals that are over 1 m long (Rickwood, 1981). If the essential minerals can be distinguished with the unaided eye then the rock may be described in the following way: (1) when the mean diameter of the grains is less than 1 mm the rock is fine-grained; (2) when it is between 1 and 5 mm it is medium-grained; (3) when between 5 and 30 mm it is coarse-grained; and (4) when greater than 30 mm the rock is very coarse-grained.

The terms used to describe the shape of minerals are: (1) euhedral (Greek, *eu* = well), for crystals bounded by straight crystal faces; (2) anhedral (Greek, *an* = without), for crystals with irregular margins; and (3) subhedral (Latin, *sub* = under or close to) for crystals with shapes intermediate between euhedral and anhedral. Mineral shape is usually controlled by the stage at which it crystallised from a magma. When determining the order of crys-tallisation the following guidelines (Rosenbusch, 1882) may be of some value: (1) when one mineral is surrounded by another the enclosing mineral is generally younger; (2) minerals that form early are more likely to be euhedral than minerals that crystallise later; (3) if both large and small minerals occur together the larger ones probably started to crystallise first; and (4) the order of crystallisation usually follows Bowen's (1928, p. 60) reaction series.

In equigranular rocks all the essential minerals are of nearly the same size. Rocks in which these miner-als are of markedly different sizes are inequigranular. The special terms allotriomorphic (Greek, *allotrios* = foreign; *morphe* = shape), or aplitic (after the rock aplite), are used to describe anhedral equigranular textures. Rocks with subhedral equigranular textures are described as having hypidiomorphic (Greek, *hyp* = less than; *idios* = own; *morphe* = shape), or grani-tic (see Plate 2.1a) textures. The special term for euhedral equigranular textures is panidiomorphic (Greek, *pan* = all; *idios* = own; *morphe* = shape).

In the most common inequigranular texture large crystals are enveloped in a groundmass of smaller crystals, or glass. This texture is called porphyritic and it is typical of most volcanic and subvolcanic rocks (see Plate 2.1b). The large crystals are called phenocrysts (Greek, *phaeno* = conspicuous). Originally the term porphyry was used to describe a rock (*porphyrites lapis*) quarried at Djebel Dokham in Egypt. This rock contains phenocrysts of feldspar embedded in a fine reddish purple groundmass. It was named porphyry (Greek, *porphyreos* = purple) because of its purple colour. In recent times the term porphyritic has been used to describe the texture of any rock with a similar texture to the original rock from Egypt. When phenocrysts are embedded in a glassy matrix the texture is designated vitrophyric (Latin, *vitrum* = glass). Many porphyritic rocks have a felsophyric texture consisting of phenocrysts set in a groundmass of intergrown quartz and feldspar. In other rocks the phenocrysts are gathered into distinct clusters. This texture is called cumulophyric (Latin, *cumulus* = heap).

If one or more minerals appear inside a larger mineral the texture is called poikilitic (Greek, *poikilos* = spotted). The most common type of poikilitic texture arises when large crystals of a clinopyroxene, such as augite, enclose smaller laths of plagioclase. This is the ophitic (Greek, *ophis* = a serpent) texture (see Plate 2.1c). It is particularly common in medium-grained basaltic dyke and sill rocks (dolerites). The name ophitic refers to the snake-like markings on the basaltic dyke rocks, or ophites, of the Pyrenees in southwestern Europe. In some granitic rocks, including very coarse-grained pegmatites, quartz and alkali feldspar crystallise simultaneously producing a type of poikilitic texture that displays a regular pattern of quartz blebs enclosed in feldspar. Such rocks are often called graphic granites and they are said to have graphic (Greek, *grapho* = I write) or micrographic textures. When a similar texture is found in basic rocks it is called myrmekitic (Greek, *myrmeke* = an ant). It typically contains intergrowths of quartz and plagioclase. In many basalts, laths of plagioclase form an open framework texture with roughly triangular spaces between the plagioclase crystals. If glass fills these spaces the texture is called intersertal (Latin, *inter* = between; *serere* = to weave), but if minerals fill the spaces the texture is called intergranular.

Lavas often display an alignment of elongate crystals. This directive texture is typical of trachytes. The various fine-grained rocks that have this texture are described as having a trachytic texture (see Plate 2.1d). A similar texture in medium-grained or coarse-grained rocks, such as nepheline syenite, is called a trachytoid texture. Some volcanic glasses contain rounded or spherical masses composed of acicular crystals that radiate from a central point. Rocks displaying these features are described as having spherulitic textures. Rhyolitic volcanic glasses with a high water content may develop concentric fractures called perlitic (German, *perl* = pearl) cracks. The various granular textures of many plutonic rocks are difficult to interpret as these rocks have usually undergone extensive recrystallisation during long periods of cooling and re-equilibration (McBirney and Hunter, 1995). The nature of these textures will be examined when considering the rocks from layered igneous complexes and the granitic rocks (see Sections 6.12 and 9.6).

2.5   Modal classifications

Streckeisen (1976 and earlier as Anon 1973) introduced the QAPF modal classification for the IUGS Subcommission on the Systematics of Igneous Rocks. It uses five components, or end members. Only four are shown on the QAPF double triangle as illustrated in Fig. 2.2. The components are Q = quartz or a high temperature form of silica, A = alkali feldspar, P = plagioclase (An05–100) and scapolite [(Na,Ca,K)$_4$[Al$_3$(Al,Si)$_3$Si$_6$O$_{24}$] (Cl,CO$_3$, SO$_4$,OH)], F = the feldspathoid group of minerals, and M = mafic and related minerals. Included in the M component are the accessory minerals, amphiboles, carbonates (primary), iron-titanium oxides, melilite group, micas, olivines and pyroxenes. The QAPF classification excludes all rocks that contain 90 per cent or more of the M component. These rocks are designated the ultramafic rocks, and are classified using the relative abundances of their mafic minerals as depicted in Figs. 2.3 and 2.4.

The colour index of a rock is the percentage of dark coloured minerals that it contains. Some petrologists use this index to define the ultramafic rocks. For example, Wyllie (1967, p. 1) defined the ultramafic rocks as having a colour index greater than 70. Irrespective of what classification one uses it is probably a sensible practice to use the term melanocratic (Greek, *melos* = black) to describe all rocks with a

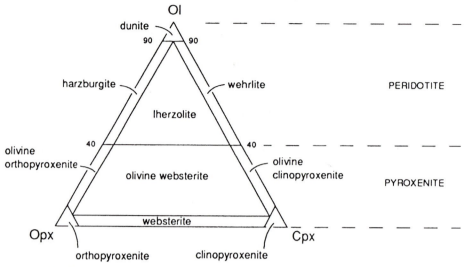

Fig. 2.3 Modal classification and nomenclature of some ultramafic rocks based on the proportions of olivine (Ol), orthopyroxene (Opx) and clinopyroxene (Cpx). After Anon (Streckeisen *et al.*), 1973, Figure 2a.

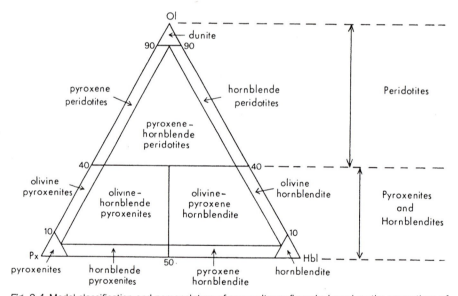

Fig. 2.4 Modal classification and nomenclature of some ultramafic rocks based on the proportions of olivine (Ol), pyroxene (Px) and amphibole (Hbl). After Anon (Streckeisen *et al.*), 1973, Figure 2b.

colour index greater than 70. Colour index is sometimes equated with the M component. This is not correct as the colour index (M') is equal to the M component minus minerals such as muscovite, apatite and the primary carbonates that are usually considered colourless minerals (Le Maitre *et al.*, 1989, p. 6). The gabbroic rocks all plot within the P corner of the QAPF double triangle. They include gabbro (Pl + Cpx +/− Ol), norite (Pl + Opx +/− Cpx), anorthosite (Pl), amphibole gabbro (Pl + Cpx +/− Am) and gabbronorite (Pl + Cpx + Opx). A systematic method of classifying these rocks is provided by Fig. 2.5. Figures 2.6 and 2.7 depict the IUGS Subcommission's modal and chemical classifications of volcanic rocks (Streckeisen, 1978, Fig. 1; Le Bas *et al.*, 1986, Fig. 2).

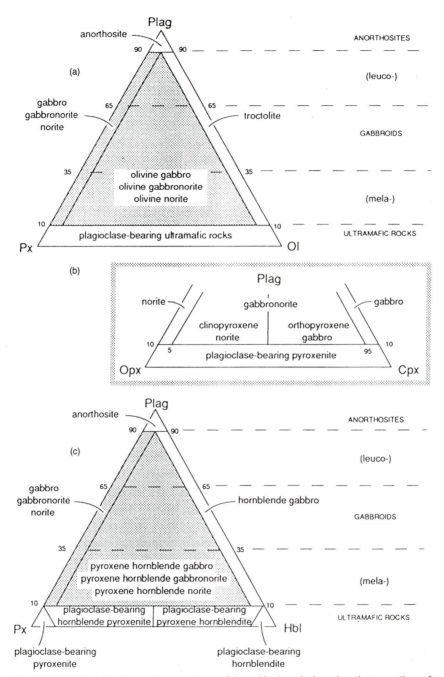

Fig. 2.5 Modal classification and nomenclature of the gabbroic rocks based on the proportions of (a) plagioclase (Pl), pyroxene (Px) and olivine (Ol), (b) plagioclase (Pl), orthopyroxene (Opx) and clinopyroxene (Cpx), and (c) plagioclase (Pl), pyroxene (Px) and amphibole (Hbl). After Streckeisen, 1976, Figure 3.

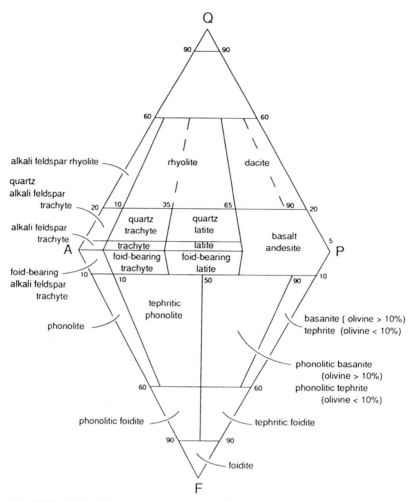

Fig. 2.6 Modal classification and nomenclature of volcanic rocks. After Streckeisen, 1978, Figure 1.

Cumulate rocks

Cumulates are magmatic rocks formed by the accumulation of crystals within a magma (Wager *et al.*, 1960, p. 73). In the simplest circumstances they are produced by the accumulation of early-formed minerals that either sink or float in their host magma. In any comprehensive classification of igneous rocks it is important to include a cumulate category because (1) they can simplify the nomenclature of many plutonic complexes and (2) they prompt petrologists to confront uncertainties about the origin of the monomineralic rocks and some picritic rocks. According to McBirney and Hunter (1995, p.

115) the term cumulate has evolved over the years, and it now usually means any coarse-grained igneous rock with a composition that does not correspond to a natural magma. The textures of cumulate rocks, like those from large plutonic bodies, are frequently difficult to interpret because they have undergone extensive recrystallisation during long periods of cooling and re-equilibration.

The cumulus concept is particularly valuable in naming the rocks found in layered intrusions, such as the Skaergaard Intrusion of East Greenland (Wager and Brown, 1968, pp. 11–244). Most cumulate rocks are either ultramafic, anorthositic, or foidolitic rocks. Feldspathoids are usually less dense than their host magmas and so tend to accumulate

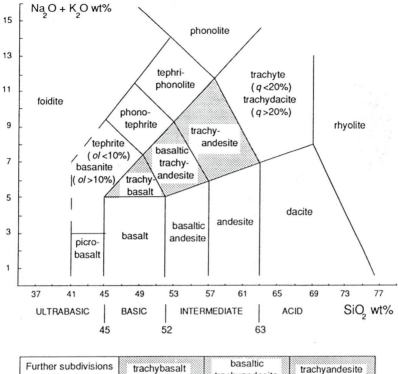

Fig. 2.7 Chemical classification and nomenclature of volcanic rocks using the total alkali versus silica (TAS) diagram. Adapted from Le Bas *et al.*, 1986, Figure 2, and Le Maitre *et al.*, 1989, p. 28.

Further subdivisions of shaded fields	trachybasalt	basaltic trachyandesite	trachyandesite
$Na_2O - 2.0 \geq K_2O$	hawaiite	mugearite	benmoreite
$Na_2O - 2.0 \leq K_2O$	potassic trachybasalt	shoshonite	latite

at the top of magma chambers. When evaluating the possible cumulate or non-cumulate origin of ultramafic rocks one is forced to ponder the question of the systematic position in a comprehensive classification of the rocks of the upper mantle. They are unlike normal plutonic rocks because of the moderately high temperatures that pervade the upper mantle and keep them in a near-magmatic, instead of a normal plutonic, milieu. In this environment their compositions change as the result of both the periodic upward removal of magma, and the gradual infiltration of volatiles and other components degassing from greater depths. Perhaps all rocks that solidify in the upper mantle should be grouped together as a special class of rocks.

2.6 Chemical diversity of magmas

Igneous rocks and magmas usually contain more than one phase. In holocrystalline rocks all the phases are minerals, in volcanic rocks one frequently finds both minerals and glass, whereas in magmas the dominant phase is usually a liquid. If one wishes to classify these materials in a similar way one is required to investigate their chemical compositions. More than 99 per cent of the common igneous rocks and magmas are made up of the major components SiO_2, TiO_2, Al_2O_3, Fe_2O_3, FeO, MnO, MgO, CaO, Na_2O, K_2O, P_2O_5, H_2O^+ and CO_2. If scatter plots of various pairs of these oxides are prepared from a

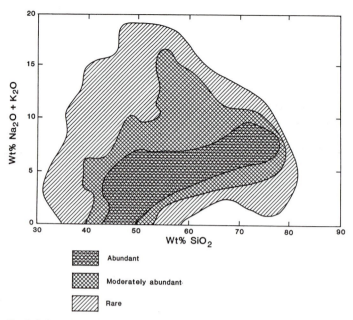

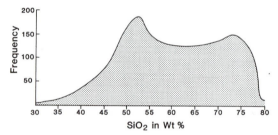

Fig. 2.8 A total alkali silica (TAS) diagram showing the relative abundance of igneous rocks. Adapted from Le Maitre, 1976, p. 599.

Fig. 2.9 Frequency distribution of silica for all igneous rocks. Adapted from the trailblazing work of Richardson and Sneesby, 1922, p. 306.

large database of igneous rocks (Le Maitre, 1976, p. 99) it is revealed that the oxides form a single, gradational series, with significant clustering of data-points in the subalkalic field that extends from basalt to rhyolite via basaltic andesite, andesite and dacite (see Fig. 2.7).

In most igneous rocks silica is the most abundant oxide by weight. Fig. 2.8 shows that in the common igneous rocks silica usually varies in abundance between 43 per cent and 80 per cent. When considering these data it is important to remember that silicon makes up only 0.2 per cent of the volume of the Earth's crust, whereas oxygen makes up 92 per cent of the volume. In 1922 Richardson and Sneesby cri-

tically examined Washington's (1917) major element database of selected analyses of igneous rocks and found that silica had a bimodal frequency distribution with maxima at 52.5 per cent and 73.0 per cent (see Fig. 2.9). When evaluating their figure one must recognise that Washington's database was biased. This was mainly because it lacked data on mid-oceanic ridge basalts. If this bias is corrected the new data shifts the primary maximum to a lower silica value in the basic field. Chayes (1975 and 1979) constructed a silica frequency distribution curve for the Cenozoic volcanic rocks. This curve is unimodal, with a strong positive skew (see Fig. 2.10). Enormous databases, such as IGBADAT, are now readily available to petrologists who wish to pursue what Chayes (1979) labelled electronic computation and bookkeeping in igneous petrology.

When the abundance data for plutonic rocks (Le Maitre, 1976) are examined one discovers that over half contain more than 63 per cent silica. This supports Daly's (1914 and 1933) pioneering research into the abundance of igneous rocks in the accessible parts of the Earth. He estimated that basalt was the most abundant volcanic rock, and granite (sensu lato) was the most abundant plutonic rock. A simple silica frequency distribution curve is misleading, because (1) it fails to separate the different sets of

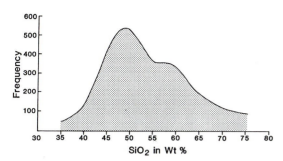

Fig. 2.10 Frequency distribution of silica for over 10 000 chemical analyses of Cenozoic volcanic rocks. Adapted from Chayes, 1979, p. 18.

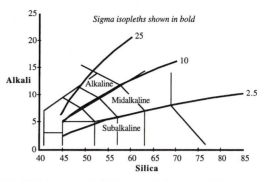

Fig. 2.11 A total alkali silica diagram showing the areas occupied by the rocks of the subalkaline, midalkaline and alkaline magmatic lineages. This diagram also shows sigma isopleths. Sigma is an index devised by Rittmann (1957, p. 34) and is equal to $(Na_2O+K_2O)^2/(SiO_2-43)$.

data from different magmatic lineages (see Fig. 2.11), and (2) silica by itself does not provide an evenly spaced measure of magmatic differentiation within the various lineages. In the basic field (45–52 per cent silica) one has to visualise the basaltic, trachy-basaltic and phonotephritic rocks, with some common cumulate rocks (e.g. anorthosites and pyroxenites) as all being stacked on top of one another. At higher silica values the data for the andesites, boninites, icelandites, trachyandesites and phonolites are combined if one only considers their silica values.

In a comprehensive evaluation of the relative abundance of the different types of igneous rocks one has to consider whether abundance has changed through time. Theoretically one would expect change because magmatic and appendant processes are directly related to heat and its transference and dissipation. In the Hadean, seas of lava-magma may have been common. According to Abbott *et al.* (1994, p. 13 847) the mean upper mantle temperature changed from 1437°C ± 40°C in the middle Archean to 1272°C ± 7°C in the Phanerozoic. During all geologic time some magmatic materials have been generated at higher than normal temperatures. These materials were probably generated by hot spot activity. In the middle Archean hot spot activity was probably responsible for the eruption of the non-cumulate komatiitic lavas of the Barberton Greenstone Belt in northeastern South Africa. These lavas contain up to 29 per cent MgO and they probably erupted at temperatures of about 1580°C (Nisbet *et al.*, 1993, p. 305). Many geochemical and tectonophysical changes are reported to have occurred during the transition from the Archeozoic to the Proterozoic (Condie, 1989, p. 14). The Cenozoic is noted for its particularly high abundance of alkaline rocks (see Fig. 2.11).

Condie (1989, p. 2) has suggested that alkaline rocks were uncommon before 1.0 giga-years, and rare, or absent, prior to 2.5 giga-years.

2.7 TAS diagrams and sigma isopleths

The total alkali versus silica, or TAS, variation diagram provides a uniform and apparently simple method of using major element abundance data to classify magmas and magmatic rocks (Le Maitre *et al.*, 1989, p. 28; Fig. 2.7). In the past this diagram has been used to illustrate a wide range of petrogenetic trends (e.g. Tomkeieff, 1937 and 1953; Tilley, 1950; Kuno *et al.*, 1957; Rittmann, 1962 and Saggerson and Williams, 1964). In 1984 the IUGS Subcommission on the Systematics of Igneous Rocks used a TAS diagram as the foundation of their chemical classification of volcanic rocks (Fig. 2.7). The idea of using this type of diagram in the classification of volcanic rocks is not new, nor can it be regarded as the ultimate classification of all volcanic rocks. It does not readily separate the various types of basaltic rocks that are frequently discussed in the petrological literature.

The Subcommission used the TAS diagram to define basalt (sensu lato) and then employed the presence or absence of normative nepheline to decide whether the rock was an alkali or subalkali basalt. In the systematic classification of basalts it is probably just as important to acknowledge the presence of normative quartz in the silica-oversaturated, subalkali basalts. It is usually important to establish whether the subalkali basalts are: (1) high-Al varieties, such as the calc-alkali basalts of island arcs or continental arcs (Nockolds and Le Bas 1977, p. 312); (2) low-K varieties, such as normal mid-oceanic ridge basalts; (3) high-Mg varieties where MgO is greater than 18 per cent; and (4) the ordinary subalkalic basalts, such as the continental flood basalts that usually lack the special chemical features characteristic of the first and third types of basalt.

Fig. 2.11 is a TAS diagram traversed by lines of constant sigma value, or sigma isopleths. Sigma is an index devised by Rittmann (1957, p. 34). It is equal to $(Na_2O + K_2O)^2/(SiO_2 - 43)$, and it provides an elegant way of defining the principal igneous lineages on a TAS diagram. Rocks of the subalkaline suite have sigma values of less than 2.5; sigma values of between 2.5 and 10 define the midalkaline suite; whereas sigma values in the range 10 to 25 separate the alkaline suite. Fig. 2.11 illustrates how the sigma five isopleth can separate the magmatic trends that normally evolve towards rhyolite from those that normally evolve towards phonolite. Normal foiditic rocks, particularly the nephelinites (Le Bas *et al.*, 1992, p. 18), occupy an area of high sigma values (sigma 25 to ∞) and a contiguous area in which sigma values are variable but negative.

A large part of the negative sigma region is occupied by rocks enriched in non-silicate phases (e.g. chromitite or carbonatite). Perhaps igneous rocks that contain either negative sigma values or values greater than 25 should be regarded as special rocks with chemical compositions that set them apart from the normal igneous rocks. The consequences of installing sigma isopleths on TAS diagrams are explored in Fig. 2.11. Sigma isopleths 2.5, 10 and 25 are perceived as tracing the boundary between the principal magmatic lineages used in the TAS classification of volcanic rocks. Fig. 2.12 illustrates both positive and negative sigma isopleths. It also shows the relationship between sigma isopleths and the field occupied by the foiditic rocks. The latter occupy a special position on the TAS diagram. Some are derived from unique primary magmas, whereas others contain accumulations of alkali-rich feldspathoids.

In the preparation of TAS diagrams attention is focussed on SiO_2, Na_2O and K_2O and all other major element information is discarded. The use of these diagrams should be regarded as only the first stage in processing and interpreting the major element abundance data on magmas and magmatic rocks. This initial processing is important as it

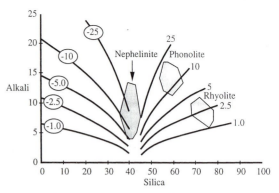

Fig. 2.12 A total alkali versus silica diagram showing both positive and negative sigma isopleths.

enables magmas and rocks to be given provisional names, and perhaps more importantly installed in an appropriate magmatic lineage. Particular magmatic lineages are usually easier to define when their compositions plot outside the congested basic field (silica = 45 per cent to 52 per cent). Le Maitre *et al.* (1989, p. 55) acknowledged this phenomenon when they defined the calc–alkali basalts as those basalts 'associated with rocks of the basalt-andesite-dacite suite of the orogenic belts and island arcs'. There is a need to establish a method of tracing magmatic lineages through the basic field. This is difficult because the depiction of magmatic differentiation is compressed in this area of the TAS diagram. Rocks as diverse as komatiites, lherzolites, peridotites, picrites and the common cumulate rocks all plot alongside the various types of basalt and gabbro. In Fig. 2.11 the subalkaline field is defined by using the 2.5 sigma isopleth. This boundary bisects the basalt field. If it is used to define the subalkalic basalts it ipso facto sets a lower sigma value for the basalts of the midalkaline suite. Perhaps the term midalkaline basalt should be used in petrology. It would include all basalts and trachybasalts that plot between the sigma isopleths of 2.5 and 10. The midalkaline suite is epitomised by trachyte (see Section 10.1).

Fig. 2.13 is a TAS diagram that shows the approximate location of selective normative quartz and normative nepheline isopleths for normal igneous rocks. The isopleths divide the basalts into three groups, the silica oversaturated (quartz normative), silica saturated and the silica undersaturated (foid normative). In 1984 Le Maitre (p. 250) found that 40.4 per cent of the basalts in his large database were oversaturated, 48.2 per cent saturated and only 11.4 per cent

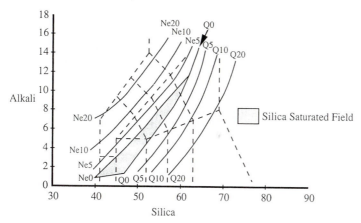

Fig. 2.13 A total alkali silica diagram showing the distribution of typical normative values for quartz and nepheline in the main sequence of igneous rocks.

undersaturated. When Figs. 2.13 and 2.11 are compared one discovers that the silica oversaturated
basalts have sigma values of less than 2.5 and are
subalkalic. All the silica undersaturated basaltic
rocks have sigma values greater than 2.5 and usually
greater than 5. The latter rocks are also alkali basalts
as they contain normative feldspathoids. Most saturated basalts plot on or near the 2.5 isopleth and
usually contain more normative hypersthene than
olivine. Mid-oceanic ridge basalts are typical of this
group. The few saturated basalts that have higher
sigma values usually contain more olivine than
hypersthene. In the past these rocks have usually
been called transitional basalts. In the field they typically crop out in close association with midalkaline
intermediate rocks.

Traditionally the common subalkalic rocks are
divided into tholeiitic and calc-alkalic lineages, but
before one can compare individual rocks from different lineages one has to factor in an index that can
account for changes in major element composition
due to magmatic differentiation. This index is
required because major element abundances are
related to both the chemical composition of a particular lineage, and the amount by which the composition of a particular sample has been modified by
magmatic differentiation. In the subalkalic basalt
field silica is a poor index of differentiation because
during this stage of differentiation the removal of the
early-formed crystals usually results in large
decreases in MgO, but only a minor change in
SiO_2 abundance (see Section 2.8). Magnesium
oxide, or an index that incorporates this oxide, is

usually used to quantify the amount of differentiation experienced by a particular tholeiitic basalt. In
Figs. 2.14 and 2.15 (SiO_2–MgO) is used as an index
of differentiation. This index is normally positive.
Negative values are only encountered in some rare
non-silicate rocks. Figure 2.14 is a (SiO_2–MgO) versus Al_2O_3/TiO_2 diagram and it provides a simple
method of separating subalkalic basic and intermediate rocks into tholeiitic and calc-alkalic lineages.
Figure 2.15 enables one to separate the normal
MOR basalts from the transitional basalts. In 1972
Chayes (p. 291) revealed that if the MOR basalts are
excluded most of the other basalts of Cenozoic age
are either oversaturated or undersaturated in silica.
If plotted on a Q-Di-Hy-Ol-Ne diagram the data-
points for these basalts cluster around two maxima
centred at $Q_{10}Di_{45}Hy_{45}$ and $Ne_{20}Di_{50}Ol_{30}$ as
depicted in Fig. 2.16.

TAS classification of plutonic rocks

Perhaps the classification of plutonic rocks should be
approached in a fresh way. One could use the successful TAS classification of volcanic rocks to
attempt to classify the plutonic rocks (Middlemost,
1994). If this is done one finds the following pairs of
equivalent rocks: (1) gabbro and basalt; (2) diorite
and andesite; (3) granodiorite and dacite; (4) foid
syenite and phonolite; (5) syenite and trachyte; and
(6) granite and rhyolite. Tonalite is particularly interesting as it plots astride the 63 per cent silica boundary on the conventional TAS diagram. One can either

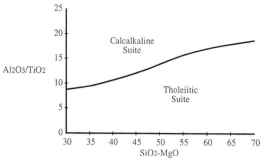

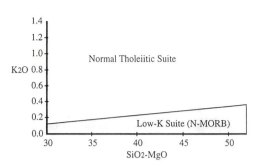

Fig. 2.14 Al$_2$O$_3$/TiO$_2$ versus SiO$_2$–MgO diagram used to separate most subalkalic basic and intermediate rocks into tholeiitic and calc-alkaline lineages.

Fig. 2.15 AlK$_2$O versus (SiO$_2$–MgO) diagram used to separate the low-K suite from the normal tholeiitic suite.

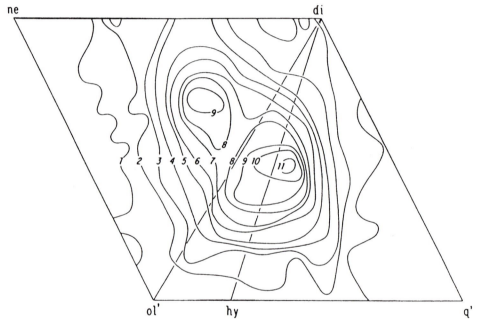

Fig. 2.16 A Q-Di-Hy-Ol-Ne diagram showing the bimodal abundance of Cenozoic basalts, excluding the mid-oceanic ridge (MOR) basalts. Adapted from Chayes (1972, p. 291).

make a special compartment for tonalite between 61 per cent and 65 per cent silica, or one can omit it, and let the classification pass directly from diorite to granodiorite. In the plutonic rocks there is no direct equivalent of basaltic andesite. It would be apposite to call such a plutonic rock gabbroic diorite. The names suggested for the plutonic rocks that occupy the same fields on the TAS diagram as trachybasalt, basaltic trachyandesite and trachyandesite, are monzogabbro, monzodiorite and monzonite, respectively (see Fig. 2.17). In the silica undersaturated fields the

plutonic equivalents of tephrite, phonotephrite and tephriphonolite would be foid gabbro, foid monzodiorite and foid monzosyenite. Many plutonic rocks of the peridotite group (e.g. the harzburgites) occupy the picrobasalt field, but some overlap into the low alkali metals part of the basalt field (e.g. the lherzolites), or the low silica side of the picrobasalt field (e.g. the dunites) (Table 2.2). It is proposed to apply the name picrogabbro to the plutonic rocks that occupy the same field as the picrobasalts in the volcanic classification. Peridotite may be regarded as

Table 2.2 Major element compositions of dunite, harzburgite, komatiite, lherzolite, peridotite and picrite.

	Dunite	Harzburgite	Lherzolite	Peridotite	Komatiite	Picrite	Picrobasalt
SiO_2	41.04	43.73	45.43	44.51	45.98	47.43	43.55
TiO_2	0.1	0.28	0.45	0.66	0.32	1.66	2.03
Al_2O_3	1.95	2.57	4.39	4.46	6.58	9.1	10.72
Fe_2O_3	3.85	6	5.15	3.8	1.69	2.67	4.31
FeO	10.05	7.09	7.44	6.93	9.18	9.09	8.58
MnO	0.76	0.16	0.18	0.43	0.19	0.17	0.21
MgO	40.66	36.34	30.31	32.9	28.98	20.07	18.08
CaO	1.08	3.18	5.68	5.32	6.38	7.75	10.24
Na_2O	0.21	0.34	0.59	0.52	0.57	1.55	1.44
K_2O	0.09	0.15	0.27	0.36	0.1	0.33	0.51
P_2O_5	0.21	0.14	0.12	0.11	0.02	0.19	0.33
Total	100	100	100	100	100	100	100
H_2O^+	4.59	4	1.07	3.91	–	0.28	–
H_2O^-	0.25	0.24	0.03	0.31	–	–	–
CO_2	0.43	0.09	0.08	0.3	–	0.06	–
	Le Maitre, 1976, p. 637	Le Maitre, 1976, p. 636	Le Maitre, 1976, p. 634	Le Maitre, 1976, p. 635	BVSP, 1981, pp. 14–15	BVSP, 1981, pp. 166–167	Le Maitre, 1984, p. 250

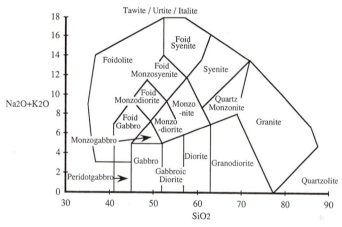

Fig. 2.17 Chemical classification and nomenclature of plutonic rocks using the total alkali versus silica diagram. Adapted from Middlemost (1994, p. 220).

the plutonic equivalent of picrite. To take the idea of equivalence a step further, one might define peridotite as a picrogabbroic or gabbroic rock that contains more than 18 per cent MgO and less than 2 per cent total alkali metals.

2.8   Variation diagrams

Bowen (1928, pp. 92–124) devoted a whole chapter of his famous book to the liquid line of descent and variation diagrams. According to him the principal purpose of a variation diagram is to provide a rapid method for depicting the chemical relations between rocks in a comagmatic suite. He regarded it as a tool for exploring the type and quantity of phases removed from an evolving magma to produce the sequence of rocks in a comagmatic suite. His term liquid line of descent refers to the gradual change in the composition of the main, or only, liquid phase in a magma as chemical constituents are removed by the nucleation and growth of crystals. Some meticulous petrologists claim that variation diagrams display crude liquid lines of descent. They would suggest that during the construction of a large volcanic cone the extrusive products are derived from many similar, often subparallel, liquid lines of descent. This means that the points on a variation diagram depicting samples collected from this volcano are likely to represent a mingling of many separate, probably similar, liquid lines of descent.

Variation diagrams may plot elements, oxides, groups of elements, groups of oxides, ratios of elements or ratios of oxides, against each other. Such diagrams are often used to test hypothetical magmatic differentiation processes. They are particularly useful at drawing attention to anomalous rocks that do not belong to a particular comagmatic suite. Current views on how they can be used in the evaluation, interpretation and presentation of geochemical data have recently been evaluated by Rollinson (1993, pp. 66–84).

A few parameters that are often used in the construction of variation diagrams include: (1) various oxides against silica (usually called Harker Diagrams, after Harker, 1909); (2) oxides against MgO; (3) oxides against Mg-number [$100 \times MgO/(MgO + FeO^*)$], where FeO^* may be FeO, total Fe as FeO, or FeO calculated from a fixed Fe_2O_3/FeO ratio; (4) oxides against Mg' value, i.e. $Mg/(Mg + Fe^{2+})$; (5) oxides against more complex parameters, such as the Larsen Index, i.e. $1/3 \ SiO_2 + K_2O (FeO + MgO + CaO)$, Larsen (1938, p. 506), or a modified Larsen Index $(1/3 \ Si + K) - (Ca + Mg)$, Nockolds in Nockolds and Allen (1953, p. 116); (6) oxides against the Differentiation Index, $Q + Or + Ab + Ne + Lc$ (using CIPW normative minerals), after Thornton and Tuttle (1960, p. 664); (7) oxides against the Crystallisation Index, $An + Di + Fo + Mg$ spinel after Poldervaart and Parker (1964, p. 281); or (8) oxides against the Solidification Index, $100 \ MgO/(MgO + FeO + Fe_2O_3 + Na_2O + K_2O)$, after Kuno and others (1957, p. 194).

Triangular variation diagrams use three components that are recast so that they add up to 100. Common triangular diagrams include: (1) normative abundance of albite–orthoclase–anorthite; (2) Na–K–Ca in weight per cent; and (3) AFM diagrams, where A = $Na_2O + K_2O$, F = $FeO + Fe_2O_3$ and M = MgO in weight per cent. It is interesting that Irvine and Baragar (1971) used an AFM diagram to separate tholeiitic rocks from calc-alkaline rocks. The disadvantages associated with triangular variation diagrams are (1) they do not display absolute

values and (2) when trace or minor elements are used, small glitches in chemical analyses are greatly exaggerated.

When evaluating the data plotted on variation diagrams it is important to consider the constant sum effect. This arises because the sum of the major oxides is 100 per cent. In igneous rocks silica normally varies between 44 per cent and 76 per cent and the sum of the other oxides ranges between 56 per cent and 24 per cent. Some negative correlation is thus likely to occur between silica and the other oxides. Harker diagrams that display a wide range of silica values normally show a decrease in scatter towards the silica-rich end. The impression of greater coherence among the silica-rich rocks should be regarded as a misleading artifact.

Harker diagrams should be regarded as a special type of ternary diagram, with the three end members being: (1) silica; (2) the designated oxide and (3) the sum of the remaining oxides. These diagrams are typically used to portray the evolutionary sequence in suites of comagmatic rocks. Studies of tholeiitic suites, such as the materials from the Alae lava lake, Hawaii (Wright and Fiske, 1971), show that during the early stages of fractional crystallisation silica values remain constant. This is because the liquidus phases are Mg-rich pyroxenes and calcic-plagioclase that have silica contents similar to their host magma. In alkali basalts the liquidus phase is usually olivine and it has a much lower silica content than the host magma (forsterite = 42.7 per cent silica, fayalite = 29.5 per cent silica). Silica is thus a useful index of magmatic differentiation for the alkalic and midalkalic basalts, intermediate rocks and silicic rocks, but it is of only limited value in portraying the evolution of tholeiitic basalts. Variation diagrams that use MgO as abscissa are more useful in displaying differentiation trends in tholeiitic basalt range of compositions, but of much less value in portraying such trends in the more silicic rocks.

Variation diagrams are also used to simulate graphically the addition or subtraction of phases from a magma. Such diagrams readily illustrate the chemical changes that take place when a single phase, such as olivine, is added or removed from a magma. They usually become difficult to interpret when more than one phase has to be added or removed simultaneously. Least-squares mass balance mixing programs, such as XLFRAC (Stormer and Nicholls, 1978) and GENMIX (Le Maitre, 1981), can solve such problems. XLFRAC compares the major

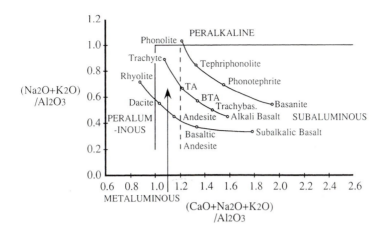

Fig. 2.18 This (Na$_2$O+K$_2$O)/Al$_2$O$_3$ versus (CaO+Na$_2$O+K$_2$O)/Al$_2$O$_3$ diagram, in molecular proportions, reveals how alkalinity changes in the subalkalic, midalkalic and alkalic suites during normal magmatic differentiation. TA = trachyandesite and BTA = basaltic trachyandesite.

element compositions of two magmas, and calculates the weight of the various phases that have to be added or subtracted to change the initial or more primitive magma into the more differentiated magma. If the initial magma was a calc-alkali basalt and some olivine, pyroxene and plagioclase was removed one is likely to move towards or produce a basaltic andesite magma. When using a mixing program one should ascertain: (1) that the mixing produces a low concentration of residual chemical components; (2) that the model makes petrological sense, and (3) that the model uses fewer phases, that is less than there are chemical components in the system. If the latter requirement is not met, all the models one tries are likely to appear feasible.

Another group of diagrams that are coupled with the standard variation diagrams depict both experimentally determined phase boundaries and major element rock data. They are essentially phase diagrams that help petrologists discover how major element data from natural rocks can be related to experimental systems that simulate natural systems like the basalt system. The simple basalt system and the basalt tetrahedron are evaluated in Section 7.3. A full account of various chemical plots on phase diagrams is given in Rollinson's (1993, pp. 84–101) excellent book on using and understanding geochemical data.

2.9 Alkalinity, the peralkaline tendency and alkaline rocks

In the feldspar group of minerals and the common feldspathoids (nepheline group and leucite) the molecular ratio of Al$_2$O$_3$ to (Na$_2$O + K$_2$O + CaO) equals 1:1. This means that in a cooling magma any excess, or deficit, in alumina with respect to this ratio must influence the compositions of the other crystallising phases. Magmas in which the molecular proportion of Al$_2$O$_3$ exceeds (Na$_2$O + K$_2$O + CaO) are called peraluminous (Shand, 1950, p. 228). The excess alumina usually results in the crystallisation of minerals such as biotite, corundum, garnet (almandine or spessartite), muscovite, topaz or tourmaline. Many rhyolites and granites are peraluminous (see Chapters 8 and 9).

Peralkaline magmas and rocks also have extreme compositions. In them the molecular proportion of Al$_2$O$_3$ is less than Na$_2$O and K$_2$O combined (see Fig. 2.18). The excess in total alkalis is usually apportioned to the soda-bearing pyroxenes (aegirine) and amphiboles (arfvedsonite, riebeckite), or to aenigmatite or eudialyte. In other minerals the deficiency in alumina results in it being replaced by iron. The phonolites and their plutonic equivalents are normally peralkaline rocks. Most basic and intermediate magmas and rocks contain just enough alumina and alkali metals to form feldspars or feldspars plus feldspathoids and they are called subaluminous (Fig. 2.18). Magmas and rocks that are transitional between the subaluminous and peraluminous groups are called metaluminous. They usually contain moderately aluminous minerals such as epidote, hornblende and/or melilite. Rocks of this type often contain a mixture of mafic phases, such as hornblende and biotite. Dacites that contain hornblende or both hornblende and biotite, are typical metaluminous rocks.

Fig. 2.18 shows that on cooling and differentiating the common subalkaline picrobasaltic and basaltic magmas tend to move out of the subaluminous field through the metaluminous field and into the peraluminous field. Midalkaline magmas have a stronger peralkaline tendency and usually terminate in the metaluminous field close to its border with the peralkaline field. Midalkaline trachytes are particularly interesting as they are often metaluminous in bulk composition, but contain a peralkaline mesostasis that may contain aegirine, arfvedsonite or riebeckite. In alkaline magmas (sigma 10 to 25) the peralkaline tendency is strong. Fig. 2.18 shows that when alkaline magmas evolve they generally migrate directly from the subaluminous field into the peralkaline field.

The phonolites and foid syenites are the usual end-members of the alkaline suite. They are chemically distinct with high absolute values of alumina and alkali metals. They share these chemical characteristics with some less common feldspathoid-cumulate rocks (see Chapter 11). The anorthosites (sensu stricto) are also characterised by high alumina values (about 26 per cent), but they generally have moderately low alkali metal contents (about 3.8 per cent). As a group the Archeozoic anorthosites are usually anorthite-rich and contain over 30 per cent Al_2O_3 and 17 per cent CaO (Ashwal, 1993, p. 34; see Section 6.13).

During the magmatic differentiation of normal subalkaline, midalkaline and alkaline magmas (sigma 0–25) CaO abundance falls from about 10 per cent in the basaltic field to about 2 per cent in the most evolved magmas. High CaO values are only found in a variety of rare igneous rocks, such as the calciocarbonatites (about 50 per cent CaO; see Chapter 12). As one would expect, rocks with high clinopyroxene contents have high CaO abundances. These rocks vary greatly in their alkali metal contents if one considers the range from websteritic clinopyroxenite to clinopyroxene-rich leucite gabbro. Alkaline rocks rich in CaO usually contain rare minerals such as melanite, melilite or even wollastonite. A rock with

the outlandish name of uncompahgrite (pyroxene + melilite) is likely to contain 29 per cent CaO.

Fig. 2.8 shows the abundance of alkali metals in the common igneous rocks. Trachytes/syenites, tephriphonolites/foid monzosyenites and the phonolites/foid syenites are the only common rocks that contain more than 10 per cent total alkali metals. The mean Na_2O plus K_2O value for the phonolites/feldspathoid syenites is about 14 per cent. Juvitic or potassic varieties may contain over 9 per cent K_2O, whereas the mariupolitic or sodic varieties may contain over 11 per cent Na_2O (Nockolds, 1954, p. 1024). The feldspathoid cumulates also have remarkably high total alkali abundances. Italites contain about 18.6 per cent K_2O, whereas the average sodalitites and urtites contain 17.9 and 14.12 per cent Na_2O respectively (Nockolds, 1954, pp. 1028–1029). Other rare potassium-rich rocks are the lamproites. According to Mitchell and Bergman (1991, p. 296) the average phlogopite lamproite from the Leucite Hills of Wyoming contains 11.6 per cent K_2O.

2.10   Iron, titanium and manganese

Iron is an essential chemical component of the common magmas and igneous rocks. The average basalt contains about 11 per cent total iron recalculated as FeO. This value decreases to about 2.5 per cent total iron as FeO in the average rhyolite, and 4.2 per cent total iron as FeO in the average phonolite (Le Maitre, 1984, pp. 250–255). While total iron decreases, the proportion of Fe_2O_3 to FeO increases through the range basalt to rhyolite, or basalt to phonolite. The mean Fe_2O_3:FeO ratio for fresh basalt is about 0.2. Slightly higher values are typical of basaltic andesite, trachybasalt and basanite. The ratio approaches unity in the dacitic, trachyandesitic and tephriphonolitic fields and often exceeds one in the rhyolitic, trachytic and phonolitic fields. In pristine rocks these values may be lower, but the trend towards a regularly increasing Fe_2O_3:FeO ratio with magmatic differentiation is well established. In addition there is a trend for the ratio to be higher in the alkaline rocks.

The presence of metallic iron in the core probably means that the deep mantle lacks phases containing ferric oxide, and detailed studies of upper mantle xenoliths suggest that the redox state of the upper mantle is closer to the wustite-iron buffer than the quartz-fayalite-magnetite buffer (Ulmer et al., 1987,

p. 5). If one examines the relative abundance of Fe_2O_3 in igneous rocks one finds that it has a remarkably high coefficient of variation. As the Fe_2O_3:FeO ratio has a powerful influence on the abundance and nature of the CIPW normative minerals calculated for igneous materials, one has to be cautious when using norms in the classification of even slightly altered rocks. Basalts with the same major element compositions, besides different Fe_2O_3:FeO ratios, may be: (1) nepheline and olivine normative when the Fe_2O_3:FeO ratio is 0.2; (2) olivine and hypersthene normative when it is 0.5; and (3) hypersthene and quartz normative when the ratio is 1.2. These changes in composition arise because in CIPW norm calculations every molecular unit of Fe_2O_3 joins with a molecular unit of FeO to form magnetite, releasing a molecular unit of SiO_2; thus as the Fe_2O_3:FeO ratio rises more SiO_2 becomes available to form hypersthene and/or quartz. Standardised Fe_2O_3:FeO ratios may be used in norm calculations, but Chayes (1973, p. 673) and others have observed that adjustments of this kind tend to blur the important distinction between observation and inference.

Within the subalkalic field the members of the tholeiitic suite usually have the highest total iron abundances. This observation was used by Kuno (1968, p. 149) and Irvine and Baragar (1971) to separate the tholeiitic suite from the calc alkaline suite. When magmatic differentiation produces residual magmas enriched in iron this tendency is usually called the Fenner Trend. Subalkalic magmas are likely to follow this trend if they differentiate in a closed system with the Fe_2O_3:FeO ratio remaining low so preventing the early precipitation of significant amounts of magnetite (Osborne, 1962).

Foiditic magmas usually contain somewhat high total iron contents (total iron as FeO = 12.3 per cent; Le Maitre, 1984, p. 255). Rocks that contain normative foids also tend to be more oxidised and have higher Fe_2O_3:FeO ratios than rocks that are free of foids (Chayes, 1973, p. 672). The igneous rocks with the highest iron contents are the magnetitites and nelsonites (see Section 6.14). The former rocks occur, either as volcanic rocks such as the Laco flows of the highlands of northern Chile (Haggerty, 1970, p. 329), or as cumulus plutonic units interbedded with other cumulate rocks in layered complexes such as the Bushveld Complex of South Africa (Reynolds, 1985). Nelsonites are igneous rocks that are essentially composed of iron-titanium oxides and apatite. Many magnetitites have high

titanium contents as their essential mineral is titanomagnetite.

In the common igneous rocks the abundance of TiO_2 is normally less than 2.5 per cent, but basanites and tephrites tend to have higher mean values (2.9 per cent and 3.4 per cent respectively). Titanium dioxide is like total iron as it usually decreases with magmatic differentiation. In 1965 Chayes (p. 128) showed that calc alkaline basalts contain significantly less TiO_2 than normal tholeiitic and midalkaline basalts. In more recent times this discovery has been used to develop geochemical discriminant diagrams that attempt to separate basalts from various tectonic environments (see Section 2.15). The high TiO_2 contents of the mare basalts are examined in Section 2.12.

Phosphorus is strongly enriched in rare apatite-magnetite-olivine-phlogopite rocks, or phoscorites, from some carbonatite complexes, such as the Phalaborwa Complex of South Africa (see Chapter 12). In the common igneous rocks manganese usually correlates with ferrous iron, and the mean Mn/Fe ratio is about 0.017. The average picrobasalt contains 0.2 per cent MnO, whereas the average rhyolite contains about 0.06 per cent of this oxide. Alkaline rocks often have high iron and manganese contents.

Volatile components

In the IUGS subcommission's comprehensive classification of igneous rocks (Le Maitre *et al.*, 1989) the lamproites, kimberlites and lamprophyres are regarded as special rocks (see Table 2.3). They are different from the common igneous rocks because they have high abundances of water, carbon dioxide and the minor elements barium, strontium, sulphur, fluorine and chlorite. Their chemical compositions are normally revealed by their modes that characteristically contain primary carbonates and hydrous phases (Streckeisen, 1978, p. 10). Because water and carbon dioxide are important components in the lamprophyric rocks their major element analyses are usually not recalculated on a water and carbon dioxide free basis (Rock, 1991, p. 82). If one compares the major element composition of the lamprophyric rocks (see Table 2.3) with the common igneous rocks one discovers that most so-called calc alkaline lamprophyres plot in the midalkaline field. Such rocks might be called midalkaline lamprophyres. Most of the alkaline lamprophyres and lamproites plot in the alkaline field

Table 2.3 Mean major element compositions of lamprophyres, lamproites, kimberlites and carbonatites.

	Calc-alkaline lamprophyres	Alkaline lamprophyres	Ultramafic lamprophyres	Kimberlites	Lamproites	Calcio-carbonatite	Magnesio-carbonatite	Ferro-carbonatite	Natro-carbonatite
SiO_2	51.08	42.3	32.33	33.75	50.25	2.69	3.6	4.55	0.04
TiO_2	1.1	2.89	3.1	2.29	4.03	0.15	0.33	0.41	0.06
Al_2O_3	14.02	13.64	6.71	4.13	7.47	1.05	0.98	1.41	0.07
Fe_2O_3/Fe_2O_3t	8.2lt	11.94t	13.61t	10.90t	7.67t	2.22	2.39	7.21	0.29
FeO	0	0	0	0	0	1	3.9	5.11	0
MnO	0.13	0.2	0.22	0.16	0.49	0.51	0.95	1.6	0.11
MgO	7.01	7.07	15.02	25.22	11.21	1.78	14.96	5.86	0.44
CaO	7.01	10.25	14.01	9.93	4.72	48.57	29.91	31.74	13.3
Na_2O	2.7	2.99	1	0.44	0.64	0.29	0.29	0.38	30.19
K_2O	3.1	1.99	1.9	1.23	7.18	0.26	0.28	0.38	7
P_2O_5	0.6	0.74	1	0.79	1.28	2.08	1.89	1.91	0.96
BaO	0.12	0.1	0.12	0.1	0.85	0.34	0.64	3.15	1.01
SrO	0.08	0.12	0.11	0.08	0.15	0.85	0.69	0.85	1.35
H_2O^+	2.4	3.09	3.5	5.8	3.15	0.75	1.19	1.21	4.31
CO_2	2	1.99	6.51	4.92	0.49	36.23	36.55	29.77	33.05
SO_3	0.3	0.5	0.5	0	0.1	0.87	1.07	4.01	2.48
F	0.11	0.18	0.32	0.24	0.32	0.29	0.31	0.44	2.33
Cl	0.01	0.03	0.04	0.02	0	0.08	0.07	0.02	3.03
Total	100	100	100	100	100	100	100	100	100
Total volatiles	4.83	5.79	10.86	10.98	4.06	38.22	39.2	35.45	45.19
	Rock, 1991, p. 78	Rock, 1991, p. 78	Rock, 1991, p. 78	Rock, 1991, p. 78	Rock, 1991, p. 78	Woolley & Kempe, 1989, p. 8	Woolley & Kempe, 1989, p. 8	Woolley & Kempe, 1989, p. 8	Dawson, 1989, p. 269, (1+2+3)/3

with some data-points overlapping into the area where the sigma values become negative. As one would expect the ultramafic lamprophyres typically plot in the area of negative sigma values. These rocks frequently contain phases such as melilite or primary carbonates and the presence of these minerals lowers the general silica content of the rock.

2.11   Interstellar clouds, meteorites, meteoroids and meteors

In the cosmos as a whole most rocks are condensates that have accreted in interstellar clouds. Such clouds are probably broadly similar to the primordial Solar Nebula. Condensation petrology is the branch of petrology that describes the formation of phases from gas–crystal reactions in the absence of melting at extremely low pressures. The condensation process generates cosmic dust and it frequently accretes into larger bodies such as meteoroids, planetoids or planets. In these larger bodies the rocks are the products of accretion, modified by impact melting, magmatic activity and a range of other transforming processes that generate metamorphic and even sedimentary rocks.

Orthodox petrography does not have a suitable vocabulary for naming rocks generated by the accretion of interstellar condensates. Such rocks are neither igneous, sedimentary nor metamorphic, but they are probably the most common type of rock in the cosmos. Another question that should be considered is whether the rocks generated by impact melting are authentic igneous rocks. They are not the products of normal igneous processes and their chemical compositions may fall outside the range typical of igneous rocks. It is recommended that these impactites should be regarded as a separate and unique group of rocks. Impactites are typically glassy, or finely crystalline, rocks produced by the fusion of surface or near-surface materials.

The countless, small, natural, solid objects that move in interplanetary space are called meteoroids. They are usually less than a hundred metres in diameter. A meteorite is a meteoroid, or a fragment of a meteoroid, that is recovered from the surface of a planet. The smallest meteorites are called micro-meteorites. They are often so small and they dissipate heat so easily that they pass through a planetary atmosphere without vaporising. Typically micro-meteorites are less than $120\,\mu m$ in diameter.

In the simplest classification meteorites are divided into irons (siderites), stony-irons (siderolites or litho-siderites) and stones (aerolites). This classification has been superseded because: (1) it disregards the considerable diversity in composition within these three divisions; (2) it often unites materials of dissimilar character and history; and (3) it separates some meteorites that are closely related to one another (Sears and Dodd, 1988, p. 8). For example, the two main types of stony-irons contain about 50 per cent metal phases and 50 per cent silicate phases, yet the two types of meteorites evolved in entirely different environments.

Irons

Irons consist almost entirely of nickeliferous iron with minor amounts of carbon, sulphur and phosphorus (Wasson, 1985, pp. 76–99). Over 40 different minerals have been identified in these meteorites, but their essential phases are two iron-nickel phases that are called kamacite (the low nickel alpha-phase) and taenite (the high-nickel gamma-phase). Historically these meteorites have been classified using both structural and chemical criteria. The simplest system is based on the structures they contain. Irons that contain up to 6 per cent nickel contain only kamacite, and individual samples are usually part of a single crystal. Because kamacite is a cubic phase, and a cube is a hexahedron (a closed form with six equivalent faces), such irons are usually called hexahedrites. When polished these meteorites often reveal parallel lines called Neumann lines. Research has shown that these lines are twin planes and their presence suggests shock deformation.

The most abundant and diverse group of irons contain between 6 and 17 weight per cent nickel. They contain both kamacite and taenite. When polished and etched they display regular octahedral intergrowth patterns, called Widmanstätten structures. These meteorites are called octahedrites and their structure results from the growth of kamacite lamellae along four sets of planes within the higher temperature taenite phase. Irons that contain more than about 17 per cent nickel are usually called ataxites. The term is derived from the Greek words *a* meaning without and *taxis* meaning arrangement. Ataxites are usually so fine grained that it is difficult to recognise any structure, but petrographic investigations may reveal Widmanstätten structures on a microscopic scale.

The chemical classification of irons is essentially based on their gallium, germanium and nickel contents. This classification was originally simple as it recognised only four groups, but the number has now grown to 12, with five or more subgroups in each of these main groups (Wasson, 1985, pp. 169–171). Logarithmic plots of germanium versus nickel and germanium versus gallium are now often used in the classification of irons. It is likely that the irons formed in the primordial Solar Nebula and small local changes in pressure and temperature influenced the abundance of gallium and germanium and these initial values have not been significantly modified by later processes.

Stony-irons

Stony-iron meteorites are rare. They contain nickel-iron and silicate phases in about equal proportions. Examples of these meteorites include the pallasites and the mesosiderites. The former contains large olivine crystals enclosed in nickel-iron, while the latter are breccias made up of fragments of silicates, such as pyroxene and plagioclase, cemented together by a nickel-iron phase. The term pallasite is also used to describe some rare iron-rich plutonic rocks from the Earth that contain more iron phases than silicate phases.

Stones and chondrites

The stony meteorites account for more than 90 per cent of the meteorites that have been observed to fall onto the surface of the Earth. They are divided into the abundant ultramafic chondrites and the less abundant achondrites. When chondrites are found on the surface of a planet they usually can be separated from other silicate rocks by the presence of small (about 1 mm in diameter) spherical inclusions called chondrules (Greek *chondros*, meaning a grain). Chondrules usually have a radiating texture and are typically composed of olivine and orthopyroxene. The following evidence supports the idea that the chondrites provide unrivalled insights into the conditions that prevailed in the early Solar System.

1 Except for the highly volatile elements such as hydrogen and helium, the chondrites have chemical compositions that are extremely similar to the solar corona (See Fig. 2.19). As the Sun contains more than 99 per cent of the mass of the Solar System there is no doubt that its chemical composition is similar to that of the primordial Solar Nebula.

2 The age of these meteorites (4.55 giga-years) is similar to the age suggested for the Earth and the other planets of the Solar System.

3 Many chondrites contain non-equilibrium assemblages that seem to have changed little since their component phases accumulated.

4 Some chondrites contain isotopes that are considered to have accumulated soon after their synthesis.

5 The isotopes of some elements are present in ratios that are totally unlike those encountered in any other known materials from the Solar System (Sears and Dodd, 1988, p. 29).

Some chondrites consist almost entirely of chondrules, whereas others are mainly or entirely made up of matrix. The term chondrite is now applied to any meteorite with a composition similar to that of the Sun excluding the highly volatile elements. One should keep in mind that there is no single group of meteorites that has a bulk composition identical with the bulk composition of the Earth (McDonough and Sun, 1995, p. 224).

Based on minor differences in major element abundance (magnesium/silicon, calcium/silicon, titanium/silicon and aluminium/silicon ratios) and mineralogy, the chondrites are divided into nine classes (see Table 2.4). The abundant H (high-iron), L (low-iron) and LL (low-iron and low-metal) classes are grouped together as the ordinary chondrites. If these ordinary chondrites were to be classified as normal igneous rocks they would be feldspathic harzburgites (see Fig. 2.3). Two classes of chondrite have modes dominated by the magnesium silicate mineral, enstatite. These enstatite chondrites (EH and EL) are rare and bizarre materials that contain so little oxygen that all their iron occurs either as a native metal or as troilite. Even elements like calcium, titanium and chromium occur combined with sulphur instead of oxygen. In some of these extraordinary meteorites sodium and potassium occur in sulphides. The rare but important carbonaceous chondrites are divided into four classes (CI, CM, CO and CV).

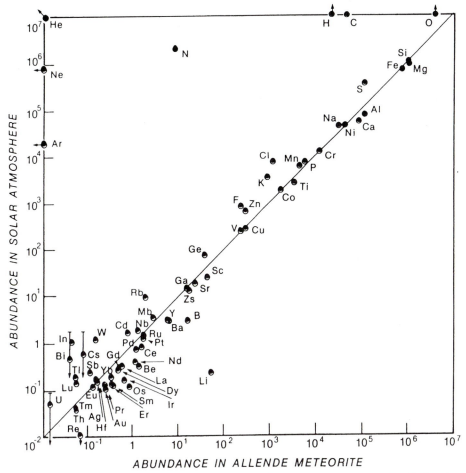

Fig. 2.19 Comparing the abundance of 69 elements in the Allende CV chondritic meteorite with the solar corona. All chemical data is relative to one million silicon atoms. After Wood, 1990, p. 243.

Most of the chondrites have been metamorphosed within their parent bodies. Samples that show no metamorphism are said to belong to petrologic type 1. The petrologic type number rises as the grade of metamorphism increases. The carbonaceous chondrites are either unmetamorphosed, or weakly metamorphosed (petrologic types 1, 2 or 3), and they usually consist of collections of chondrules and unmelted aggregates, embedded in a fine-grained, volatile-rich matrix. Type-1 samples and the matrices of the other carbonaceous chondrites carry the unambiguous data on the early history of the Solar System. All the dark-coloured, class-I samples belong to the petrologic type-1. They are physically and chemically the most primitive types of meteorites, and mainly consist of low-temperature phases that

condensed out of the Solar Nebula. They are called I after the Ivuna in Tanzania, where the type specimen fell to Earth. CI chondrites do not contain chondrules, but contain an abundance of low temperature, hydrous phases. Some samples contain 20 per cent water. They also may contain phases such as serpentine group minerals and epsonite ($MgSO_{4.7}H_2O$). The only high temperature phase found in these chondrites is olivine. It usually forms small, isolated, euhedral crystals.

Most petrologists would agree that chondrules were once molten or partly molten and that they once existed, at least briefly, as independent entities (Grossman and Wasson, 1983, p. 89). The most likely setting for their formation was the Solar Nebula, but the nature of the source of heat required for the

Table 2.4 Parameters used in classifying the chondrites. After Wasson and Kallemeyn, 1988, p. 537.

Chondrite group	Al/Si normalised ratio	Mg/Si normalised ratio	Ni/Si normalised ratio	Zn/Si normalised ratio	FeOx/FeOx + MgO molecular per cent	delta oxygen 17 parts/ thousand	delta oxygen 18 parts/ thousand
CV	1.34	1	0.85	0.25	35	−3	1
CO	1.07	0.97	0.87	0.21	33	−4	0
Cm	1.1	0.97	0.92	0.48	43	1	7
CI	1	1	1	1	45	9	17
H	0.8	0.89	0.94	0.092	17	3	4.2
L	0.78	0.84	0.64	0.088	22	3.5	4.6
LL	0.75	0.84	0.53	0.08	27	3.8	4.9
EH	0.58	0.68	1.04	0.49	0.05	2.9	5.7
EL	0.67	0.81	0.69	0.03	0.05	2.9	5.7

Table 2.5 Chemical composition of selected basaltic meteorites. After B.V.S.P., 1981, p. 221.

Name	Sioux County	Cachari	Haraiya	Macibini	Stannern	Mean
SiO_2	49.03	48.26	48.69	49.32	49.33	48.93
TiO_2	0.62	0.63	0.56	0.72	0.96	0.7
Al_2O_3	12.84	12.85	12.49	12.12	12.34	12.53
Cr_2O_3	0.35	0.32	0.32	0.42	0.28	0.34
FeO	18.58	19.1	19.25	18.31	17.92	18.63
MnO	0.56	0.59	0.59	0.54	0.5	0.56
MgO	7.11	7.14	7.11	8.37	6.36	7.22
CaO	10.35	10.25	10.26	9.97	10.58	10.28
Na_2O	0.45	0.51	0.5	0.48	0.6	0.51
K_2O	0.04	0.05	0.03	0.05	0.08	0.05
P_2O_5	0.09	0.08	0.1	0.11	0.13	0.1
Total	100.02	99.78	99.9	100.41	99.08	99.85

melting process is still being debated. Some possible sources of heat include lightning, heating in shock fronts created during the collapse of the Solar Nebula, chemical energy, heating events associated with the evolution of the primitive Sun, and impact between small objects in space (Taylor *et al.*, 1983, p. 273). It is now generally agreed that most of the ultramafic meteorites were derived from asteroids or comets.

Achondrites

Achondrites are stony meteorites of non-solar composition. As their name suggests they normally do not contain chondrules. They are usually more coarse-grained than the chondrites, contain no nickel-iron phases, and look like the rocks commonly found on the surface of the Earth. These meteorites are rare but very diverse in their properties and in their inferred histories. According to Sears and Dodd (1988, p. 9) they have all experienced processing within a planet and this is likely to have obliterated all the phases generated in the Solar Nebula. Some samples provide pointers to the size and type of igneous processes that operated in the early Solar System.

Some achondrites, such as the eucrites and shergottites, have basaltic compositions, whereas others, such as the howardites and mesosiderites, contain a mixture of clasts that include basaltic fragments. Eucrites are the most abundant type of basaltic meteorite (see Table 2.5). They look like terrestrial basalts and are difficult to separate from terrestrial basalts, yet by some curious quirk their name is derived from the Greek word *eukritos* that means easily distinguished. Eucrites are essentially composed of two minerals, calcium-rich plagioclase and calcium-poor pyroxene. They contain no water and their iron is reduced. Their main claim to distinction is that they are the oldest known basalts. Their mean

Table 2.6 Mean chemical composition of mare basalts. After Heiken *et al.*, 1991, pp. 449–453.

	Apollo 11	Apollo 12	Apollo 15	Apollo 17	Luna 16	Luna 24
SiO_2	40.46	44.88	46.68	39.03	44	46
TiO_2	10.41	3.62	1.97	11.94	4.77	1.02
Al_2O_3	10.08	8.93	10.21	9	13.83	12.2
FeO	19.22	20.48	19.87	18.89	18.7	21.58
MgO	7.01	10.64	8.76	8.54	6.4	7.4
CaO	11.54	9.81	10.63	10.82	11.82	12.2
Na_2O	0.38	0.25	0.3	0.39	0.53	0.23
K_2O	~0.09	0.07	0.05	0.05	0.12	–
Total	99.1	98.68	98.57	98.66	100.17	100.63

age is 4.54 giga-years. It is usually postulated that they originally erupted onto the surface of a large asteroid. Remote sensing has shown that basaltic rocks crop out on the surface of the large belt asteroid, Vesta. In contrast to the eucrites, the rare SNC meteorites (shergottites, nakhlites and chassignites) are far younger (1.3 to 0.18 giga-years). A comparative study of the isotopic compositions of the noble gases and nitrogen in the Martian atmosphere and the isotopes of these gases trapped in shergottite has provided clear evidence supporting the idea that the SNC meteorites are impact ejecta derived from the surface of Mars (Bogard *et al.*, 1984; Robinson and Wadhwa, 1995). Several achondrites from Antarctica are composed of a breccia that contains clasts of anorthosite. These meteorites are believed to come from the Moon or less likely Mercury (see Section 6.13).

2.12   Diversity among lunar and other extraterrestrial rocks

When viewed from Earth the surface of the Moon is readily divided into light-coloured, densely-cratered highland areas called the terra and dark-coloured, moderately smooth, level, low-lying plains called the mare. Basaltic rocks cover the mare but they occupy only 17 per cent of the total surface area of the Moon. When they erupted between 3.9 and 3.2 giga-years ago they were particularly fluid and flowed long distances. Individual lavas flowed up to 600 km on slopes of only one in a hundred. The various volcanic landforms found on the Moon show that the mare lavas erupted quietly and rapidly, without significant explosive activity. Volcanic craters and domes are rare. Sinuous rilles are common and are

usually regarded as open lava channels that were diverted by obstacles and generally followed preexisting depressions.

The Moon is particularly important in our review of the magma/igneous rock system because it provides excellent data on the first billion and a half years in the evolution of the Earth–Moon system. A formidable amount of information has been collected on the petrography and chemistry of the lunar rocks. Fortunately these data have been systematically summarised by Heiken *et al.* (1991). A typical mare basalt has the following chemical criteria that set it apart from the normal terrestrial basalts: (1) a dearth of volatiles and no detectable water; (2) low levels of oxidation, as suggested by the presence of metallic iron in many lunar samples; (3) low to very low silica contents; (4) high to very high titanium and iron contents; and (5) low potassium and sodium values. Four of the six mare basalt analyses displayed in Table 2.6 contain less than 45 per cent silica. Using chemical criteria these rocks would be classified as picrobasalts and not basalts (sensu stricto). Modally they do resemble normal basalts as their essential minerals are augite and plagioclase.

The enigma of discovering a rock with the modal composition of a basalt but the chemical attributes of a picrobasalt can be explained if one considers the composition of the plagioclase found in the mare basalts. A typical plagioclase from a Hawaiian tholeiitic basalt contains about 52 per cent anorthite that corresponds to about 66 per cent silica (BVSP, 1981, p. 182), whilst a typical plagioclase from a mare basalt contains 91 per cent anorthite that corresponds to only 46.4 per cent silica (Heiken *et al.*, 1991, p. 163). Since the publication of the book on the Basaltic Volcanism Study Project in 1981 the term basalt has collected yet another meaning. It is now used for all basic lavas (sensu lato), from sources other than the Earth, that play the same petrological role or occupy the same petrological niche as the basalts of Earth.

The origin of the very high (see Table 2.6) titanium contents in some mare basalts is a particularly interesting problem. A clue to solving this problem is supplied by the insight that on Earth anorthosite bodies are frequently associated with large concentrations of iron-titanium oxide minerals (Ashwal, 1993, p. 157). Anorthosites are abundant in the lunar highlands where they usually contain less than 0.15 per cent TiO_2. If the lunar anorthosites evolved from picrobasaltic or basaltic magmas they must have shed large amounts of TiO_2, probably as the minerals armalcolite or ilmenite. The aberrant titanium-rich lavas probably evolved from a magma contaminated by reaction with, or incorporation of, cumulate rocks with a high-titanium content. If the high-titanium (greater than 9 per cent) mare basalts were contaminated by cumulates, it is clearly not appropriate to use the TAS diagram for their classification. If one wished to assume a neutral position on the origin of these lavas one might regard all volcanic rocks that contain less than 52 per cent SiO_2, more than 7.5 per cent TiO_2 and more than 16 per cent total iron as FeO as special rocks. Such rocks might be called apolloites. This discussion steers one's attention back to the problem of classifying cumulates and it is indisputable that the simplest way of doing this is to use modal criteria.

The lunar highlands are densely covered by impact craters that abut and overlap to form a continuously cratered surface. About 25 per cent of the samples collected from the highlands have textures that superficially resemble those of basaltic lavas. Detailed examination reveals that many of these rocks were generated by hypervelocity impacts. Three components normally occur in the surface rocks of the lunar highlands. The components are ANT, KREEP and a meteoritic component. ANT is an acronym for the rocks of the anorthosite-norite- troctolite suite (see Fig. 2.4), whereas KREEP is an acronym for materials enriched in potassium (K), the lanthanoids (rare earth elements or REE), phosphorus (P) and the other incompatible elements. Igneous textured KREEPy rocks have seldom been recognised in the lunar samples returned to Earth as the samples are mainly basalts. A very few coarser grained samples that display intergrowths of quartz and potassium feldspar have been described (Heiken *et al.*, 1991, p. 217). Outcrops of anorthosite account for the light colour of the lunar highlands and trails of anortho-

site fragments account for the light coloured 'rays' that radiate out from many impact sites. Plagioclase (An 95–97) is normally the only essential phase in the lunar anorthosites.

Soon after fragments of anorthosite had been recovered from the Moon the idea of a magma-lava ocean was introduced to account for their origin. It was postulated that in the pre-Nectarian Period the outer layer of the Moon was partly or completely molten and an anorthositic crust formed as flotation cumulates in a high temperature magma-lava ocean. The complementary ferromagnesian phases sank into the ocean and they eventually formed the source materials for the younger mare basalts. Perhaps the most compelling evidence in support of a magma-lava ocean is the global distribution of anorthosite on the lunar surface. The speculation about a deep magma-lava ocean is reinforced by the absence of any complementary ferromagnesian cumulates down to the depths excavated by the large hypervelocity impacts. The great age (4.44 giga-years) of the anorthosites and the geochemical uniformity of the KREEPy materials also bolster the magma-lava ocean model. According to Ashwal (1993, p. 329) geochemical mass balance calculations suggest that the initial magma-lava ocean was 250 km deep. Not all petrologists support this idea. Longhi and Ashwal (1985) proposed a two-stage model for the origin of the lunar anorthosites. During the first stage a series of overlapping layered bodies intruded into the primitive lunar crust. The second stage was triggered by a pulse of heating that mobilised the less dense anorthositic layers and enabled them to move upwards and congeal as a high-level layer.

Most petrologists were surprised when KREEP was discovered because it contains high concentrations of elements that are usually typical of differentiated residual magmas. KREEP probably represents the final 2 per cent of residual liquid that evolved during the formation of the thick anorthositic highland crust. Potassium, uranium and thorium are all likely to be concentrated in the KREEPy materials near the lunar surface (Taylor, 1982, p. 243).

2.13 Pyroclastic rocks and their classifications

Pyroclastic rocks and tephra are products of explosive volcanic activity and their classification is based on particle size. These rocks include volcanic air-fall,

Lithophile elements	
Refractory	Be, Al, Ca, Ti, V*, Sr, Y, Zr, Nb, Ba, REE, Hf, Ta, Th and U
Transitional	Mg, Si and Cr*
Moderately volatile	Li, B, Na, K, Mn*, Rb and Cs*
Highly volatile	F,Cl, Br, I and Zn
Siderophile elements	
Refractory	Mo, Ru, Rh, W, Re, Os, Ir and Pt
Transitional	Fe, Co, Ni and Pd
Moderately volatile	P, Cu, Ga, Ge, As, Ag, Sb and Au
Highly volatile	Tl and Bi
Chalcophile elements	
Highly volatile	S, Se, Cd, In, Sn, Te, Hg and Pb
Atmophile elements	
Highly volatile	H, He, C, N, O, Ne, Ar, Kr and Xe

At high pressure these elements may develop siderophile behaviour and partition into the core.

Fig. 2.20 Geochemical classification of the elements. After McDonough and Sun, 1995, p. 225.

flow and surge deposits, lahars and a variety of fragmental materials that occur in diatremes, dykes and vents. Air-fall deposits consist of pyroclasts that move through the air while falling out of a volcanic eruption column. Flow deposits are pyroclasts that move over the ground as hot, particulate flows with a high proportion of particles to gas. Surge deposits are pyroclasts that are carried in ground hugging, dilute particulate flows. Lahar is an Indonesian word that describes a variety of water-supported flows of volcaniclastic materials that are propelled down the flanks of volcanoes. Lahars may form as the direct result of volcanic activity, such as an eruption taking place through a crater lake, or they may be triggered by heavy rain or local earthquake activity.

Pyroclastic deposits and tephra are assemblages of clastic materials that contain more than 75 per cent by volume of pyroclasts. It is now standard usage to call these materials pyroclastic rocks when they are consolidated but tephra when they are unconsolidated. The various types of pyroclasts and pyroclastic rocks are classified according to their size (Le Maitre *et al.*, 1989, pp. 7–9). Bombs are pyroclasts with a mean diameter that exceeds 64 mm and a surface shape that reveals that they were at least partly molten during their formation and subsequent transport. Pyroclastic rocks that contain clasts in this size range are usually called agglomerates. Blocks are

similar in size to bombs, but they have an angular to subangular shape. This shows that they were solid when erupted. Rocks composed of blocks are called pyroclastic breccias or block tephras. Pyroclasts of any shape with a mean diameter between 64 and 2 mm are called lapilli. Rocks containing pyroclasts in the lapilli range are called lapilli tuffs or lapilli tephras. Ash grains are pyroclasts of any shape that have a mean diameter of less than 2 mm. Rocks that contain pyroclasts in the ash range are called tuff or ash. Tuffites are rocks that contain both clastic and pyroclastic fragments and the clastic fragments exceed 25 per cent of the volume of the rock (Le Maitre *et al.*, 1989, p. 9).

2.14 Distribution of trace elements between coexisting phases

In igneous petrology trace elements are usually defined as either the non-essential elements in a particular rock-forming mineral or all elements other than the major elements. The major elements are oxygen, silicon, aluminium, iron, magnesium, calcium, sodium and potassium, and they are present in the Earth's crust in amounts greater than 1 per cent (see Table 1.3). Phosphorus, sulphur, chlorine, titanium, chromium and manganese are either called minor elements or trace elements. Fig. 2.20 contains a recent geochemical classification of the elements. They are divided into the lithophile, siderophile, chalcophile and atmophile groups (McDonough and Sun, 1995, p. 225). The elements that are usually found in oxide minerals and to a lesser extent as halides, and are particularly abundant in the crustal rocks, are known as the lithophiles or rock-loving elements. Other elements that are often found in the crust in combination with non-metals such as sulphur, selenium and arsenic, instead of the oxide minerals, are known as the chalcophiles. Elements that are considered to concentrate in the core are known as the siderophiles or iron-loving elements. The atmophiles are elements that mainly occur in volatile form in the atmosphere or the oceans.

Goldschmidt (1937, p. 660) provided the first systematic account of the distribution of trace elements in minerals. His ideas developed from his research into the structure of minerals and he attempted to relate the abundance of trace elements in minerals to their ionic size and ionic charge. He proposed three empirical rules. Rule one asserts that ions of similar radii and of the same charge will enter a crystal in amounts proportional to their concentration in the magma. For example, rubidium (Rb^+) of ionic radius 0.147 nm is said to be camouflaged by potassium (K^+) of ionic radius 0.133 nm. Rule two asserts that if two ions have the same charge the ion of smaller radius will be incorporated preferentially into a growing crystal. Magnesium (Mg^{2+}) of ionic radius 0.066 nm is usually concentrated in the early formed olivines and pyroxenes in preference to Fe^{2+} of ionic radius 0.074 nm. Rule three asserts that an ion of the same radius but of higher charge than another will be incorporated preferentially into a growing crystal. Calcium (Ca^{2+}) of ionic radius 0.099 nm enters the plagioclase lattice more readily than Na^+ of ionic radius 0.097 nm (cf. Henderson, 1982, p. 125). These rules explain the abundance and distribution of many trace elements found in igneous rock-forming minerals. An excellent example of the first rule in action is the camouflaging of hafnium (ionic radius 0.079 nm) by zirconium (ionic radius 0.080 nm) in the mineral zircon (see Table 2.1). The alkali metal rubidium often concentrates in late-forming potassium feldspars and/or micas, as predicted by the second rule. An example of the third rule in action is the capture of Ba^{2+} of ionic radius 0.155 nm by K^+ in the potassium feldspars.

Over the years Goldschmidt's Rules have been modified to increase their usefulness. The first major modification was by Ringwood (1955), who showed that it was important to consider the electronegativities of substituting ions. Electronegativity (expressed in eV) is the capacity of an atom in a molecule or a crystal to attract extra electrons. Ringwood proposed that if two elements have similar radii and the same charge, the one with the lower electronegativity will be preferentially concentrated in the minerals that crystallise first. This rule is generally useful for elements that differ in electronegativity by more than 0.1. Ringwood also discussed how the formation of complexes may modify the distribution of trace elements in a cooling and crystallising magma. The small highly charged ions that tend to form complexes usually play an important role during the late, or pegmatitic, stage in the crystallisation of magmas.

When investigating the distribution of trace elements between coexisting magmatic phases, it is usually convenient to use a partition coefficient (k) to express this relationship. These distribution coefficients, or Nernst distribution coefficients, describe the equilibrium distribution of a trace element between a mineral and its host magma (Henderson, 1982, pp. 88–97). Such coefficients are usually obtained by determining the chemical composition of both the mineral and the igneous matrix that surrounds it. Ideally these coefficients should be determined as functions of temperature, pressure, oxygen fugacity and bulk composition. In practice most coefficients are not well defined in terms of all these parameters (see Table 2.7). For example, Green and Pearson (1986) have shown experimentally how lanthanoids (rare earth elements) partitioning, between sphene (titanite) and a coexisting silicate liquid, changes when temperature, pressure or liquid composition change.

In 1954 Neumann, Mead and Vitaliano established that when magma crystallises, at constant temperatures and pressures the trace elements are distributed between the magma and the solid phases in a predictable manner (see Fig. 2.21). If, for example, the partition coefficient (k) equals two, then the trace element will concentrate in the mineral in preference to the magma and there is a concomitant drop in the concentration of the trace element in the residual magma. If k equals 0.5 then the trace element concentrates in the magma and there is an accompanying increase of the element in the residual magma relative to the original magma. Fig. 2.21 shows the relative changes in element concentration for a series of ideal trace elements during the crystallisation of a magma. Partition

Table 2.7 Approximate mineral/matrix partition coefficients for selected elements in basaltic rocks. Adapted from Henderson, 1982, p. 91.

Element	Olivine	Orthopyroxene	Clinopyroxene	Amphibole	Plagioclase
K	0.0007	0.015	0.03	0.6	0.17
Ca	0.03	0.3	–	2.05	–
Cr	2.1	~10	8.4	~0.85	–
Mn	1.2	~1.4	0.9	0.05	–
Co	3.8	~3	1.2	–	–
Ni	14	~5	2.6	–	–
Zn	~0.7	–	~0.41	~0.56	–
Rb	0.006	0.025	0.04	0.25	0.1
Sr	0.01	0.015	0.14	0.57	1.8
Ba	0.006	0.013	0.07	0.31	0.23
La	–	–	~0.08	0.27	0.14
Ce	0.009	0.02	0.34	0.34	0.14

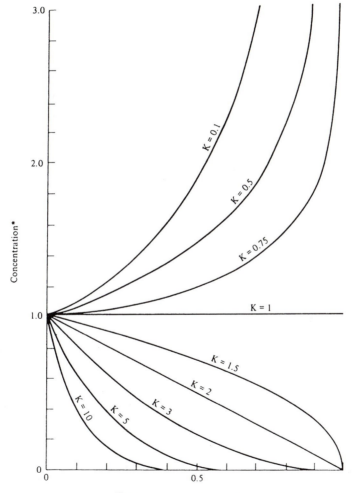

*Relative concentration of an element in the residual melt relative to the original melt.

Fig. 2.21 Equilibrium distribution of individual components (trace elements) between minerals and magma. After Neumann, Mead and Vitaliano, 1954, p. 94.

Table 2.8 Mean Ti, Zr, Y, Nb and Sr values in p.p.m. for lavas from nominated tectonic structural settings. After Pearce and Cann, 1973, p. 293.

Magma type	Ti	Zr	Y	Nb	Sr
Mid-oceanic ridge basalt	8 350	92	30	5	131
Volcanic arc tholeiitic basalt	5 150	52	19	1.5	207
Volcanic arc calc-alkali basalt	5 400	106	23	2.5	375
Oceanic island basalt	16 250	215	29	32	438
Continental basalt	15 150	215	29	20	460

coefficients are now widely used in petrogenetic modelling, and as a semi-quantitative guide to element distribution.

2.15   Trace elements and discrimination diagrams

At the present stage in the Earth's history volcanic activity is limited to a few discrete tectonic settings. In the early 1970s an investigation was launched to find out whether geochemical criteria could be used to discover the tectonic setting of altered volcanic rocks. Cann (1970) examined basaltic rocks from the ocean floor and claimed that the abundances of the trace elements rubidium, strontium, yttrium, zirconium, niobium and titanium from these rocks was unaffected by secondary processes including low-grade metamorphism. Pearce and Cann (1971) then collected data on the abundance of the trace elements titanium, zirconium and yttrium from a variety of tectonic settings, such as (1) basaltic rocks from the ocean floor; (2) tholeiitic basalt lavas from Hawaii; (3) alkali basalts from Flores in the Azores in the mid-Atlantic Ocean; (4) basaltic andesites and andesites from Japan and New Zealand; and (5) a special type of island arc tholeiitic basalt from Japan. The results of their research were encouraging and they (1971, p. 343) concluded that it was possible to use chemical discriminant diagrams to assign tectonic settings to volcanic rocks of unknown origin.

In 1973 Pearce and Cann prepared a larger geochemical database (see Table 2.8). It was used to evaluated trace element data on basalts from four major tectonic settings: (1) the mid-oceanic ridges; (2) volcanic arcs; (3) oceanic islands; and (4) continental areas. They (p. 292) were unable to separate the oceanic island from the continental basalts. This forced them to merge these rock types and propose a 'within plate' category. They also argued that the yttrium/

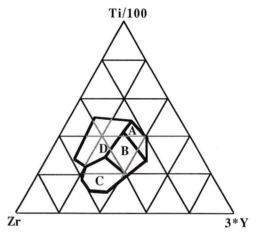

Fig. 2.22 Ternary diagram (Ti/100, Zr and 3*Y) that is used to discriminate between plate continental and oceanic island basalts (field D), the mid oceanic ridge basalts (field B), the volcanic arc tholeiitic basalts (fields A and B), and the volcanic arc calc-alkali basalts (fields B and C). After Pearce and Cann, 1973, p. 295.

niobium ratio offered a ready method for finding out the alkalinity of altered basalts because the yttrium/niobium ratio decreased with increasing alkalinity (p. 297). Their 1973 paper also contains the now well known Ti/100–Zr–Y*3 geochemical discriminant diagram (see Fig. 2.22). This diagram includes scaling factors for titanium and yttrium. Scaling factors provide a means of bringing datapoints into the centre of a graph or diagram without significantly distorting their relative positions. In the past 25 years the Pearce and Cann chemical discriminant diagram and a series of other broadly similar diagrams such as the Nb*2–Zr/4–Y (Meschede, 1986), the TiO_2–MnO*10–P_2O_5*10 (Mullen, 1983) and the Hf/3–Th–Ta (Wood *et al.*, 1979), have been widely used.

Many investigators have questioned the immobility of the so-called immobile trace elements. Trace element mobility is normally controlled by: (1) the nature of the changes in mineral assemblage induced by superficial alteration or low grade metamorphic processes, and (2) the chemical and physical characteristics of the fluid phase that triggers alteration. As a general rule the large cations of small charge, or low field strength incompatible elements, are mobile, whereas the small highly charged cations, or high field strength elements, are immobile. Other petrologists have questioned the ability of chemical discriminant diagrams to predict the tectonic settings of magmatic rocks (Vallance, 1974; Thompson *et al.*, 1980; Holm, 1982; Philpotts, 1985; Arculus, 1987; Wang and Glover, 1992). Some petrologists imply that the positions occupied by samples plotted on discrimination diagrams normally have more to do with their magmatic history than their tectonic setting. Early criticism by Vallance was based on his investigation of the chemical composition of a gradationally altered series of basaltic rocks from Bombay in India. He found titanium, zirconium and yttrium to be moderately stable but not absolutely immobile. This was because during alteration no single phase acted as a repository for all these elements. He concluded that ratios that use titanium, zirconium and yttrium can only serve as reliable discriminants in rocks that contain phases that can stabilise these elements during the alteration process. Arculus (1987) found that even if investigations are restricted to fresh volcanic rocks of known tectonic setting some samples will have ambiguous trace element signatures. He also showed that some volcanic suites which evolved in an unambiguous tectonic setting have trace element signatures that straddle more than one field on the various chemical discriminant diagrams.

Pearce (1987) acknowledged the shortcomings of these diagrams and introduced a more complex expert system to improve his rate of success at predicting the tectonic setting of old volcanic rocks. An expert system is a computer system that uses software to store and apply the knowledge of experts in a particular field. This enables the user of the system to draw on this expertise when making decisions. In his prototype expert system, Pearce has attempted to define a series of 'tectonomagmatic environments', and the geochemical, geological, mineralogical and petrological characteristics of lavas from these environments. This approach for detecting the tectonic setting of ancient volcanic suites seems to have great potential for further development and fine-tuning.

The trace elements commonly used in chemical discrimination diagrams (e.g. Ti, Y, Zr, Nb, Hf, Ta and Th) are all high-field-strength trace elements. Their abundance is strongly contingent on the stabilisation of suitable refractory phases during the partial melting process, and this usually depends on the composition of the magma and the prevailing pressure and temperature. This suggests that tectonic activity is normally only one of the many factors that govern the major and trace element compositions of magmas. Arculus (1987, p. 9) has warned that the uncritical acceptance of a trace element signature as diagnostic of a particular tectonic setting is likely to result in confusion.

Much of the early enthusiasm for chemical discriminant diagrams was probably misplaced. It was based on an over-optimistic view of the immobility of the so-called immobile trace elements, and a simplistic view of the often complex processes that generate magma. It is probably sensible to regard chemical discriminant diagrams as one of a battery of chemical and petrographic techniques that may provide clues to the tectonic setting of magmatic rocks. Many magmatic rocks such as the continental flood basalts have complicated evolutionary histories and one must consider both the initial source of the magma and the tectonic controls that operated during its subsequent evolution. Saunders *et al.* (1992, p. 49) have shown that continental flood basalts may reveal geochemical evidence of plume affinities, lithospheric affinities or crustal contamination. The geochemical characteristics of these various types of continental flood basalts are readily shown on normalised, multi-element diagrams, or spidergrams.

Before concluding this brief review of chemical discrimination diagrams, it is apposite to consider another group of discrimination diagrams that use the compositions of clinopyroxenes to reveal the magmatic affinities of slightly altered volcanic rocks. This method is based on two assumptions: (1) that the compositions of clinopyroxenes vary according to the chemical composition of their host magma, or lava; and (2) that the cores of the clinopyroxene phenocrysts in many metabasalts and altered volcanic rocks remain chemically unaltered. In the discrimination diagrams introduced by Leterrier and others (1982) the elements sodium, aluminium, calcium, titanium and chromium are used to figure out the tectonic settings of the altered rocks being investigated.

Spidergrams

Spidergrams are widely used as they enable petrologists to compare speedily the abundance of a

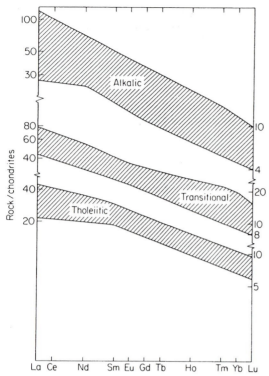

Fig. 2.23 A chondrite-normalised lanthanoid diagram showing various lavas from Loihi Seamount, Hawaii. After Clague, 1987, p. 244.

Table 2.9 Recommended chondrite values for lanthanoid normalisation. After Boynton, 1984, p. 91.

La	lanthanum	0.31
Ce	cerium	0.808
Pr	praseodymium	0.122
Nd	neodymium	0.6
Sm	samarium	0.195
Eu	europium	0.0735
Gd	gadolinium	0.259
Tb	terbium	0.0474
Dy	dysprosium	0.322
Ho	holmium	0.0718
Er	erbium	0.21
Tm	thulium	0.0324
Yb	ytterbium	0.209
Lu	lutetium	0.0322

not easily fit into the crystal structures of the early-formed minerals. They are continuously enriched in the residual magma. This explains why these elements typically concentrate in late-stage granitic pegmatites. In petrogenesis the incompatible elements are sometimes defined as those elements that do not readily fit into the structural sites of the essential minerals of the upper mantle.

Several different types of spidergrams have been proposed with variations in both the order in which the elements are portrayed and in the normalisation standards. Elements that are frequently used include barium, potassium, rubidium, thorium, tantalum, niobium, lanthanum, cerium, strontium, neodymium, phosphorus, samarium, zirconium, hafnium, titanium, gadolinium, yttrium and ytterbium. They may be normalised to chondritic values (see Table 2.9), but they are often normalised to the composition of the primary magma for a particular suite of rocks or a hypothetical primordial upper mantle composition. The order in which the elements are plotted is arbitrary, but the usual order is one that gives a moderately smooth graph for the elements found in a typical mid-oceanic ridge basalt (Wood *et al.*, 1979).

The lanthanoids

The recommended name for the group of elements that range from atomic number 57 (lanthanum) to 71 (lutetium) is the lanthanoids. In petrology these important trace elements are often known as the

broad spectrum of elements from different rocks. These diagrams evolved from the well-known, chondrite-normalised lanthanoid abundance diagrams (see Fig. 2.23). The expression chondrite-normalised data means that the elemental abundance values for a rock are divided by standard values obtained from carbonaceous chondrites (see Table 2.9). As has already been noted these meteorites are normally considered the most primitive material in the Solar System. If element abundance values for a particular rock are normalised using these chondritic data the computation removes fluctuations in elemental abundance initially produced by the processes of nucleogenesis. The computation also reduces the extreme range in elemental abundance found in most sets of raw data.

Spidergrams usually contain the lanthanoids plus some of, or all, the other incompatible elements. The incompatible elements have been defined in various ways. In a cooling and crystallising picrobasaltic or basaltic magma ions of the incompatible elements do rare earth elements, or REE. Traditionally the lanthanoids with low atomic numbers are called the light rare earths (LREE) and those with higher atomic numbers are called the heavy rare earths (HREE). The lanthanoids usually form stable 3^+ ions of similar, but slightly different, size. As the charge on the nucleus increases across the lanthanoid group, it pulls the various subshells closer to the nucleus. This results in the radii of their 3^+ ions decreasing as their atomic number increases, thus causing an effect called the lanthanoid contraction. The lanthanum ion (La^{3+}) has a radius of 0.106 nm, whereas the heavier lutetium ion (Lu^{3+}) has a radius of 0.085 nm. In some magmatic processes these differences in ionic size induce the lanthanoids to become fractionated relative to one another and this enables petrologists to use the lanthanoids to help unravel the processes responsible for the origin and evolution of some magmatic rocks. The chemical characteristics of the lanthanoids are essentially similar. This is because the differences in their electronic structures chiefly involve their inner electrons and not their outer electrons, and it is the latter that determine their chemical behaviour.

Cerium and europium are of special interest to petrologists because they do not always occur in the 3^+ oxidation state. The former sometimes occurs in the 4^+ oxidation state and in many magmatic environments europium ion has a charge of 2^+. The conundrum, of why the mare basalts have negative europium anomalies, was solved when it was realised that divalent europium had been preferentially incorporated into the plagioclase of an earlier formed suite of plagioclase cumulate rocks (see Sections 6.13 and 7.12). It is now recognised that positive and negative europium anomalies are usually produced by the addition or removal of feldspars. To a lesser extent other minerals, like clinopyroxene, garnet, hornblende, orthopyroxene and sphene (titanite), can also cause europium anomalies. Garnet also plays another important role as it often seems to control the abundance pattern of the lanthanoids in primary magmas. When such magmas display extreme depletion of the heavy lanthanoids relative to the light ones it is usually claimed that this pattern is a result of the presence of garnet in the source rocks. Felsic and foiditic rocks often display remarkable concentrations and fluctuations in their lanthanoid abundance patterns. These changes are usually produced by fluctuations in the abundance of accessory minerals such as allanite, apatite, monazite, sphene (titanite) and zircon.

2.16 Hierarchical and simple classifications

The many recommendations of the IUGS subcommission on the systematics of igneous rocks can be integrated to form a broad hierarchical classification (see Figs 2.24 and 2.25). Rocks belonging to the pyroclastic group, the carbonatitic group, the lamprophyric group, the melilitic group and the charnockitic group are separated from and classified before the other plutonic and volcanic rocks. The pyroclastic rocks and tephra belong to a special group of rocks because they are products of explosive volcanism. Their textures and distribution patterns are distinct from the other magmatic rocks. The separation of the carbonatitic, lamprophyric and melilitic rocks from the common magmatic rocks implicitly recognises that, just as there is a main sequence of stars illustrated by the well-known Hertzsprung-Russell diagram, there is also a main sequence of igneous rocks. These abundant main sequence rocks grade from picrobasalt and basalt to the strongly differentiated rhyolites and phonolites. The charnockitic rocks are special because they evolved in a distinctive metamorphic environment where enstatite (hypersthene) crystallised. Cumulate rocks are special because they lie outside, yet are usually complementary to, the main sequence of igneous rocks. Most cumulates excluding some feldspathoid-cumulates can be divided into feldspar-depleted or feldspar-enriched rocks. The former have low-Al_2O_3 contents (less than 6 per cent), whereas the latter have high-Al_2O_3 contents (more than 20 per cent). These chemical characteristics set cumulates apart from the rocks of the main sequence. When thinking about this chapter it is probably important to keep in mind that when one describes a building one does not have to name every brick in the structure.

The public is fascinated by the awesome behaviour of the volcano-magmatic system, yet they usually know so little about it. Many young people are captivated by volcanic activity and the great size and apparent homogeneity of granitic batholiths. Their inquisitiveness about the volcano-magmatic system is often suffocated by the many technical names used in igneous petrology. Petrological nomenclature must be radically pruned and reformed if the public is to enjoy the delights of igneous petrology. The minimum requirement is a straightforward classification of magmas and magmatic rocks that is internally

SIMPLE HIERARCHICAL CLASSIFICATION OF IGNEOUS ROCKS

SPECIAL IGNEOUS ROCKS

Cumulate Rocks
including
Ferromagnesian Cumulates, Feldspar Cumulates and Feldspathoid Cumulates

High-magnesium Rocks
including
Boninite, Komatiite, Meimechite and Picrite

Igneous-looking Rocks
including
Charnockitic Rocks, Eclogite, Glimmerite, Gneiss and Migmatite

Lamprophyric Rocks
including
Carbonatite, Kimberlite and Lamproites

Mantle Rocks
including
Depleted Mantle Rock, Fertile Mantle Rock or Hypothetical Mantle Rock (pyrolite)

Melilitic Rocks
excluding
Melilite-bearing Lamprophyres

Non-silicate Rocks
excluding
Carbonatite and Cumulate Rocks

Pyroclastic Rocks
classified using
Clast Size and Major Element Composition

Foidolites and Foidites
including
Sodic, Potassic and Ultrapotassic Rock-types

MAIN SEQUENCE IGNEOUS ROCKS

Subalkaline Series *	Midalkaline Series *	Alkaline Series
Picrobasalt/Basalt	Picrobasalt/Basalt	Picrobasalt/Basanite
to	to	to
Rhyolite (Granite)	Trachyte/Trachydacite (Syenite/Quartz Monzonite)	Phonolite (Foid Syenite)

Fig. 2.24 Simple hierarchical classification of igneous rocks. This system of classification separates the common, or main sequence, igneous rocks from the other, or special, igneous rocks that are normally classified separately. Adapted from Le Maitre and others, 1989, and Middlemost, 1991, p. 76.

consistent and can be used as a mastering net to enclose and link the various parts of the volcano-magmatic system.

The various igneous magmas and rocks can be grouped into associations that keep recurring in space and time. These associations are the key to developing a simplified classification of igneous materials. This approach is important because it separates igneous petrology from its natural history past and repositions it as a physical science. As a branch of physical science, one would expect igneous petrology to be committed to reclassifying its subject-matter in the light of new discoveries. Rocks can no longer be regarded as haphazard assemblages of minerals that

Alkaline Series

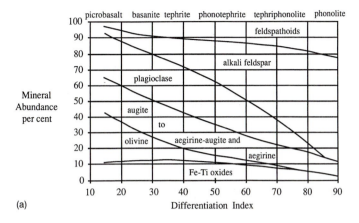

(a)

Midalkaline Series

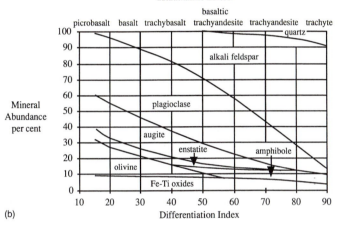

(b)

Subalkaline Series

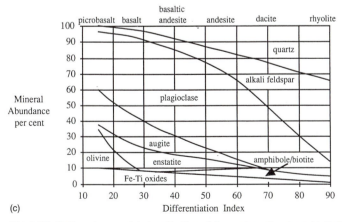

(c)

Fig. 2.25 Rudimentary modal classification of the major volcanic rocks that belong to the subalkaline, midalkaline and alkaline lineages. Note how the modal compositions of the rocks of each lineage become more distinct as their differentiation index increases.

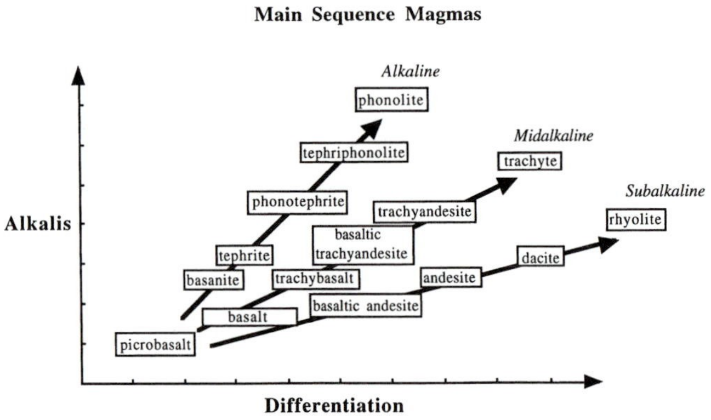

Main Sequence Magmas

Fig. 2.26 Diagram depicting the principal types of magmas that belong to the subalkaline, midalkaline and alkaline suites.

need to be classified. They are products of complex, dynamic processes and fascinating materials that carry information about their genesis and their standing in the grand planetary volcano-magmatic system.

Magma is the pivotal concept in igneous petrology and the naming of magmas should be an important component in any system developed to classify igneous materials. This stratagem also will ensure that the planned classification is sympathetic to the needs of petrogenesis. When attempting to assemble a practical classification of the principal magmas and igneous rocks it is important to define: (1) the primitive field; (2) the end members in the principal liquid lines of descent; and (3) the rare materials that lie outside this main sequence of igneous rocks. These simple goals are readily met by the TAS diagram. In addition sigma isopleths can be superimposed on this diagram to provide a successful way of defining the principal magmatic lineages. Rocks of the subalkaline suite have sigma values less than 2.5, the midalkaline suite is bound by sigma values of between 2.5 and 10, and the alkaline suite is defined by sigma values in the range 10 to 25. The principal types of magmas, plutonic rocks and volcanic rocks can all be defined on appropriately modified TAS diagrams (see Figs 2.7, 2.11, 2.17 and 2.26).

Materials that lie outside the main sequence field have high (greater than 25) or negative sigma values. Foiditic/foidolitic magmas and rocks, and magmas and rocks enriched in non-silicate phases typically plot outside the main sequence field. Many of these rocks are cumulates. In any practical classification of igneous rocks it is desirable to insert a cumulate category because this enables one to simplify the procedures used to classify many foidolitic rocks also the

rocks found in layered igneous complexes. The rare rocks that contain less silica than the picrobasalts and basanites are likely to contain non-silicate phases such as carbonates or Cr-Fe-Ti oxides and silicates with exceptionally low-silica contents. The latter include olivine, some feldspathoids and melilite (see Chapter 11).

References

Abbott, D., Burgess, L. and Longhi, J., 1994. An empirical thermal history of the Earth's upper mantle. *Journal of Geophysical Research*, 99 (B7), pp. 13 835–13 850.

Anon, 1973. Plutonic Rocks: Classification and nomenclature recommended by the IUGS Subcommission on the Systematics of Igneous Rocks. *Geotimes*, 18 (10), pp. 26–30.

Arculus, R.J., 1987. The significance of source versus process in the tectonic controls of magma genesis. *Journal of Volcanology and Geothermal Research*, 32 (1–3), pp. 1–12.

Ashwal, L.D., 1993. *Anorthosites*. Springer-Verlag, Berlin, 422 pp.

Basaltic Volcanism Study Project (BVSP), 1981. *Basaltic Volcanism on the Terrestrial Planets*. Pergamon Press, Inc., New York, 1286 pp.

Bogard, D.D., Nyquist, L.E. and Johnson, P., 1984. Noble gas contents of shergottites and implications for the Martian origin of SNC meteorites. *Geochimica et Cosmochimica Acta*, 48, pp. 1723–1739.

Bowen, N.L., 1913. Melting phenomenon of plagioclase feldspar. *American Journal of Science*, Fourth Series, 38, pp. 207–264.

Bowen, N.L., 1928. *The evolution of the igneous rocks.* Princeton University Press, Princeton, New Jersey, 334 pp.

Boynton, W.V., 1984. Cosmochemistry of the rare earth elements: meteorite studies. In Henderson, P. (editor) *Rare Earth Element Geochemistry*, pp.63–114. Elsevier, Oxford, 510 pp.

BVSP, 1981. See Basaltic Volcanism Study Project.

Cann, J.R., 1970. Rb, Sr, Y, Zr and Nb in some ocean floor basaltic rocks. *Earth and Planetary Science Letters*, 10, pp. 7–11.

Chayes, F., 1949. On correlation in petrography. *Journal of Geology*, 57, pp. 239–254.

Chayes, F., 1956. *Petrographic modal analysis.* John Wiley and Sons, Incorporated, New York, 113 pp.

Chayes, F., 1965. Titania and alumina content of oceanic and circumoceanic basalt. *Mineralogical Magazine*, 34, pp. 126–131.

Chayes, F., 1972. Silica saturation in Cenozoic basalt. *Philosophical Transactions of the Royal Society of London*, A. 271, pp. 285–296.

Chayes, F., 1973. A relation between silica saturation and oxidation state in class II, (ol + hy) normative basalts. *Carnegie Institution of Washington Yearbook* 72, pp. 671–673.

Chayes, F. 1975. Statistical Petrology. *Carnegie Institution of Washington Yearbook* 74, pp. 542–550.

Chayes, F., 1979. Electronic computation and book-keeping in igneous petrology. *Episodes* (IUGS Newsletter), 1979 (1), pp. 16–19.

Clague, D.A., 1987. Hawaiian alkaline volcanism. In Fitton, J.G. and Upton, B.G.J. (editors), *Alkaline Igneous Rocks*, pp. 227–252, Geological Society of London Special Publication Number 30, 568 pp.

Clarke, A.M., 1993. *Hey's Mineral Index: Mineral Species, Varieties and Synonyms.* Natural History Museum Publications, Chapman and Hall, London, 852 pp.

Clarke, D.B., 1992. *Granitoid Rocks.* Chapman and Hall, London, 283 pp.

Condie, K.C., 1989. Geochemical changes in Basalts and Andesites across the Archean-Proterozoic Boundary: Identification and Significance. *Lithos*, 23, pp. 1–18.

Daly, R.A., 1914. *Igneous Rocks and Their Origins.* McGraw-Hill Book Company, New York, 598 pp.

Daly, R.A., 1933. *Igneous Rocks and the Depths of the Earth.* McGraw-Hill Book Company, New York, 598 pp.

Dawson, J.B., 1989. Sodium carbonatite extrusions from Oldoinyo Lengai, Tanzania: Implications for carbonatite complex genesis. In Bell, K. (editor) *Carbonatites, genesis and evolution*, pp. 255–277, Unwin Hyman, London, 618 pp.

Deer, W.A., Howie, R.A. and Zussman, J., 1992. *An Introduction to the Rock-Forming Minerals.* Second Edition, Longman Scientific and Technical, Harlow, U.K., 696 pp.

Goldschmidt, V.M., 1911. Die Gesetze der Mineralassosiation vom Standpunkt der Phasenregel. *Zeitschrift für Anorganische Chemie*, 71, pp. 313–322.

Goldschmidt, V.M., 1937. The principles of distribution of chemical elements in minerals and rocks. *Journal of the Chemical Society, London*, pp. 655–673.

Green, T.H. and Pearson, N.J., 1986. Rare-earth element partitioning between sphene and coexisting silicate liquid at high pressure and temperature. *Chemical Geology*, 55, pp. 105–119.

Grossman, J.N. and Wasson, J.T., 1983. The compositions of chondrules in unequilibrated chondrites: An evaluation of models for the formation of chondrules and their precursor materials. In King, E.A. (editor), *Chondrules and Their Origin*, pp. 88–121, Lunar and Planetary Institute, Houston, USA, 377 pp.

Haggerty, S.E., 1970. The Laco magnetite flow, Chile. *Carnegie Institution of Washington Yearbook*, 68, pp. 329–330.

Harker, A., 1909. *The Natural History of Igneous Rocks.* Macmillan Company, New York, 384 pp.

Heiken, G., Vaniman, D. and French, B.M., 1991. *Lunar Sourcebook: A User's Guide to the Moon.* Cambridge University Press, Cambridge, U.K., 736 pp.

Heinrich, E.W., 1966. *The Geology of Carbonatites.* Rand McNally and Company, Chicago, USA, 555 pp.

Henderson, P., 1982. *Inorganic Geochemistry.* Pergamon Press, Oxford, 353 pp.

Holm, P.E., 1982. Non-recognition of continental tholeiites using Ti-Y-Zr diagram. *Contributions to Mineralogy and Petrology*, 79, pp. 308–310.

Irvine, T.N. and Baragar, W.R.A., 1971. A guide to the chemical classification of the common volcanic rocks. *Canadian Journal of Earth Sciences*, 8, pp. 523–548.

Kuno, H., 1968. Origin of andesite and its bearing on the island arc structure. *Bulletin Volcanologique*, series 2, volume 32, pp. 141–176.

Kuno, H., Yamasaki, K., Iida, C. and Nagashima, K., 1957. Differentiation of Hawaiian Magmas. *Japanese Journal of Geology and Geography*, 28 (4), pp. 179–218.

Larouziere, F.D. de, 1989. *Dictionnaire des roches d'origine magmatique*. Editions du BRGM, Orleans, France, 188 pp.

Larsen, E.S., 1938. Some new variation diagrams for groups of igneous rocks. *Journal of Geology*, 46, pp. 505–520.

Le Bas, M.J., Le Maitre, R.W., Streckeisen, A. and Zanettin, B., 1986. A chemical classification of volcanic rocks based on the total alkali-silica diagram. *Journal of Petrology*, 27, pp. 745–750.

Le Bas, M.J., Le Maitre, R.W. and Woolley, A.R., 1992. The construction of the total alkali-silica chemical classification of volcanic rocks. *Mineralogy and Petrology*, 46, pp. 1–22.

Le Maitre, R.W., 1976. Chemical variability of some common igneous rocks. *Journal of Petrology*, 17 (4), pp. 589–637.

Le Maitre, R.W., 1981. GENMIX – a generalised petrological mixing model program. *Computers and Geoscience*, 7, pp. 229–247.

Le Maitre, R.W., 1984. A proposal by the IUGS Subcommission on the Systematics of Igneous Rocks for a chemical classification of volcanic rocks based on the total alkali silica (TAS) diagram. *Australian Journal of Earth Sciences*, 31 (2), pp. 243–255.

Le Maitre, R.W., Bateman, P., Dubek, A., Keller, J., Lameyre, J., Le Bas, M.J., Sabine, P.A., Schmid, R., Sorensen, H., Streckeisen, A., Woolley, A.R. and Zanettin, B. (editors), 1989. *A Classification of Igneous Rocks and Glossary of Terms: Recommendations of the IUGS Subcommission on the Systematics of Igneous Rocks*. Blackwell Scientific Publications, Oxford, 193 pp.

Leterrier, J., Maury, R.C., Thonon, P., Girard, D. and Marchal, M., 1982. Clinopyroxene composition as a method of identification of the magmatic affinities of paleo-volcanic series. *Earth and Planetary Science Letters*, 59, pp. 139–154.

Longhi, J. and Ashwal, L.D., 1985. Two-stage models for lunar and terrestrial anorthosites: Petrogenesis without a magma ocean. Proceedings of the Fifteenth Lunar and Planetary Science Conference. *Journal of Geophysical Research*, 90, pp. C571–C584.

MacKenzie, W.S., Donaldson, C.H. and Guilford, C., 1982. *Atlas of Igneous Rocks and their Textures*. Longman, Harlow, U.K., 148 pp.

Mason, B., 1992. Victor Moritz Goldschmidt: Father of Modern Geochemistry. *The Geochemical Society*, Special Paper Number 4, San Antonio, Texas, 184 pp.

McBirney, A.R. and Hunter, R.H., 1995. The cumulate paradigm reconsidered. *The Journal of Geology*, 103, pp. 114–122.

McDonough, W.F. and Sun, S., 1995. The composition of the Earth. *Chemical Geology*, 120, pp. 223–253.

McPhie, J., Doyle, M. and Allen, R.L., 1993. *Volcanic Textures: A guide to the interpretation of textures in volcanic rocks*. CODES, University of Tasmania, Australia, 196 pp.

Meschede, M., 1986. A method of discriminating between different types of mid-ocean ridge basalts and continental tholeiites with the Nb-Zr-Y diagram. *Chemical Geology*, 56, pp. 207–218.

Middlemost, E.A.K., 1991. Towards a comprehensive classification of igneous rocks and magmas. *Earth-Science Reviews*, 31, pp. 73–87.

Middlemost, E.A.K., 1994. Naming materials in the magma/igneous rock system. *Earth-Science Reviews*, 37, pp. 215–224.

Mitchell, R.H. and Bergman, S.C., 1991. *Petrology of Lamproites*. Plenum Press, New York, 447 pp.

Mitchell, R.S., 1985. *Dictionary of Rocks*. Van Nostrand Reinhold Company, New York, 188 pp.

Morimoto, N., Fabries, J., Ferguson, A.K., Ginzburg, I.V., Ross, M., Seifert, F.A. and Zussman, J., 1988. Nomenclature of pyroxenes. *Mineralogical Magazine*, 52, pp. 535–550.

Mullen, E.D., 1983. $MnO/TiO_2/P_2O_5$: a minor element discriminant for basaltic rocks of oceanic environments and its implications for petrogenesis. *Earth and Planetary Science Letters*, 62, pp. 53–62.

Neumann, H., Mead, J. and Vitaliano, C.J., 1954. Trace element variation during fractional crystallisation as calculated from the distribution law. *Geochimica et Cosmochimica Acta*, 6, pp. 90–99.

Nisbet, E.G., Cheadle, M.J., Arndt, N.T. and Bickle, M.J., 1993. Constraining the potential temperature of Archean mantle: A review of the evidence from komatiites. *Lithos*, 30, pp. 291–307.

Nockolds, S.R., 1954. Chemical composition of some igneous rocks. *Bulletin of the Geological Society of America*, 65, pp. 1007–1032.

Nockolds, S.R. and Allen, R., 1953. The geochemistry of some igneous rock series, Part 1. *Geochimica et Cosmochimica Acta*, 4, pp. 105–142.

Nockolds, S.R. and Le Bas, M.J., 1977. Average calc-alkali basalt. *Geological Magazine*, 114, pp. 311–312.

Osborne, E.F., 1962. Reaction series for subalkalic igneous rocks based on different oxygen pressure conditions. *American Mineralogist*, 47, pp. 211–226.

Pearce, J.A., 1987. An expert system for the tectonic characterisation of ancient volcanic rocks. *Journal of Volcanology and Geothermal Research*, 32 (1–3), pp. 51–65.

Pearce, J.A. and Cann, J.R., 1971. Ophiolite origin investigated by discriminant analysis using Ti, Zr and Y. *Earth and Planetary Science Letters*, 12, pp. 339–349.

Pearce, J.A. and Cann, J.R., 1973. Tectonic setting of basic volcanic rocks using trace element analyses. *Earth and Planetary Science Letters*, 19, pp. 290–300.

Philpotts, J.A., 1985. Pearce-Cann discriminant diagrams applied to eastern North America Mesozoic diabase. *United States Geological Survey Circular Number 946*, pp. 114–117.

Pitcher, W.S., 1993. *The Nature and Origin of Granite*. Blackie Academic and Professional, Glasgow, 321 pp.

Poldervaart, A. and Parker, A.B., 1964. The crystallization index as a parameter of igneous differentiation in binary variation diagrams. *American Journal of Science*, 262, pp. 281–289.

Reynolds, I.M., 1985. The Nature and Origin of Titaniferous Magnetite-Rich Layers in the Upper Zone of the Bushveld Complex: A Review and Synthesis. *Economic Geology*, 80, pp. 1089–1108.

Richardson, W.A. and Sneesby, G. 1922. The frequency distribution of igneous rocks. *Mineralogical Magazine*, 19, pp. 303–313.

Rickwood, P.C., 1981. The largest crystals. *American Mineralogist*, 66, pp. 885–907.

Ringwood, A.E., 1955. The principles governing trace element distribution during magmatic crystallisation: Parts 1 and 2. *Geochimica et Cosmochimica Acta*, 7, pp. 189–202 and 242–254.

Rittmann, A., 1957. On the serial character of igneous rocks. *The Egyptian Journal of Geology*, 1 (1), pp. 23–48.

Rittmann, A., 1962. *Volcanoes and Their Activity*. Interscience Publishers, New York, 305 pp.

Rittmann, A., 1973. *Stable Mineral Assemblages of Igneous Rocks*. Springer-Verlag, Berlin, 262 pp.

Robinson, M. and Wadhwa, M., 1995. Messengers from Mars. *Astronomy*, 23, 8, pp. 44–49.

Rollinson, H.R., 1993. *Using Geochemical Data: Evaluation, Presentation, Interpretation*. Longman Scientific and Technical, Harlow, U.K., 352 pp.

Rock, N.M.S., 1991. *Lamprophyres*. Blackie, Glasgow, Scotland, 285 pp.

Rosenbusch, H., 1882. Ueber das Wesen der körnigen und porpyrischen Struktur bei Massengesteinen. *Neues jahrbuch*, 2, pp. 1–17.

Saggerson, E.P. and Williams, L.A.J., 1964. Ngurumanite from southern Kenya and its bearing on the origin of rocks in the northern Tanganyika alkaline district. *Journal of Petrology*, 5 (1), pp. 40–81.

Saunders, A.D., Storey, M., Kent, R.W. and Norry, M.J., 1992. Consequences of plume-lithosphere interactions. In Storey, B.C., Alabaster, T. and Pankhurst, R.J. (editors), *Magmatism and the causes of continental break-up*. Geological Society of London, Special Publication Number 68, pp. 41–60.

Sears, D.W.G. and Dodd, R.T., 1988. Overview and classification of meteorites. In Kerridge, J.F. and Matthews, M.S., *Meteorites and the early Solar System*, pp. 3–31. The University of Arizona Press, Tucson, USA, 1269 pp.

Shand, S.J., 1917. A system of petrography. *Geological Magazine*, Volume 4, Decade 6, pp. 463–469.

Shand, S.J., 1950. *Eruptive Rocks*. Fourth edition, Thomas Murby and Company, London, 488 pp.

Stormer, J.C. and Nicholls, J., 1978. XLFRAC: A program for interactive testing of magmatic differentiation models. *Computers and Geoscience*, 4, pp. 143–159.

Streckeisen, A., 1976. To each plutonic rock its proper name. *Earth-Science Reviews*, 12, pp. 1–33.

Streckeisen, A., 1978. IUGS Subcommission on the Systematics of Igneous Rocks: Classification and Nomenclature of Volcanic Rocks, Lamprophyres, Carbonatites and Melilite Rocks: Recommendations and Suggestions. *Neues Jahrbuch fur Mineralogie*, Stuttgart, Abhandlungen, 143, pp. 1–14.

Streckeisen, A. and Le Maitre, R.W., 1979. A chemical approximation to the QAPF classification of the igneous rocks. *Neues Jahrbuch fur Mineralogie*, Stuttgart, Abhandlungen, 136 (2), pp. 169–206.

Taylor, G.J., Scott, E.R.D. and Keil, K., 1983. Cosmic setting for chondrule formation. In King, E.A. (editor), *Chondrules and their origin*, pp. 262–278. Lunar and Planetary Institute, Houston, USA, 377 pp.

Taylor, S.R., 1982. *Planetary science: A lunar perspective*. Lunar and Planetary Institute, Houston, 481 pp.

Taylor, S.R., 1992. *Solar System Evolution: A New Perspective*. Cambridge University Press, Cambridge, U.K., 307 pp.

Thompson, R.N., Morrison, M.A., Mattey, D.P., Dickin, A.P. and Moorbath, S., 1980. An assessment of the Th-Hf-Ta diagram as a discrimination for tectonomagmatic classification of crustal contamination of magmas. *Earth and Planetary Science Letters*, 50, pp. 1–10.

Thornton, C.P. and Tuttle, O.F., 1960. Chemistry of igneous rocks, 1, Differentiation Index. *American Journal of Science*, 258, pp. 664–684.

Tilley, C.E., 1950. Some aspects of magmatic evolution. *Quarterly Journal of the Geological Society of London*, 106, pp. 37–61.

Tomkeieff, S.I., 1937. Petrochemistry of the Scottish Carboniferous-Permian igneous rocks. *Bulletin Volcanologique*, series 2, volume 1, pp. 59–87.

Tomkeieff, S.I., 1953. Petrochemistry and petrogenesis. In *The Tectonic Control of Igneous Activity*, pp. 24–27, Inter-University Geological Congress, Department of Geology, The University of Leeds, U.K.

Tomkeieff, S.I., Walton, E.K., Randall, B.A.O., Battey, M.H. and Tomkeieff, O., 1983. *Dictionary of Petrology*. Wiley, New York, 680 pp.

Ulmer G.C., Grandstaff, D.E., Weiss, D., Moats, M.A., Buntin, T.J., Gold, D.P., Hatton, C.J., Kadic, A., Koseluk, R.A. and Rosenhauer, M., 1987. The mantle redox state; an unfinished story? *Geological Society of America Special Paper 215*, pp. 5–23.

Vallance, T.G., 1974. Spilitic degradation of a tholeiitic basalt. *Journal of Petrology*, 15, pp. 79–96.

Wager, L.R., and Brown, G.M., 1968. *Layered Igneous Rocks*. Oliver and Boyd, Edinburgh, 588 pp.

Wager, L.R. Brown, G.M. and Wadsworth, W.J., 1960. Types of Igneous Cumulates. *Journal of Petrology*, 1, pp. 73–85.

Wang, P. and Glover, L., 1992. A tectonics test of the most commonly used geochemical discriminant diagrams and patterns. *Earth-Science Reviews*, 33, pp. 111–131.

Washington, H.S., 1917. *Chemical analyses of rocks published from 1884 to 1913 inclusive, with a critical discussion of the character and use of analyses*. A revision and expansion of Professional Paper 14. United States Geological Survey, Professional Paper 99.

Wasson, J.T., 1985. *Meteorites: Their Records of early Solar-System History*, W.H. Freeman and Co., New York, 267 pp.

Wasson, J.T. and Kallemeyn, G.W., 1988. Compositions of chondrites. *Philosophical Transactions of the Royal Society of London*, Series A, 325, pp. 535–544.

Wedepohl, K.H., 1969. Composition and Abundance of Common Igneous Rocks. In Wedepohl, K.H. (editor) *Handbook of Geochemistry*, pp. 227–249, Volume 1, Springer-Verlag, Berlin, 442 pp.

Wood, D.A., Joron, J.L. and Treuil, M.l., 1979. A reappraisal of the use of trace elements to classify and discriminate between magma series erupted in different tectonic settings. *Earth and Planetary Science Letters*, 45, pp. 326–336.

Wood, J.A., 1990. Meteorites. In Beatty, J.K. and Chaikin, A. (editors) *The New Solar System*, Third Edition, pp. 241–250. Cambridge University Press, Cambridge, UK, 326 pp.

Woolley, A.R. and Kempe, D.R.C., 1989. Carbonatites: Nomenclature, average chemical composition and element distribution. In Bell, K. (editor) *Carbonatites, genesis and evolution*, pp. 1–14. Unwin Hyman, London, 618 pp.

Wright, T.L. and Fiske, R.S., 1971. Origin of the differentiated and hybrid lavas of Kilauea Volcano, Hawaii. *Journal of Petrology*, 12 (1), pp. 1–65.

Wyllie, P.J., 1967. Ultramafic and ultrabasic rocks; Petrography and petrology. In Wyllie, P.J. (editor) *Ultramafic and Related Rocks*, pp. 1–7. John Wiley and Sons, Incorporated, New York, 464 pp.

CHAPTER 3

Heat flow and planetary interiors

What were the range and distribution of internal temperature in the newborn Earth?... How far has convection aided pure conduction in the wastage of heat?

Daly, 1933, p. 227.

3.1 Heat and its transport

On a global scale the Earth operates like a complex, self-regulating heat pump. Thermal instabilities in the mantle produce mechanical movements that interact with and help drive the lithospheric plates. Seismic tomography provides a technique for charting the complex thermal patterns within the mantle. A budget of the heat lost from the mantle can be prepared if one develops a thermal model that simulates the processes of ocean-floor spreading, ocean-crust cooling, subduction, and intraplate volcanism. Oceanic lithosphere is now being subducted at velocities that vary between 10 and 110 km per million years. During this process the bulk density of the descending crust increases and this helps boost the rate of subduction. The continual dispatch of cold lithospheric materials into the upper mantle is a particularly important method of cooling the upper mantle. The complementary processes of water cooling submarine lava flows and subduction provide a remarkable cooling mechanism that is unique to the Earth.

If one wishes to understand the various dynamic processes that operate within a planet one has to try to predict the changing temperature patterns within the planet through time. The heat that reaches the surface of the planets comes from the Sun and from the interior of the planets. On Earth and the inner planets most of the heat comes from the Sun. In striking contrast Jupiter is currently losing about twice as much heat as it traps from the Sun. Its atmosphere is a large bottom-heated system like the Earth's mantle. Venus has a mean surface temperature of 475°C. This high temperature is the result of a huge input of solar energy, combined with the operation of a runaway greenhouse effect. On Earth most

of the Sun's heat is either reflected or reradiated out into space. Solar energy interacts with the atmosphere, biosphere and hydrosphere. It helps to propel the global weather system that plays an essential part in the alteration and erosion of surface rocks. At the present stage in the Earth's history solar heating has little influence on the behaviour of the planet's magma/igneous rock system. In the distant future this will change because the Sun will transform into a red giant star and provide enough heat to cause extensive melting of the surface layers (see Section 13.5).

Within the planets heat is transported by advection, convection, radiation and thermal conduction. Convection is the main heat transfer system through the mantle and outer core of the Earth. Most of the mantle behaves like a high-viscosity liquid and heat is transported by convection, or the actual movement of part of this body. In the simplest hypothetical system, mantle materials meet a source of heat, such as the outer core, become hotter and expand. Being less dense they rise and their place is taken by colder materials. In the Earth's mantle the pattern of heat transfer is complex. Some heat that flows from the core is likely to perturb the thermal boundary layer separating it from the mantle. This is likely to generate narrow plumes of rising liquid. Such hot plumes are usually called mantle plumes or thermal plumes. In a recent review paper on the core–mantle boundary region Loper and Lay (1995, p. 6410) have stated that the thermal boundary layer at the base of the mantle is the most likely source of buoyant material to feed mantle plumes. The present heat flux from the core to the mantle is probably about three terawatts (3×10^{12} W).

For many years geophysicists have argued about whether the lower and upper mantle belong to either a single convective system or to two almost independent systems. Recent computer simulations suggest that both hypotheses are probably partially correct. Convective flow in the mantle is likely to be complex and unstable. It will oscillate between the two principal types of circulation. Usually the materials in the lower and upper mantle will convect as independent systems, but occasionally the build-up of cooler materials at the base of the upper mantle will reach a critical mass and plunge down through the boundary layer. This will generate a countercurrent of hot lower mantle material that spurts into the upper mantle.

Thermal conduction is the transmission of heat from an area of higher to an area of lower temperature. This process results from the interaction of atoms or molecules that contain different amounts of energy. The rate of heat conduction is normally proportional to the temperature gradient and it is fastest when the gradient is large. Thermal conduction is an essential means of heat transference through the rocks of the lithosphere and inner core.

Radiation describes the direct transfer of heat from a source by electromagnetic radiation. One can feel molten lava radiating heat but within the Earth radiation is unlikely to be an important method of transferring heat. Advection is usually regarded as the process by which materials and heat are transferred horizontally from one place to another. In tectonic discussions advection usually means the lateral movement of mantle material, but heat is sometimes said to be advected when hot materials are lifted up into a cooler environment by tectonic activity.

At a very early stage in the evolution of the Earth there was a separation of phases of different density to form the core and mantle. This process liberated a huge amount of gravitational energy that produced an enormous amount of heat. Part of this heat remains stored in the core. It is released as latent heat of solidification as the molten outer core crystallises. This heat helps drive the convection system in the outer core, and the latter is responsible for generating the Earth's magnetic field. Convection in the outer core was probably more active during the early history of the Earth (Elder, 1976, p. 81).

During this early stage in planetary development heat was also being generated from the kinetic energy possessed by impacting planetoids and planetesimals. One such impact was the hypervelocity event that created the Moon. Within the Earth's interior heat has always been produced by the decay of radioactive isotopes. Currently isotopes of the elements thorium,

uranium and potassium generate most of this radioactive heating. In the early Solar System these, and other radioactive isotopes, would have been available at much higher concentrations. The present continental crust contains about 9130 ppm potassium, 3.5 ppm thorium and 0.91 ppm uranium (see Table 1.3). In the mantle these elements are close to two orders of magnitude less abundant (see Table 3.1). Isotopes of potassium are particularly interesting because only potassium 40 is radioactive, and it now comprises only 0.01 per cent of ordinary potassium. The common isotopes are potassium 39 (93.10 per cent) and potassium 41 (6.88 per cent). If one estimates the present radioactive heat-generating capacity of the materials in the crust and upper mantle one discovers that thorium and uranium generate significantly more heat than potassium. Other radioactive isotopes with short half-lives were probably more important during the early stages in the evolution of the planets. The radioactive disintegration of thorium 232, uranium 235, uranium 238 and potassium 40 accounts for about half the heat currently escaping from the interior of the Earth.

The short-lived radionuclides include aluminium 26, manganese 53, palladium 107, iodine 129, samarium 146 and plutonium 244, and they are often called the extinct radionuclides. Iodine 129 has often been studied. It has a half-life of 16 million years and its prior existence can be detected in some meteorites by looking for an excess of its beta-decay radiogenic daughter xenon 129. The Earth's atmosphere contains about 7 per cent more xenon 129 than predicted from geochemical models of the primordial composition of the atmosphere. This is regarded as unambiguous evidence of the presence of radiogenic xenon generated by iodine 129 (Podosek and Swindle, 1988, p. 1111).

The boundary between the Earth's mantle and core is probably the most fundamental interface within the planet. It separates unlike materials with very different properties. Density and seismic velocity differences across this boundary are, for example, much larger than those across the interface between the mantle and crust. The lowermost mantle, or D'' layer, is as heterogeneous as the crust (see Sections 1.3 and 3.4). Geodynamic modelling and ultra-high pressure laboratory simulations provide a guide for estimating the temperature at the transition between the molten outer core and the solid inner core. These data can then be used to calculate the temperature at the core–mantle interface and even at the Earth's centre. The pressure and temperature at the centre

Table 3.1 Major element compositions of selected upper mantle rocks.

Number	1	2	3	4	5
SiO_2	44.48	44.2	45.83	45.1	45
TiO_2	0.1	0.13	0.09	0.2	0.15
Al_2O_3	3.34	2.05	1.57	4.6	3.3
Cr_2O_3	0.46	0.44	0.32	0.3	0.44
FeO total	8.72	8.29	6.9	7.9	8
MnO	0.13	0.13	0.11	0.1	0.13
NiO	0.26	0.28	0.29	0.2	0.25
MgO	39.24	42.21	43.41	38.1	39.8
CaO	3.14	1.92	1.16	3.1	2.6
Na_2O	0.27	0.27	0.16	0.4	0.34
K_2O	0.03	0.06	0.12	0.02	0.02
P_2O_5	0.01	0.03	0.04	0.02	nd
H_2O^+	nd	nd	nd	nd	nd
CO_2	0.03	nd	nd	nd	nd
Total	100.21	100.01	100.00	100.04	100.03
	BVSP, 1981, p. 288	Maaloe & Aoki, 1977, p. 165	Carswell, 1980, p. 135	Ringwood, 1975, p. 188	Taylor & McLennan, 1985, p. 261

1 = mean major element composition of eight spinel lherzolite xenoliths from Kilbourne Hole, New Mexico (BVSP, 1981, p. 288), USA; 2 = mean major element composition of 384 spinel lherzolite xenoliths (Maaloe and Aoki, 1977, p. 165) ; 3 = mean major element composition of 61 garnet lherzolite (sensu lato) xenoliths (Carswell, 1980, p. 135; analyses were recalculated on a water and carbon dioxide free basis); 4 = major element composition of the average mantle pyrolite (Ringwood, 1975, p. 188); and 5 = estimated major element composition of the upper mantle (Taylor and McLennan, 1985, p. 261).

of the Earth are estimated to be about 364 giga-pascals and 6600°C. At the core–mantle boundary the pressure is probably about 136 giga-pascals, and according to Yoneda and Spetzler (1994) the temperature just above the D'' layer is about 2225°C.

Six processes seem to dominate the present thermal behaviour of the Earth's mantle. They are: (1) generation of heat by the decay of radioactive isotopes; (2) movement of heat from the core to the mantle; (3) transport of heat by convection; (4) movement of magma and other fluids from the mantle into the crust; (5) the subduction of cold lithospheric materials into the upper mantle; and (6) conductive transfer of heat into the crust (see Fig. 3.1).

Viscosity

Viscosity is a most important property of magmas and convecting fluids. It can be regarded as a measure of the ability of a fluid to offer internal resistance to flow and it is often regarded as a type of internal friction. Viscosity is particularly important during the initial segregation, ascent, emplacement and differentiation of a magma. It even influences the rate at which elements diffuse through a magma. Viscosity generally decreases with increasing temperature and with increasing pressure (see Plate 3.1). It is usually low in magmas with low silica contents. Scarfe and Hamilton (1980, p. 319) have shown experimentally that at 1300°C rhyolites are three orders of magnitude more viscose than comagmatic basaltic materials. X-ray and neutron diffraction studies of silicate melts show that they display long-range ordering, or polymerisation. This means that they contain linked silicon-oxygen tetrahedra and as the silica content of a magma increases these structures grow in size and complexity. The viscosity of a magma or a lava is also enhanced when the crystal content increases. For example once the crystal content of a lava exceeds about 55 per cent there is a large increase in apparent viscosity (Pinkerton and Stevenson, 1992).

Minor changes in the volatile content of a magma may have a significant effect on its viscosity, polymerisation and density. Water is a particularly important volatile in the generation of magmas because it lowers the melting temperature of silicate minerals. Experimental studies show that the solubility of water in a silicate melt normally increases with increasing pressure until a maximum is reached.

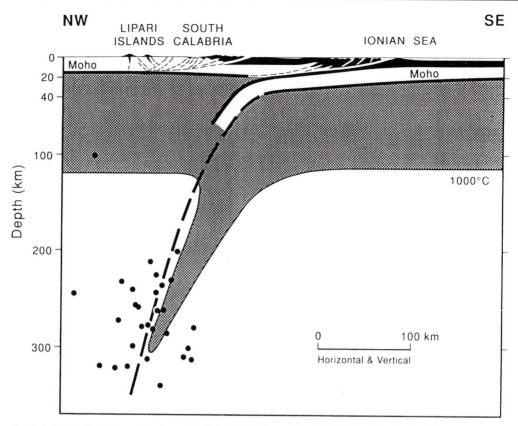

Fig. 3.1 Schematic section across the active Calabrian Arc in southern Italy. The rocks in the shaded area are cooler than 1000°C. This indicates that the subduction of lithospheric rocks is an important mechanism for cooling the upper mantle. The dots display the positions of seismic foci of deep earthquakes beneath the Aeolian Archipelago. After Blundell, Freeman and Mueller, 1992, p. 212.

Such solubility is usually greater in silicic magmas than in mafic magmas. For example at one giga-pascal and between 1000°C and 1200°C a rhyolite melt can dissolve 21 per cent water, while under the same physical conditions a basaltic melt dissolves only 14 per cent water (Sood, 1981, p. 143). At low pressures carbon dioxide is not particularly soluble in silicate melts, but at pressures above one and a half giga-pascals it is highly soluble. It frequently occurs in alkaline magmas generated at these higher pressures.

Flow of magma

To understand how lavas and magmas flow one has to delve into the physics of deformable materials. Nicholls (1990) has reviewed the mathematical operators and concepts used to formulate the equations that describe fluid flow. In the simplest model of the volcano-magmatic system small pockets of molten material segregate in the source region, and then coalesce to produce batches of magma. These batches either move along fractures if the surrounding rocks deform by fracture or move by diapirism if the surrounding rocks deform by flowing. The flow of lava is different and this is mainly because lava is not covered by a superincumbent load of rock. In a planet such as the Earth one would expect transition stages between each of these flow regimes.

Tait and Jaupart (1990) have reviewed the mathematics of the physical processes that operate when convective systems develop in magma chambers. They show that the equations used to describe these processes are complex, but rewarding because they provide realistic models that use standard fluid dynamical processes. In simple physical terms convection is caused by the action of gravity on density gradients. These gradients arise as the result of spatial variations in fluid composition, temperature and differences in the abundance of solid phases or bubbles. In the past 25 years consistent methods have been formulated for predicting the changing densities of silicate liquids during fractional crystallisation (Bottinga and Weill, 1970; Ghiorso *et al.*, 1994).

Any review of convection processes in large bodies of magma would be incomplete unless it recorded that the temperatures at which magmas begin to crystallise are related to pressure. This means that in a layer of magma crystallisation is likely to begin at a higher temperature in its basal layer than in its upper layer. In a thick layer of basaltic magma one might expect the initial crystallisation process to be more rapid at the floor than near the roof of the magma chamber. In a purely thermal model with the cooling of a simple liquid one would expect an unstable upper boundary layer, a convecting interior and a stagnant conductive lower boundary layer. Most magmas have complex compositions and complex cooling histories, and as they evolve their density is likely to change thus introducing the likelihood of convective instability in the lower boundary layer. All the outer boundary layers are likely to retain steep temperature, density and viscosity gradients. Because viscosity increases as temperature decreases, steep thermal gradients tend to induce steep gradients in viscosity. The higher viscosity of the cooler outer boundary layer, particularly the upper layer, dampens out minor instabilities and encourages compositional zoning.

Magma chambers are likely to be periodically replenished by new batches of magma. If the new magma is more dense than the resident magma it will pond at the base of the magma chamber and its momentum will govern its initial mixing characteristics. In the less likely situation when the incoming magma is less dense than the resident magma a buoyant plume will evolve and its attributes depend on the momentum of the incoming magma, and the density and viscosity contrasts between the two magmas.

3.2 Heat flow and melting in the Hadean and Archean Eras

Many petrologists have tried to analyse the thermal history of the Earth. In the Hadean Era (see Section 1.4) it was probably covered by a magma/lava ocean. This body of molten rock-forming material may have extended down to a depth of 400 km. Below this depth it is likely that the ambient pressure would induce the magma to crystallise. Until the end of the Hadean Era the magma/lava ocean is likely to have been agitated by vigorous convection currents that helped cool it, but these currents would also have overturned and destroyed any large body of proto-crust that formed on its surface.

In the early Archean the Earth developed its first solid crust. Abbott *et al.* (1994) have investigated the liquidus temperatures of many basaltic and picrobasaltic rocks that erupted at various times extending back to the Isuan or early Archean. Their data suggests that the mean liquidus temperature of these lavas has decreased from $1437 \pm 40°C$ in the middle Archean to $1272 \pm 7°C$ in the Phanerozoic. They also estimate that the mean upper mantle temperature has declined from $1618 \pm 55°C$ in the middle Archean to $1380 \pm 10°C$ in the Phanerozoic. These calculations assume that the komatiites have always evolved from atypical magmas that equilibrated at higher temperatures than was normal for the period when they erupted (see Section 7.4). Perhaps these high temperature magmas were transported to the surface in the high temperature central zones of mantle plumes.

The Kaapvaal Craton of southern Africa and the Western Shield of western Australia (see Figs 3.2 and 3.3) are the largest regions on Earth that retain rocks that can provide a coherent record of the early Archean. It has been suggested that for part of the Archean these cratons were joined to form a supercraton (De Wit and Hart, 1993, p. 313). The present Kaapvaal Craton covers about 1.2 million km^2. It is a block of lithosphere that varies in thickness between 170 and 350 km. At the surface it is mainly composed of metamorphosed and metasomatised mafic to ultramafic rocks (greenstones) and leucocratic tonalites. De Wit and Hart (1993) have used data from the Kaapvaal Craton to propose that the development of the earliest continental crust was probably triggered by the sinking of most of the primordial mid-oceanic ridge system below the level of the sea. This event introduced a new heat exchange process between hot igneous materials and the hydrosphere, and led to the development of vigorous hydrothermal systems that caused widespread metasomatism of the submarine crust. Such hydrothermal processes are likely to have dominated the early Archean environment resulting in widespread metasomatic alteration of the various basaltic and picrobasaltic rocks that

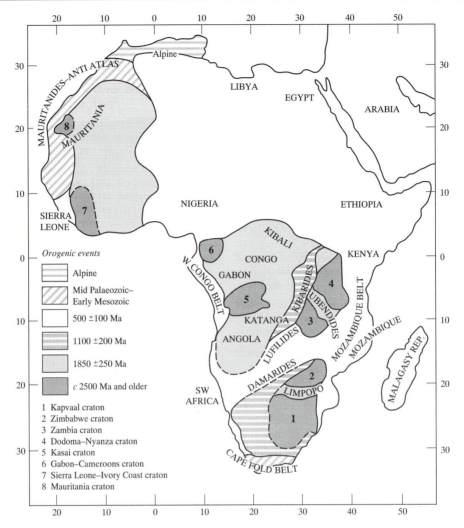

Fig. 3.2 Generalised map of the major structural and time units of south and west Africa showing the major cratons. After Clifford, 1971, p. 316 and Brown and Mussett, 1981, p. 189.

erupted beneath the primitive oceans. All the main petrographic units found in the present oceanic crust have been recorded from the 3.5 giga-year-old Jamestown Ophiolite Complex that crops out in the east-central part of the Kaapvaal Craton. Most of these rocks show unequivocal evidence of pervasive hydrothermal alteration.

De Wit and Hart (1993) postulate that the initial Archean continents formed as the result of a process they call regional intraoceanic obduction. The process began with the thrusting and stacking of slices of metasomatised oceanic lithosphere. Once this ma-

terial had been thrust into a thick pile, its weight triggered subsidence that promoted the partial melting of these hydrated materials resulting in the generation of dacitic magma. Much of this magma crystallised at depth where it produced leucocratic tonalites. Convergence between continental nuclei caused further magmatic activity. These processes culminated in the collision and amalgamation of miscellaneous tonalite-greenstone terrains. Collision between these proto-continents was also likely to trigger more partial melting and the generation of large volumes of rhyolitic magma.

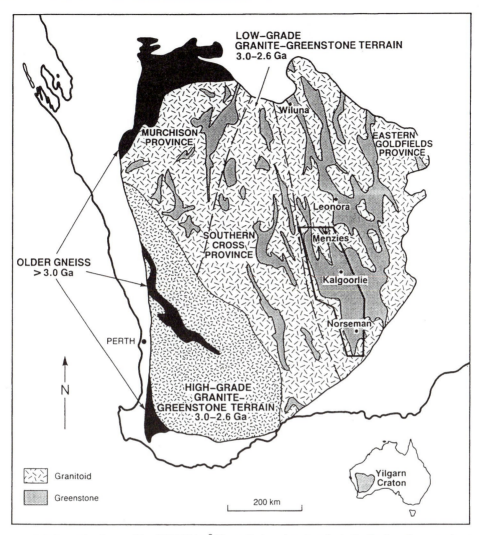

Fig. 3.3 Generalised map of the 650 000 km^2 Yilgarn Craton of western Australia. The boxed area encloses the Kalgoorlie and Norseman Terranes that are well known for their rich gold and nickel mineralisation. After Morris, Barnes and Hill, 1993, p. 2.

There has been much discussion about when modern-style plate tectonics and subduction began to operate within the Earth. An investigation of seismic reflections from an Archean collision suture that lies between the Abitibi Greenstone Belt and the Opatica Belt in the Superior Province of Canada has revealed evidence of subduction being an active process about 2.69 giga-years ago (Calvert *et al.*, 1995). This late Archean collision suture appears as a steeply dipping seismic reflector that extends down to a depth of about 65 km.

3.3 Core rocks and iron meteorites

The radius of the Earth's core is about 3485 km. It is the most inaccessible part of the planet and the only properties that can be measured directly are some of its physical properties. Estimates of its chemical composition are partly constrained by these physical properties, but most geochemical models are augmented by data collected from iron meteorites (see Section 2.11). At a depth of about 2885 km within the

Earth the silicate mantle passes into a core of much higher mean atomic weight (see Section 1.3). As S-waves are not transmitted through the outer core this major shell of the Earth is assumed to be liquid. The inner core, below about 5145 km, is considered to be a solid. From thermodynamic considerations analogy with the phases found in iron meteorites and from high pressure experimental studies it is concluded that the metallic iron-rich core is likely to contain significant amounts of phosphorus, cobalt, nickel, silver, the platinum group elements and gold.

Detailed investigations of the density of the core show that it is lower than would be expected for an iron-nickel alloy held in the pressure range prevailing in the core. This means that the core probably contains between 6 and 10 per cent of some lighter element such as sulphur, oxygen or silicon. In Section 1.3 it was suggested that the core was likely to be composed of 86.3 per cent iron, 7.4 per cent nickel, 5.9 per cent sulphur and 0.4 per cent cobalt, with traces of other elements including potassium. Recently Allegre and others (1995, p. 522) have suggested that the composition of the core is considerably different and contains 79.39 per cent iron, 7.35 per cent silica, 4.87 per cent nickel, 4.10 per cent oxygen, 2.30 per cent sulphur, 7790 ppm chromium, 5820 ppm manganese, 3690 ppm phosphorus and 2530 ppm cobalt.

It is enlightening to examine the relative abundances of pairs of elements that behave similarly during crystallisation in silicate magmas, but have different partition coefficients when in the presence of silicate materials and an immiscible metallic phase. An immiscible metallic phase of this sort probably existed prior to and during the core-forming process and it may still exist in the D'' layer at the base of the Earth's mantle. Titanium and phosphorus both respond as incompatible elements during the crystallisation of silicate magmas, though phosphorus is also strongly partitioned into iron-rich metallic phases. The titanium/phosphorus ratio in modern basalts is about ten, yet in the Solar System as a whole this ratio is less than one. This difference probably reflects the strong partitioning of phosphorus into the core. If the core formed during the Archean then the picrobasaltic and basaltic rocks of this era should have titanium/phosphorus ratios that increased with time. Such a change has not been found. This probably shows that the core-forming process was completed in the Hadean. Sun (1987, p. 79) has postulated a date of 4.6 giga-years for the core-forming event on Earth.

Many geochemists believe that after core formation the Earth received a further 1 per cent of chondritic material (see Section 2.11). A late addition of undifferentiated material would help account for the abundance of siderophile elements and platinum group elements in the crust and upper mantle. Because the crustal abundance of iridium is remarkably low it has been possible to use this element and other elements that are usually locked away in the core to support the hypothesis that a large asteroid or comet hit the Earth circa 60 million years ago at the changeover from the Cretaceous to the Tertiary period. The various sedimentary and impact rocks deposited at this time often contain anomalously high amounts of iridium, platinum and gold (Alvarez et al., 1979).

Geochemical models of the Earth usually estimate that most of the planet's supply of sulphur is stored in the core. This is most fortunate for the well-being of the common life-forms, because if sulphur were more abundant at the surface it probably would generate a toxic environment (Broecker, 1985, pp. 135–138). The transference of most of the Earth's iron into the core has also had a profound influence on the chemical composition and mineralogy of the mantle and crust.

Iron meteorites

Most iron meteorites are considered fragments of the cores of planetoids or asteroids (Wasson, 1985, p. 76) and rhenium 187–osmium 187 radiometric dating has shown that they formed at about the same time as the Earth (Luck and Allegre, 1983). When evaluating how they formed one has to consider how their parental planetoid obtained enough heat to enable differentiation to go to the core-forming stage. In small differentiated planetary bodies the initial source of heat had to be large, because: (1) heat is normally generated at a rate proportional to the volume of a planetoid; (2) the amount of heat available to most planetoids was greatest during the early stages in their evolution and has since decreased with time; and (3) heat is lost more rapidly from small bodies with a high ratio of surface area to volume. In a planetoid of normal chondritic composition (see Section 2.11) the core-forming processes would not be active below a temperature of about 1350°C. The most likely sources of heat available to a differentiating planetoid are long-lived radionuclides, short-lived radionuclides, electric currents induced by the immersion of a planetoid in an intense solar wind, radiation from a super-luminous phase of the Sun and conversion of kinetic energy into heat during the collisions associated with the accretionary process. It is probable that the necessary heat was generated by a combination of these processes with most of it coming from the short-lived radionuclides and the electric currents induced by strong solar winds.

As we have noted in Section 2.11 the iron meteorites are usually composed of nickel-iron metallic phases, such as kamacite (low-nickel) and taenite (high-nickel), with accessory amounts of troilite (FeS), schreibersite [(Fe,Ni)$_3$P] and graphite. The iron-nickel metallic phases tend to occur in lamellar intergrowths that show up as criss-cross or triangular patterns on polished and etched surfaces. These distinctive patterns are known as Widmanstätten structures. Stony-iron meteorites usually contain almost equal amounts of iron-nickel and silicate phases. The pallasites (see Section 2.11) are a particularly interesting group of these meteorites as they comprise large olivine crystals embedded in nickel-iron. They are usually considered to have evolved near the interface between the mantle and core in a layered planetoid and are likely to be similar to the material in the D$''$ layer of the lower mantle of the Earth.

3.4   The Earth's mantle dissected

A variety of indirect methods have been used to investigate the modal and chemical composition of the upper mantle (see Section 1.3). In 1897 Wiechert used theoretical reasoning based on the density of the Earth to suggest that it had an iron-rich core surrounded by a 1200 km thick rocky mantle. The densities of the upper mantle rocks are now considered to range from 3240 kg/m^3 to 3320 kg/m^3. These values are consistent with the upper mantle being principally composed of peridotites or olivine-rich rocks (see Table 2.2). Peridotites are subdivided into dunites, harzburgites, lherzolites and wehrlites (see Fig. 2.3). Geochemists used chondritic abundance data and data on the physical properties of the mantle to devise elaborate models of the chemical composition of the mantle (cf. Ringwood, 1975). Experimental petrologists try to simulate the conditions present in the upper mantle when primary magmas are generated (cf. Holloway and Wood, 1988). Their results have produced independent models of the modal and chemical composition of the upper mantle.

Upper mantle materials are sometimes discovered at the surface. Most of them have been thrust or obducted onto, or into, the crust. They may consist of slabs of the sub-oceanic upper mantle, or smaller samples of sub-continental upper mantle. The latter are usually xenoliths that occur in volcanic or sub-volcanic rocks. These xenoliths are typically found in alkali basalts, basanites, carbonatites, kimberlites and lamproites. More data on the physical nature and composition of the upper mantle can be obtained from the study of rocks that have been emplaced directly from the upper mantle as hot quasi-solid diapirs such as the Tinaquillo Peridotite of Venezuela. In Lanzo in Italy, Lherz in France, and Beni Bousera in Morocco one can examine extraordinary marble-cake peridotites composed of strips of eclogite embedded in peridotite.

Xenoliths provide the freshest samples of the upper mantle. Some of these fragments of the upper mantle are considered to have been rapidly transported to the surface at speeds of between 0.01 and 20 m/s. If the upward movement of these xenoliths was rapid it is probable that their compositions have not been significantly modified during transit to the surface. If a particular volcanic rock carries xenoliths of mantle origin it is usually regarded as having a primary, or near primary, composition.

The spinel lherzolites are the most abundant and widespread group of xenoliths derived from the upper mantle. They usually occur in basanites. A typical spinel lherzolite from Kilbourne Hole, New Mexico contains 60–70 per cent forsteritic olivine, 10–30 per cent aluminous enstatite, 8–12 per cent aluminous chrome diopside and 3–13 per cent aluminous chrome spinel, with minor amounts of sulphides and glass (see Table 3.1).

Garnet lherzolites are usually found as xenoliths in kimberlites (see Section 11.10). They are considered typical of the material found in the sub-continental upper mantle. According to Mathias *et al.* (1970, p. 85) the southern African garnet lherzolites usually contain 56 per cent forsteritic olivine (plus altered olivine), 25 per cent aluminous enstatite, 9 per cent aluminous chrome diopside and 7 per cent chrome pyrope garnet (plus reaction rims), with a few minor phases. Some garnet lherzolite samples contain diamond and the phases included in these diamonds are estimated to have equilibrated at pressures of between 5 and 6 gigapascals and a temperature of about 1100°C.

Harzburgites are peridotites that are essentially composed of olivine and orthopyroxene but they contain less than 5 per cent clinopyroxene. Xenoliths of this composition are probably more abundant than is usually acknowledged. Dawson *et al.* (1980, p. 325) found that only 16 of the 91 published modes of garnet lherzolite (sensu lato) xenoliths from southern Africa contained more than 5 per cent clinopyroxene. Most were samples of garnet-clinopyroxene harzburgite. It is likely that in most areas the sub-continental upper mantle contains a layer of less dense harzburgite above the denser garnet lherzolite layer. The harzburgites are usually discussed in terms of two endmembers: (1) a highly refractory type and (2) a rock that contains significant amounts of the elements sodium, calcium, aluminium and chromium. With partial melting in the upper mantle harzburgites of the latter type readily yield picrobasaltic magmas.

It is interesting to consider how the various types of peridotitic xenoliths found in continental areas are related to one another. The parental rock is perceived as a fertile garnet lherzolite that contains appreciable amounts of garnet and diopside with minor amounts of accessory minerals such as phlogopite, apatite and amphibole. If various amounts of magma are extracted from this rock various refractory materials are left behind. They are likely to range in composition from garnet lherzolite that is slightly depleted in garnet, diopside and accessory phases, via garnet harzburgite, to highly refractory harzburgite or even dunite. This process is likely to change the host rock by decreasing its aluminium, calcium, titanium, sodium and potassium content, and boosting its $Mg/(Mg + Fe)$ and $Cr/(Cr + Al)$ ratios (Nixon *et al.*, 1981).

The origin of dunite in the upper mantle has recently been explained in a new way. Kelemen and others (1995, pp. 747–748) have suggested that the dunites located in the mantle sections of ophiolites form because of the dissolution of pyroxene and the concomitant precipitation of olivine during the porous flow of an olivine-saturated magma through the upper mantle. They also speculate that dunites are likely to form in the upper mantle by replacement in ductile shear zones and in various porous reactive zones around propagating cracks. The tabular bodies of dunite that are found in some ophiolites are interpreted as porous reaction zones that evolved around what had once been magma-filled fractures.

Eclogites occur as rare xenoliths in alkaline rocks. Some of these eclogites contain phases that show that they equilibrated in the upper mantle. In a few kimberlites, such as Obnazhennaya (Yakutia on the Eastern Siberian Platform), Orapa (Botswana) and Roberts Victor (South Africa), this special type of eclogite is abundant. Their essential minerals are garnet and omphacite (a green aluminous Ca-Na clinopyroxene). The garnets display considerable variation in their calcium (CaO = 3–20 per cent), magnesium (MgO = 3–22 per cent) and iron (total Fe as FeO = 8–19 per cent) contents. These eclogites usually contain a range of accessory minerals that include kyanite, corundum, coesite (high-pressure silica), diamond, graphite, rutile, sulphides, amphiboles and phlogopite. Eclogites generally have the same major element compositions as basalts and as we have already proposed they are likely to form when basaltic oceanic crust is subducted. The remarkable large bodies of eclogite with their marble-cake structures such as those from Beni Bousera in Morocco will be discussed later in this Section.

Metasomatised peridotites

The compositions of some samples of upper mantle rocks have been modified by metasomatic processes. Metasomatism usually produces phases such as phlogopite, ilmenite, rutile, sulphides (pyrrhotite, pentlandite and chalcopyrite), richterite (alkali amphibole), apatite and carbonates. The compositions and textures of these rocks show that several distinct styles of metasomatism have operated within the upper mantle. In the most conspicuous type of metasomatism new phases have crystallised in veins or as overgrowths that cover earlier formed phases. The various products of metasomatism show that the upper mantle has contained and probably still contains migrating fluids enriched in water, carbon dioxide, fluorine, sodium, phosphorus, sulphur, chlorine, potassium, calcium, titanium, manganese, iron, copper, rubidium, strontium, yttrium, zirconium, niobium, barium, the light rare earth elements, hafnium, tantalum, thorium and uranium (Bailey, 1987).

Inclusions that represent melts occur within minerals derived from the mantle. These xenoliths were transported in magmas that evolved beneath both continental and oceanic intraplate regions (Schiano *et al.*, 1994). The first type of melt is a normal silicate magma that contains 54.3–66.0 per cent SiO_2, 15.6–22.4 per cent Al_2O_3, 2.2–7.4 per cent Na_2O, 1.0–6.9 per cent K_2O, 0.1–2.2 per cent TiO_2, 0.7–5.0 per cent FeO, 0.6–4.1 per cent MgO, 1.4–11.0 per cent CaO, and more than 1000 ppm chlorine. A magma of this composition would normally be called a latite (see Fig. 2.7). The second type of inclusion is rich in carbonate phases. Because it is usually found physically connected with the silicate melt inclusions, it is likely that both inclusions were originally part of a homogeneous fluid that subsequently separated into two immiscible phases. A third type of inclusion is a carbon dioxide-rich fluid. These inclusions are interpreted as samples of the metasomatic fluids that migrate through the upper mantle. Fluids derived from a deeper source in the mantle occur as inclusions in some diamonds. Samples from African, Indian and Yakutian diamonds have been studied. They reveal a wide range of compositions that grade from carbonatites to hydrous end-members enriched in SiO_2, K_2O and Al_2O_3 (Schrauder et al., 1994; Schrauder and Navon, 1994).

Another interesting type of xenolith found in the kimberlites of southern African and Yakutian are the glimmerites. These mica-rich rocks are probably produced by metasomatic fluids. When they are essentially composed of micas, such as phlogopite or tetraferriphlogopite, they have also been called phlogopitites (Larouziere, 1989, p. 132). The various phlogopite wehrlite and phlogopite and amphibole-bearing clinopyroxenite xenoliths found in the ultrapotassic volcanic rocks of the Western Branch of the African Rift System and the Roman Magmatic Province are reviewed in Section 11.6.

In 1977 Dawson and Smith introduced the idea of the MARID-suite to describe the xenoliths that showed evidence of mantle-metasomatism. MARID is a mnemonic for the minerals, mica (phlogopite or tetraferriphlogopite), amphibole (potassic richterite), rutile, ilmenite and diopside. They are the crucial phases in this suite. Some petrologists regard mantle metasomatism as a necessary precursor and cause of alkaline magmatism. Geochemical modelling has revealed that the source rocks characteristic of most primary alkaline magmas are depleted mantle peridotites that have been selectively enriched in the incompatible elements including potassium.

of harzburgite that persists to a depth of between 100 to 120 km. Beneath the harzburgite there is a layer that contains both garnet-clinopyroxene harzburgite and garnet lherzolite. The fertile materials usually have the highest densities. Veins, ribbons and small bodies of eclogite and MARID-suite rocks occur sporadically throughout this part of the upper mantle. Many hydrous and carbonate phases occur within MARID-suite rocks. Calcite is the main carbon dioxide-bearing mineral in the harzburgites and it is probably replaced by dolomite and magnesite at greater pressures.

As the uppermost unit of the mantle beneath the oceans is normally nearer the surface than the equivalent unit beneath the continents it may contains phases, such as plagioclase, that equilibrated at lower pressures. It is postulated that in normal oceanic lithosphere there is a 20 km thick layer of harzburgite, followed by a 30 km layer of spinel lherzolite (sensu lato). At greater depths the spinel lherzolite is transformed into a garnet lherzolite assemblage.

In 1962 Ringwood coined the term pyrolite to describe the chemical composition of a hypothetical fertile upper mantle rock (see Table 3.1). Over the years he proposed several compositions for this hypothetical rock (Ringwood, 1975, pp. 180–189). The first pyrolite was a somewhat arbitrary mixture of three parts peridotite and one part basalt. His second composition was obtained by combining 99 per cent of a slightly fractionated peridotite from the Lizard area of Cornwall in England with 1 per cent nephelinite. Another pyrolite composition was obtained from mixing 83 per cent residual harzburgite with 17 per cent primitive mid-oceanic ridge basalt. Recently Allegre and others (1995, p. 518) have estimated the chemical composition of the primitive mantle (PRIMA) using ratios of major and trace elements. Their estimate is as follows: 46.12 per cent SiO_2, 37.77 per cent MgO, 7.49 per cent FeO, 4.09 per cent Al_2O_3, 3.23 per cent CaO, 0.38 per cent Cr_2O_3, 0.36 per cent Na_2O, 0.25 per cent NiO, 0.18 per cent TiO_2, 0.15 per cent MnO, and 0.034 per cent K_2O.

Petrographic models of the mantle

Samples of the upper mantle show that beneath continental areas it is mineralogically complex and chemically heterogeneous (Erlank et al., 1980, and Dawson et al., 1980). It usually has an upper layer

Simple model of the upper mantle

The upper mantle gains heat and volatile components from a convecting lower mantle that is traversed by mantle plumes. Subduction enables it to accumulate

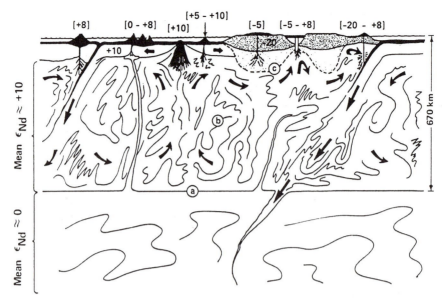

Fig. 3.4 Geochemical model of the interior of the Earth based on isotopic data from basalts. The numbers in brackets are typical eNd values of basalts erupted in the tectonic settings indicated. Within the upper mantle there are ribbons of recycled crustal material and material injected from the lower mantle. After De Paolo, 1988, p. 110.

some hydrous, altered igneous and sedimentary materials from the crust. Both abundant and rare magmas are extracted from the upper mantle. The components in these magmas are either transported out of the mantle and into the crust, hydrosphere or atmosphere, or they remain in the mantle where they congeal to form various intrusive bodies and promote local metamorphism and metasomatism.

When meditating on the abundance and distribution of elements within the Earth it is convenient to consider the planet to contain a series of geochemical reservoirs. Energy and matter is exchanged between these reservoirs. Beneath active mid-oceanic ridges both new oceanic crust and a complementary layer of depleted upper mantle forms from an initially homogeneous source; whereas magma-mixing processes generally result in the combining of two reservoirs into one unit. Since the initial differentiation of the Earth into core and mantle the growth of the continental crust has significantly depleted the upper mantle in the incompatible elements and subduction is also changing the chemical character of the upper mantle. After magma has been generated by subduction-related process the refractory remains of the layered oceanic lithosphere continues to sink through the mantle to the base of the upper convecting layer at a depth of about 670 km. The subducted

lithospheric slab is then likely to be heated, softened, deformed and drawn out to greater length by the process of streamline mixing. This produces elongate layers that have discrete geochemical, isotopic and age characteristics and a marble-cake structure. Rock complexes with this distinctive structure are found at Lanzo in Italy, Lherz in France and Beni Bousera in Morocco. These marble-cake rocks typically consist of bands of eclogite interbedded with peridotite.

Studies of chronometric nuclide pairs such as samarium-147/neodymium-143 have enabled geochemists to establish the timing of major events in the chemical evolution of the Earth. Such isotopes have been used to probe the interior of our planet because they can sense the existence of domains of different chemical composition and they can trace the origin and evolution of these domains back through time. Fig. 3.4 shows an Earth model based on such isotopic data (De Paolo, 1988, p. 110).

Lower mantle of the Earth

At depths of about 400 and 680 km below the surface there are rapid increases in seismic wave velocities and in the density of the mantle rocks. These changes occur within the transition zone that lies at a depth of between about 400 and 900 km. Phase changes have probably produced the various discontinuities found in this zone. Experimental studies by Ringwood (1979, p. 19) and others have shown that a whole series of phase changes are likely to occur in rocks of peridotitic composition at pressures corresponding to the range in depths characteristic of the transition zone. Perhaps the most fundamental of these changes is the transformation of olivine into the beta-Mg_2SiO_4 structure. This phase is about 6 per cent more dense than normal olivine and it is closely related to the structure of spinel. At depths of about 520 km the calcium silicate component in garnet transforms into a dense perovskite structure. Other phase changes are also expected to occur in the transitional zone.

In the lower mantle the rate of increase of seismic velocities and density is readily accounted for by self-compression of an essentially homogeneous material. The bulk chemical composition of the lower mantle is not known, but it is probably broadly similar to that of the upper mantle. The lowermost 200 km of the mantle abuts the core and it is known to have highly variable physical characteristics. This probably reflects thermal and chemical interaction between the rocks of this zone and the liquid core. Ito et al. (1995) have studied the reactions between molten iron and silicate melts at pressures of up to 26 GPa. They suggest that silicon and oxygen were originally important light elements in the Earth's core. With the gradual cooling and growth of a relatively pure iron inner core, silicon and oxygen have been expelled resulting in the bottom layer of the outer core becoming supersaturated in these elements. An immiscible oxide liquid is likely to develop. It forms blobs that move through the outer liquid core and they accumulate and solidify and contribute to the evolution of the D'' layer.

Deep earthquakes and the make-up of the mantle

In most parts of the world earthquakes take place at depths of less than 30 km, but in areas of active subduction earthquakes periodically occur at depths greater than 300 km. At these depths the ambient pressure is greater than 10 GPa. This means that these earthquakes are not triggered by brittle fracture. The foci of these deep earthquakes trace the outline of the slabs of cold oceanic lithosphere that are subducted. Individual earthquakes take place along the edges of the slowly resorbing lithospheric slabs.

According to Green and others (Green and Burnley, 1989; Green et al., 1990; Green, 1994) normal brittle fractures account for shallow earthquakes. At deeper levels in subduction zones the partially hydrated oceanic lithosphere is gradually dehydrated and this promotes fluid-assisted faulting that triggers earthquakes. The steady decrease in earthquake frequency between 30 and 300 km is related to the progressive heating and dehydration of the descending slab. At 300 km the interior of the slab is usually sufficiently cool to prevent olivine from transforming into the beta-Mg_2SiO_4 structure. As the outer layer of the slab is heated the metastable olivine reaches a critical temperature and converts to the beta-phase. This causes 'anticrack faulting' and deep earthquakes (Green et al., 1990). In lithospheric slabs that are particularly old and cold metastable olivine and deep earthquakes are likely to persist down to a depth of 700 km.

Brittle failure and anticrack failure are the main mechanisms for triggering shallow and deep earthquakes, respectively. When brittle rocks are stressed microcracks open parallel to the direction of compression. At depths greater than 300 km the ambient pressure inhibits dilation and instead microanticracks, or small lens-shaped bodies filled with fine-grained, beta-Mg_2SiO_4 phase, form perpendicular to the direction from which these rocks are compressed. During an earthquake the microfeatures generated by both mechanisms link to create a fault. In brittle failure the fault forms an open fracture whereas in anticrack failure the fault follows a plane linking microanticracks enriched in small crystals of beta-Mg_2SiO_4.

3.5 Mantle plumes, crustal swells and epeirogeny

The interaction of hot, petrologically complex, mantle plumes with the lithosphere is likely to produce large-scale vertical movements in the overlying crust. Generally the elevation of the oceanic floor can be related to age. Traditionally this relationship has been interpreted as signifying that regions of cooler, older oceanic lithosphere are less elevated than hotter, younger regions. According to Crough (1983) if one changed the average temperature of a 100 km thick layer of oceanic lithosphere by 600°C this would change the surface elevation by 3 km. This is approximately the shift in elevation observed between the crest of an active mid-oceanic ridge and the centre of an old oceanic basin.

All ocean basins contain depth anomalies or regions that are more elevated than one would expect from their crustal age. These anomalies usually enclose major volcanic edifices. Such broad topographic swells are found associated with most volcanic islands, seamounts or seamount chains. A typical volcanic swell is 1000–1500 km across and between 500 and 1200 m high. These swells form the most prominent positive topographic features in the oceanic basins. Their lack of correlation with plate boundaries seems to support the idea that mantle plumes and the global convective system usually operate independently.

It is essential to speculate on how these volcanic swells are supported at depth. Over the years several shallow compensation models have been proposed but they are now considered inadequate. Models that use deeper support mechanisms belong to two main categories: (1) models that claim that the support takes place within the lithosphere and (2) models that suggest that the support comes from beneath this layer. Crough (1983, p. 190) suggested that the swells located along the mid-oceanic ridge system were supported isostatically by the upwelling of asthenospheric materials, whereas normal intraplate swells are mainly supported by the reheating and expansion of lithospheric materials. Volcanic swells located in intraplate settings probably represent the surface expression of ascending mantle plumes that come from the deep mantle. If these plumes contain enough heat to produce the broad halo of heating required by the reheating model they must represent a fundamental method for transferring heat from the deep mantle to the lithosphere. While such heating is probably important in generating mantle swells it is likely that another important factor is the accumulation and ponding at the base of the lithosphere of lower density depleted mantle phases that had previously become entrained in a rising mantle plume. The huge Tharsis Dome on Mars has existed for too long to be supported by a thermal anomaly or active subcrustal convection. It is probably underlain by an enormous pillow-shaped body of low-density mantle residuum generated by the partial melting and expulsion of vast quantities of iron-rich picrobasaltic or basaltic magma (see Section 4.5).

When considering swells and domes in continental regions it is essential to recollect that the lithosphere is composed of crustal and mantle layers with very different densities; when normal picrobasaltic magma arrives at the mantle-crust interface its density is likely to exceed that of the average continental crust. It is proposed when such magma reaches this critical interface it insinuates itself along the interface between the crust and mantle and heats up the surrounding rocks. The process of inserting magma beneath a less dense layer has been called underplating. Such underplating may generate swells, domes or regional epeirogeny. The shape of the upwarp depends on the size of the mantle plume, the configuration of the sublithospheric topography encountered by the plume and local plate velocities.

During the period of extensive continental crust development that occurred in the early Archean Era many igneous rocks, particularly the silicic plutonic rocks, evolved from magmas generated by heat supplied by magmatic underplating (Kroner and Layer, 1992). As volcanic swells and domes are also found on Venus and Mars, it is likely that thermal plumes are an essential mechanism for transporting heat on all the terrestrial planets (see Section 1.5). The high surface temperature of Venus probably enables smaller and slower magma bodies to reach, or nearly reach, its surface (Sakimoto and Zuber, 1995, p. 58).

If it is assumed that the distribution of mantle plumes is nonuniform and the lithospheric plates have been moving over them since the Archean Era then it can be deduced that the older continental regions have all passed over several plumes (Crough, 1981). This suggests that the thermal and structural reworking of continental lithosphere above plume heads has played a substantial role in the evolution of the lower continental crust. The development of surface swells and domes as the result of the passage of mantle plumes is likely to assist surface erosion. Occasionally they may have been partly responsible for the commencement of a new cycle of continental rifting.

If some mantle plumes are even partly responsible for continental rifting one has to consider the question of the size of mantle plumes and whether continental crust can be broken up by mega-plumes. Perhaps such large plumes represent an early stage in the evolution of normal plumes. The Earth has experienced many episodes of flood volcanism (McDougall, 1988). During these episodes over a million cubic kilometres of lava is erupted (see Section 7.6). This is considerably greater than the volume produced by current intraplate mantle plume volcanism.

In 1981 Morgan attempted to link lavas from various Mesozoic and Cenozoic continental flood basalt provinces with currently active mantle plumes. He proposed the following pairs: (1) Siberian Platform Province and the Jan Mayen mantle plume; (2) Karoo Province and the Crozet mantle plume in the southern Indian Ocean; (3) Parana-Etendeka Province and the Tristan da Cunha mantle plume in the middle of the South Atlantic Ocean; (4) Deccan Province and the Reunion mantle plume in the Indian Ocean; (5) Thulean Province and the Iceland mantle plume; (6) Columbia River Province and the Yellowstone mantle plume; and (7) Ethiopian-Yemen Province and the Afar mantle plume. Morgan also suggested that flood basalts were extruded during the initial crust-breaching phase in the development of a mantle plume. This idea has been supported by later research (Campbell and Griffiths, 1990; Hill et al., 1992; Farnetani and Richards, 1994).

It is now usually proposed that when a new mantle plume impinges on the lithosphere it develops a tadpole-like shape, with a huge bulbous head tapering to a somewhat slender tail. During ascent newly established mantle plumes are likely to grow by entraining some of the surrounding mantle materials that they have heated up. When they emerge beneath the more rigid lithosphere the magma in the head cools (1350°C) and spreads laterally, meanwhile in the central zone hotter (1550°C), uncontaminated material continues to rise. Intrusion of a huge, possibly 2000 km in diameter and 150 km-thick, plume-head beneath a plate of continental lithosphere is likely to cause thermal erosion, heating, stretching, doming and fracturing of the overlying rocks (Hill et al., 1992, p. 187). These events are likely to cause decompression melting resulting in rapid magma generation in the plume-head.

Plume tectonics

Hill et al. (1992, p. 187) have introduced the term plume tectonics to describe the crustal doming and subsequent subsidence induced by the emplacement and cooling of thermal plumes. They emphasise that the impingement of a mega-plume beneath the continental crust is likely to generate a thermal anomaly that takes hundreds of millions of years to dissipate. They also trace the sequence of magmatic events that follow the impingement of a plume-head under a section of lithosphere that contains non-cratonic, or fertile, continental crust. Their normal magmatic sequence is: (1) eruption of a thick sequence of flood basalts; (2) concurrent eruption of lesser amounts of higher temperature volcanic rocks, such as picrites or komatiites, from the core of the dome; (3) gradual conduction of heat into the crust eventually resulting in the production of a range of magmas principally derived from the melting of crustal rocks; and (4) continued conduction of heat into the crust may result in further partial melting and the generation of a second group of more felsic, often alkaline, magmas. Hill et al. (1992) used the rocks of the Yilgarn area of southwestern Australia as an example of their magmatic cycle. An 8 km-thick sequence of picrobasalts, basalts and komatiites was extruded some 2.715 giga-years ago. The next major magmatic event (2.688 giga-years) was the eruption of a sequence of felsic volcanic rocks. Both volcanic sequences were intruded by a series of felsic plutonic rocks. Later between 2.665 and 2.660 giga-years ago these rocks were intruded by a group of granitic rocks. In the younger (193–177 mega-years) Karoo province of southern Africa the extrusion of huge volumes of basic flood lava was followed by the eruption of felsic lavas of the Lebombo belt.

The Yellowstone plume produced the Yellowstone–Snake River Plain volcanic field of Idaho, Wyoming and Montana. It is an excellent example of a currently active mantle plume located in an intraplate continental setting. During the past 16 million years the passage of the North American plate over this plume has remoulded the geology of more than 20 per cent of the western United States (Smith and Braile, 1994). An 800-km long linear chain of silicic volcanic centres traverses the Snake River Plain and terminates in the active Yellowstone Volcanic Centre. As one moves from the Yellowstone Plateau across the Snake River Plain and into southwestern Idaho one discovers a systematic decrease in elevation. On a regional scale the Yellowstone Volcanic Centre is situated on the crest of a dome some 600 m high and 600 km across. The volcanic centre is also in the middle of the most substantial geoid anomaly associated with the North American continent. It is a circular high about 1000 km in diameter. Extremely high heat flow values, more than 30 times the North American continental average, are also recorded in the Yellowstone Volcanic Centre.

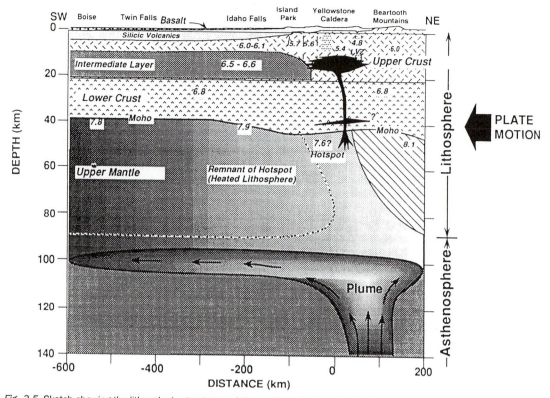

Fig. 3.5 Sketch showing the lithospheric structure and the position of a hypothetical thermal plume beneath the Yellowstone Volcanic Centre and the Snake River Plain of northwestern USA. After Smith and Braile, 1994, p. 176.

Geophysical studies of the crust beneath the Yellowstone–Snake River Plain reveal that the granitic upper crustal layer, typical of much of the North American continent, is absent or at least of greatly diminished thickness. According to Smith and Braile (1994, p. 168) the granite has been largely replaced by silicic volcanic rocks and mafic intrusions generated by the Yellowstone plume. Beneath the Yellowstone Volcanic Centre heterogeneous upper crustal rocks extend to a depth of about 10 km, whereas anomalous seismic velocities extend to depths of at least 200 km (see Fig. 3.5). In the past two million years the Yellowstone Volcanic Centre has produced three vast caldera-forming eruptions and many smaller explosive events. The enormous Huckleberry Ridge tuff deposit that erupted two million years ago contains about 2500 km^3 of silicic material.

The present surface of the Earth is embellished by many volcanic chains that appear to trace the tracks of waning mega-plumes. Some tracks are thousands of km long and probably evolved over periods of up to 200 million years. The flood lavas of the Thulean Province are particularly interesting because they erupted in both continental and oceanic settings. Clear evidence of the impingement of a juvenile mantle plume, or mega-plume, beneath an area of oceanic lithosphere is found in the lavas of the Ontong Java Plateau in the west-central Pacific. This oceanic plateau contains a huge pile of flood lavas that erupted over a period of about three million years. The lavas cover an area of approximately four million km^2, and the plateau as a whole contains about 50 million km^3 of magmatic rocks. At the time it was forming the rate of eruption was at least 15 km^3 per year. This eruption rate is similar to the present eruption rate for the whole mid-oceanic ridge system.

Another large oceanic plateau is the Kerguelen-Heard Plateau in the remote Antarctic sector of the Indian Ocean. This plateau is 2300 km long, and between 200 and 600 km wide. According to Storey *et al.* (1992, p. 33) it was originally contiguous with Broken Ridge, and together these two topographic features have a combined crustal volume of over fifty million km^3. Schlich and Wise (1992, p. 12) have suggested that the Kerguelen-Heard Plateau was constructed between 130 and 110 million years ago when the plateau was at, or near, a developing spreading centre situated between India, Australia and Antarctica (Whitechurch et al., 1992). It has been established that flood lavas erupt in both oceanic and continental settings. The eruptions that constructed the Ontong Java and the Kerguelen-Heard plateau were among the largest volcanic events recorded on Earth during the past 200 million years.

The mantle plumes of Earth can be arranged in four groups: (1) rare plumes in their mega-plume stage of evolution that release vast floods of lavas, produce significant amounts of underplating and profoundly change the atmosphere and biosphere; (2) ordinary plumes that intrude into the base of normal, or fast moving, lithospheric plates and produce crustal swells and a trail of extinct volcanoes; (3) ordinary plumes that intrude into the base of slow-moving lithospheric plates and produce clusters of volcanoes; and (4) ordinary plumes that intrude divergent plate margins and trigger abnormally productive volcanic activity (Marzocchi and Francesco, 1993). The Azores, Galapagos and Iceland are examples of abnormally productive volcanic provinces fed by mantle plumes located on or close to mid-oceanic ridges.

Characteristics of the Icelandic plume

Iceland is the most important dry land location for investigating what happens when a mid-oceanic spreading axis cuts across the convectively driven core of an active mantle plume. Ryan (1990, p. 178) has synthesised a formidable volume of geophysical, petrological and structural data to uncover the 'magmatic arterial system' beneath Iceland. In particular, he reveals the depth at which the physical characteristics ascribed to the mid-oceanic ridge terminate and are replaced by the physical features associated with the mantle plume. He has constructed a three-dimensional image of the magma system beneath Iceland. In the upper 175 km the form of the magma body is strongly influenced by the tectonic activity occurring within the mid-oceanic ridge system. Between 175 and 275 km the magma body becomes more columnar in form, and has a mean diameter of about 175 km. At greater depths the body bifurcates and

at a depth of about 375 km it forms two discrete columns.

According to Ryan (1990) a three-dimensional melt network forms in the core of the plume at a depth of about 110 km. Above this level magma is transported upwards and outwards by intergranular flow. The magma becomes concentrated at the base of the lithosphere in areas that are locally elevated. In detail the transport and storage of this magma is regulated by the local variations in the structural properties of the lithosphere. These properties are continually changing in response to changes in the stress field in the rocks along the margins of the diverging plates. Ryan (1990, p. 218) has suggested that beneath Iceland there is a conflict between: (1) the steady state extension and flow in the asthenosphere; (2) episodic compression at mid crustal levels because of magma emplacement and the generation of dyke complexes; and (3) tensile extension in the zone above the dilating dyke complexes.

The diameter of the swell associated with the Icelandic plume is about 2000 km and the maximum elevation at its centre is about 2 km. It is probable that vigorous convection in the core and head of the plume will provoke the entrainment of materials from the surrounding mantle. According to Hauri et al. (1994) a steady-state mantle plume is likely to entrain between 5 and 90 per cent of the surrounding mantle materials. Such entrainment is induced by the radial conduction of heat away from the plume core. This raises the buoyancy and lowers the viscosity of the surrounding mantle. Shreds of mantle material are also likely to be dragged up by the plume because of viscous coupling between the plume and the materials it traverses.

Thirlwall (1995) has investigated the lead isotopic characteristics of the Icelandic plume. He discovered that few, if any, modern Icelandic lavas contain a lead isotopic component similar to the isotopic material typical of mid-oceanic ridge basalts found outside the thermal influence of the Icelandic plume. These data suggest that the Icelandic plume is composed of a mixture of at least two components with isotopic characteristics that are unlike those of the normal mid-oceanic ridge basalts of the North Atlantic.

Other detailed geochemical studies of the rocks of Iceland and its bordering seas have verified the heterogeneous nature of the Icelandic plume. It is a blend of: (1) batches of incompatible element-enriched materials that are derived from an old, deep-seated source that was isolated from the convecting mantle for a long period and (2) a more abundant component, that is less enriched in the incompatible elements, and was probably added at a higher level in the mantle (Hards *et al.*, 1995, p. 1007). The high helium-3/helium-4 ratio of the Icelandic volcanic rocks supports the view that an essential part of the Iceland plume is derived from the deep mantle, possibly from near the core-mantle interface (Condomines *et al.*, 1983). This view is also supported by thermal modelling, because a large heat flux is required to maintain the extensive topographic swell associated with this plume. Geochemical and geophysical modelling of several other mantle plumes reveals that they are normally made up of a mixture of batches of old material derived from the deep mantle and a selection of other mantle materials entrained during the plumes rise to the surface.

The maximum amount of partial melting attained by an upwelling mantle plume is normally directly related to its temperature and the depth at which the final episode of decompression melting occurred. According to White *et al.* (1995, p. 1044) the temperature of the core of the Icelandic plume is about 1500°C. This is some 100°C hotter than the mantle beneath the segment of the mid-oceanic ridge that adjoins southern Iceland. The thermal influence of the Icelandic plume continues for about 1350 km to the south of the island.

In Iceland the maximum amount of partial melting occurs in the axial rift zone where the overlying lithosphere is between 15–20 km thick. Under these conditions the maximum amount of partial melting is likely to be about 20 per cent (Hards *et al.*, 1995, p. 1007). In other oceanic areas where mantle plumes are active the lithosphere is likely to be older and thicker and in such areas one can predict less partial melting than that currently taking place beneath Iceland. Hards and others (1995, p. 1008) have investigated the substantial geochemical and isotopic variation found in the Icelandic picrobasaltic and basaltic rocks. They suggest that the meagre quantities of alkali basalts are mainly derived from the partial melting of dismembered batches of the oldest materials within the Icelandic plume, whereas the more abundant tholeiitic basalts are usually generated from materials entrained during the ascent of the plume.

Hydrothermal flux through the ocean-floor

When discussing the origin of the Earth's earliest continental lithosphere it was stated that the initial oceanic crust was cooled and pervasively altered by hydrothermal activity. Such hydrothermal fluxing is still important. Wolery and Sleep (1976) estimated that the flux of hydrothermal fluid necessary to account for the low conductive heat flow currently found on mid-oceanic ridges was approximately 1.3×10^{14} kg of fluid per year. De Wit and Hart (1993, p. 326) have used three different methods to calculate the rate of hydrothermal flux and their estimates are essentially similar. They have calculated that both heat flow and hydrothermal fluxing have decreased by an order of magnitude since the Archean. The latter statement seems to imply that if the magma supply system in the Archean was broadly similar to the one presently in operation then either the ridge length, the spreading rate, or both the ridge length and the spreading rate, must have been much greater 3.5 giga-years ago.

A typical segment of the present oceanic crust has an upper layer of fragmented and pillowed lavas overlying a dyke-sill complex that contains a variety of gabbroic and cumulus rocks (see Section 7.5). Sea-water percolates through the hot, fractured rocks where it is heated. At 3000 m beneath the oceans water boils at close to 400°C and it is extremely buoyant. As this fluid circulates within the hot fractured rocks it reacts with them and the compositions of both the fluid and the host rocks change. The difference in chemical composition between sea-water and hydrothermal fluids that traverse the oceanic crust provides a straightforward measurement of mass transfer from the crust to the oceans. At high temperature the leaching process usually yields an acidic fluid enriched in silica, hydrogen sulphide and metals. Such hot, buoyant, metal-rich fluid usually exits from vents on the sea-floor at velocities of approximately one m/s. On mixing with sea-water the pH of the introduced fluid rises, and its temperature drops. This results in the precipitation of a variety of sulphide and oxide minerals, including pyrrhotite, chalcopyrite, sphalerite and amorphous iron oxides. Such exhalations are sometimes called 'black smoke' and precipitation of part of this fluid may produce 'smoker chimneys'. The latter are concentrically layered hollow spires that may be up to 7 m high with internal diameters of between 1 and 10 cm. Recent detailed studies of the abundance of hydrothermal vents on the mid-Atlantic ridge have led to the prediction that there are likely to be about 5000 active vent sites in the oceans (Franklin, 1994, p. 20).

In August 1986, when exploring the Juan de Fuca Ridge to the west of Oregon and Washington states in the United States, a huge thermal plume was discovered. The original volume of fluid in this plume was probably about 100 million m^3, at a temperature of 350°C. The sudden release of this fluid was likely to have been triggered by the catastrophic rupturing of a section of the neovolcanic zone because the hydrothermal discharge was followed by an eruption of lava. Another large hydrothermal plume was recorded on the same ridge the following year. Similar discharges have now been recorded from other regions, such as the area southwest of Samoa in the Fiji Basin. In some areas, particularly where the oceanic crust is cut by major faults, sea-water reacts with peridotitic rocks within the crust and produces serpentinites. These serpentinites are less dense and more plastic than the original peridotites, and they often intrude up towards the surface of the crust.

References

Abbott, D., Burgess, L. and Longhi, J., 1994. An empirical thermal history of the Earth's upper mantle. *Journal of Geophysical Research*, 99 (B7), pp. 13 835–13 850.

Allegre, C.J., Poirier, J-P., Humler, E. and Hofmann, A.W., 1995. The chemical composition of the Earth. *Earth and Planetary Science Letters*, 134, pp. 515–526.

Alvarez, L.W., Alvarez, W., Asaro, F. and Michel, H.V., 1979. Experimental evidence in support of an extraterrestrial trigger for the Cretaceous-Tertiary extinctions. *EOS*, 60, p. 350.

Bailey, D.K., 1987. Mantle metasomatism – perspective and prospect. In Fitton, J.G. and Upton, B.G.J. (editors), *Alkaline Igneous Rocks*, pp. 1–13, Geological Society of London Special Publication Number 30, 568 pp.

Basaltic Volcanism Study Project (BVSP), 1981. *Basaltic Volcanism on the Terrestrial Planets*. Pergamon Press, Incorporated, New York, 1286 pp.

Bottinga, Y. and Weill, D.F., 1970. Densities of liquid silicate systems calculated from partial molar volumes of oxide components. *American Journal of Science*, 267, pp. 169–182.

Broecker, W.S., 1985. *How to Build a Habitable Planet*. Eldigio Press, Palisades, New York, 291 pp.

Brown, G.C. and Mussett, A.E., 1981. *The Inaccessible Earth*. George Allen and Unwin, London, 235 pp.

Calvert, A.J., Sawyer, E.W., Davis, W.J. and Ludden, J.N., 1995. Archean subduction inferred from seismic images of a mantle suture in the Superior Province. *Nature*, 375, pp. 670–673.

Campbell, I.H. and Griffiths, R.W., 1990. Implications of mantle plume structure for the evolution of flood basalts. *Earth and Planetary Science Letters*, 99, pp. 79–93.

Carswell, D.A., 1980. Mantle derived lherzolite nodules associated with kimberlite, carbonatite and basaltic magmatism: A review. *Lithos*, 13, pp. 121–138.

Clifford, T.N., 1971. Location of mineral deposits. In Gass, I.G., Smith, P.J. and Wilson, R.C.L. (editors) *Understanding the Earth*, pp. 315–325, The Open University, UK, 355 pp.

Condomines, M, Grönvold, K., Hooker, P.J., Muehlenbachs, K., O'Nions, R.K., Oskarsson, N. and Oxburgh, E.R., 1983. He, O, Sr Nd isotopic relationships in Icelandic volcanoes. *Earth Science and Planetary Letters*, 66, pp. 125–136.

Crough, S.T., 1981. Mesozoic hotspot epeirogeny in eastern North America. *Geology*, 9, pp. 2–6.

Crough, S.T., 1983. Hotspot swells. *Annual Review of Earth and Planetary Sciences*, 11, pp. 165–193.

Daly, R.A., 1933. *Igneous Rocks and the Depths of the Earth*. McGraw-Hill Book Company, Inc., New York, 598 pp.

Dawson, J.B. and Smith, J.V. 1977. The MARID (mica–amphibole–rutile–ilmenite–diopside) suite of xenoliths in kimberlite. *Geochimica et Cosmochimica Acta*, 41, pp. 309–323.

Dawson, J.B., Smith, J.V. and Hervig, R.L., 1980. Heterogeneity in upper mantle lherzolites and harzburgites. *Philosophical Transactions of the Royal Society (London)*, A 297, pp. 323–331.

De Paolo, D.J., 1988. *Neodymium Isotope Geochemistry: An Introduction*. Springer-Verlag, Berlin, 187 pp.

De Wit, M.J. and Hart, R.A., 1993. Earth's earliest continental lithosphere, hydrothermal flux and crustal recycling. *Lithos*, 30, pp. 309–335.

Erlank, A.J., Allsopp, H.L., Duncan, A.R. and Bristow, J.W., 1980. Mantle heterogeneity beneath southern Africa: evidence from the volcanic record. *Philosophical Transactions of the Royal Society (London)*, A 297, pp. 295–307.

Elder, J., 1976. *The Bowels of the Earth*. Oxford University Press, Oxford, 222 pp.

Farnetani, C.G. and Richards, M.A., 1994. Numerical investigations of the mantle plume initiation model for flood basalt events. *Journal of Geophysical Research*, 99 (B7), pp. 13 813–13 833.

Franklin, C., 1994. Black smokers multiply on ocean floor. *New Scientist*, 144 (1948), p. 20.

Ghiorso, M.S., Hirschmann, M. and Sack, R.O., 1994. MELTS: Software for thermodynamic modelling of magmatic systems. *EOS*, 75, p. 571.

Green, H.W., 1994. Solving the paradox of deep earthquakes. *Scientific American*, 271 (3), pp. 50–57.

Green, H.W. and Burnley, P.C., 1989. A new self-organising mechanism for deep-focus earthquakes. *Nature*, 341 (6244), pp. 733–737.

Green, H.W., Young, T.E., Walker, D. and Scholtz, C.H., 1990. Anticrack-associated faulting at very high pressure in natural olivine. *Nature*, 348 (6303), pp. 720–722.

Hards, V.L., Kempton, P.D. and Thompson, R.N., 1995. The heterogeneous Iceland plume: new insights from the alkaline basalts of the Snaefell volcanic centre. *Journal of the Geological Society*, London, 152, pp. 1003–1009.

Hauri, E.H., Whitehead, J.A. and Hart, S.R., 1994. Fluid dynamics and geochemical aspects of entrainment in mantle plumes. *Journal of Geophysical Research*, 99, pp. 24 275–24 300.

Hill, R.I., Campbell, I.H., Davies, G.F. and Griffiths, R.W., 1992. Mantle plumes and continental tectonics. *Science*, 256, pp. 186–193.

Holloway, J.R. and Wood, B.J., 1988. *Simulating the Earth*. Unwin Hyman, London, 196 pp.

Ito, E., Morooka, K., Ujike, O. and Katsura, T., 1995. Reactions between molten iron and silicate melts at high pressure: implications for the chemical evolution of the Earth's core. *Journal of Geophysical Research*, 100, B4, pp. 5901–5910.

Kelemen, P.B., Shimizu, N. and Salters, V.J.M., 1995. Extraction of mid-ocean-ridge basalt from the upwelling mantle by focussed flow of melt in dunite channels. *Nature*, 375, pp. 747–753.

Kroner, A. and Layer, P.W., 1992. Crust formation and plate motion in the early Archean. *Science*, 256, pp. 1405–1411.

Larouziere, F.D. de, 1989. *Dictionnaire des roches d'origine magmatique*. Editions du BRGM, Orleans, France, 188 pp.

Loper, D.E. and Lay, T., 1995. The core–mantle boundary region. *Journal of Geophysical Research*, 100, B4, pp. 6397–6420.

Luck, J. and Allegre, C.J., 1983. 187 Re–187 Os systematics in meteorites and cosmochemical consequences. *Nature*, 302, 130–132.

Maaloe, S. and Aoki, K., 1977. The major element composition of the upper mantle estimated from the composition of lherzolites. *Contributions to Mineralogy and Petrology*, 63, pp. 161–173.

Mathias, M., Siebert, J.C. and Rickwood, P.C., 1970. Some aspects of the mineralogy and petrology of ultramafic xenoliths in kimberlite. *Contributions to Mineralogy and Petrology*, 26, pp. 75–123.

McDougall, J.D., (editor) 1988. *Continental Flood Basalts*. Kluwer Academic Publishers, Dordrecht, The Netherlands, 341 pp.

Marzocchi, W. and Francesco, M., 1993. Patterns of hot spot volcanism. *Journal of Geophysical Research*, 98 (B8), pp. 14 029–14 039.

Morgan, W.J., 1981. Hotspot tracks and the opening of the Atlantic and Indian Oceans. In Emiliani, C., *The Sea*, pp. 443–487, Volume 7, The Oceanic Lithosphere, John Wiley and Sons, New York, 1738 pp.

Morris, P.A., Barnes, S.J. and Hill, R.E.T., 1993. Eruptive environment and geochemistry of Archaean ultramafic, mafic and felsic volcanic rocks of the Eastern Yilgarn Craton. *IAVCEI Excursion Guide, Canberra, Record 1993/62*, Australian Geological Survey Organisation, 40 pp.

Nicholls, J., 1990. The mathematics of fluid flow and a simple application of problems of magma transport. In Nicholls, J. and Russell, J.K. (editors), *Modern methods of igneous petrology: understanding magmatic processes*, pp. 107–123. Reviews in Mineralogy, 24, Mineralogical Society of America, 314 pp.

Nixon, P.H., Rogers, N.W., Gibson, I.L. and Grey, A., 1981. Depleted and fertile mantle xenoliths from southern African kimberlites. *Annual Review of Earth and Planetary Sciences*, 9, pp. 285–309.

Pinkerton, H. and Stevenson, R.J., 1992. Methods of determining the rheological properties of magmas at sub-liquidus temperatures. *Journal of Volcanological and Geothermal Research*, 53, pp. 47–66.

Podosek, F.A. and Swindle, T.D. 1988. Extinct Radionuclides. In Kerridge, J.F. and Matthews, M.S. (editors), *Meteorites and the Early Solar System*, pp. 1093–1113, The University of Arizona Press, Tucson, USA, 1269 pp.

Ringwood A.E., 1962. A model for the upper mantle. *Journal of Geophysical Research*, 67, pp. 857–866.

Ringwood A.E., 1975. *Composition and Petrology of the Earth's Mantle*. McGraw-Hill, New York, 618 pp.

Ringwood, A.E., 1979. *Origin of the Earth and Moon.* Springer-Verlag, New York, 295 pp.

Ryan, M.P., 1990. The physical nature of the Icelandic magma transport system. In Ryan, M.P. (editor), *Magma Transport and Storage*, pp. 175–220, John Wiley and Sons, Chichester, 420 pp.

Sakimoto, S.E.H. and Zuber, M.T., 1995. Effects of planetary thermal structure on the ascent and cooling of magmas on Venus. *Journal of Volcanology and Geothermal Research*, 64, pp. 53–60.

Scarfe, C.M. and Hamilton, T.S., 1980. Viscosities of lavas from the Level Mountain Volcanic Centre, Northern British Columbia. *Carnegie Institution of Washington Year Book*, 79, pp. 318–320.

Schiano, P., Clocchiatti, R. and Shimizu, N., 1994. Melt inclusions trapped in mantle minerals: a clue to identifying metasomatic agents in the upper mantle beneath continental and oceanic intraplate regions. *Mineralogical Magazine*, 58 A, pp. 807–808.

Schlich, R. and Wise, S.W., 1992. The geologic and tectonic evolution of the Kerguelen Plateau: an introduction to scientific results of leg 120. *Proceedings of the Ocean Drilling Program, Scientific Results*, 120, pp. 5–32.

Schrauder, M. and Navon, O., 1994. Hydrous and carbonatitic fluids in fibrous diamonds from Jwaneng, Botswana. *Geochimica et Cosmochimica Acta*, 58 (2), pp. 761–771.

Schrauder, M., Navon, O., Szafranek, D., Kaminsky, F.V. and Gamimov, E.M., 1994. Fluids in Yakutian and Indian diamonds. *Mineralogical Magazine*, 58 A, pp. 813–814.

Smith, R.B. and Braile, L.W., 1994. The Yellowstone hotspot. *Journal of Volcanology and Geothermal Research*, 61, pp. 121–187.

Sood, M.K., 1981. *Modern Igneous Petrology.* John Wiley and Sons, New York, 244 pp.

Storey, M., Kent, R.W., Saunders, A.D., Salters, V.J., Hergt, J., Whitechurch, H., Sevigny, J.H., Thirlwall, M.F., Leat, P., Ghose, N.C. and Gifford, M., 1992. Lower Cretaceous Volcanic Rocks on Continental Margins and their relationship to the Kerguelen Plateau. *Proceedings of the Ocean Drilling Program, Scientific Results*, 120, pp. 33–53.

Sun, S.S., 1987. Chemical composition of Archean komatiites: implications for early history of the Earth and mantle evolution. *Journal of Volcanology and Geothermal Research*, 32 (1–3), pp. 67–82.

Tait, S. and Jaupart, C., 1990. Physical processes in the evolution of magmas. In Nicholls, J. and Russell, J.K. (editors), *Modern methods of igneous petrology: understanding magmatic processes*, pp. 125–152, Reviews in Mineralogy, 24, Mineralogical Society of America, 314 pp.

Taylor, S.R. and McLennan, S.M. 1985. T*he Continental Crust: Its Composition and Evolution.* Blackwell Science Publishers, Oxford, 312 pp.

Thirlwall, M.F., 1995. Generation of the Pb isotopic characteristics of the Iceland plume. *Journal of the Geological Society*, London, 152, pp. 991–996.

Wasson, J.T., 1985. *Meteorites: Their Records of Early Solar-System History.* W.H. Freeman and Company, New York, 267 pp.

White, R.S., Bown, J.W. and Smallwood, J.R., 1995. The temperature of the Iceland plume and the origin of outward-propagating V-shaped ridges. *Journal of the Geological Society*, London, 152, pp. 1039–1045.

Whitechurch, H., Montigny, R., Sevigny, J., Storey, M. and Salters, V.J.M., 1992. K-Ar and 40Ar/39Ar ages of central Kerguelen Plateau Basalts. *Proceedings of the Ocean Drilling Program, Scientific Results*, 120, pp. 71–77.

Wiechert, E., 1897. *Üeber die Massenverheilungim Innern der Erde.* Naturwissenschaft Gesellschaft Wissenschaft Gottingen, pp. 221–243.

Wolery, T.J. and Sleep, N.H., 1976. Hydrothermal circulation and geochemical flux at mid-ocean ridges. *Journal of Geology*, 84, pp. 249–275.

Yoneda, A. and Spetzler, H., 1994. Temperature fluctuations and thermodynamic properties in Earth's lower mantle: an application of the complete travel time equation of state. *Earth and Planetary Science Letters*, 126, pp. 369–377.

CHAPTER 4

Volcanism and planetary development

> ...the types of volcanic activity observed on Earth represent only a fraction
> of the array of volcanic phenomena that are possible.
>
> Carr, 1987, p. 128.

4.1 Volcanoes and volcanism

Volcanoes are both vents that extrude lavas, pyroclastic materials and gases, and landforms, often of elegant symmetry, constructed or sculpted by eruptive activity. Currently most of the Earth's active volcanoes erupt within linear belts linked to either divergent or convergent plate boundaries. Spectacular volcanic eruptions have long fascinated humankind. For example, after the awe-inspiring eruption of Hekla in Iceland in AD 1104 many pious Europeans believed that it was the gateway to Hell. To them Hekla seemed to provide proof of the existence within the Earth of an infernal region of eternal fires and suffering souls (Thorarinsson, 1970).

Francis (1993, p. 2) has defined volcanism as 'the manifestation at the surface of a planet or satellite of internal processes through the emission at the surface of solid, liquid, or gaseous products'. This stilted definition was framed to include various unfamiliar phenomena such as the release of sulphur dioxide plumes on the Jovian satellite, Io, the extrusion of geyser-like plumes on Triton, Neptune's largest satellite, and the various processes known as cryovolcanism. The first part of this term is derived from the Greek word *kryos* meaning frost or icy cold. Cryovolcanism means icy cold volcanism. It was introduced to describe the processes that generate materials with the appearance of volcanic products found on the icy satellites of the Jovian planets (Cattermole, 1989, pp. 364–369). Splendid examples of cryovolcanism are found on Jupiter's huge satellite Ganymede. It is likely that the fluid that erupts onto the surfaces of the icy satellites is a mixture of ammonia and water, or methane and water.

Volcanoes of the Earth

Most maps that purport to show the abundance and distribution of volcanoes on Earth show some 1300 volcanoes that have been active during the last 10 000 years. According to Simkin (1993, p. 429) the Earth is girdled by approximately twenty volcanic belts that account for 94 per cent of all the known historic eruptions. These volcanic belts are about 32 000 km in length. If one assumes they have a mean width of a hundred km, then they cover an area less than 0.6 per cent of the Earth's surface. Few eruptions are normally reported from under the oceans, yet mid-oceanic ridge volcanism is the most voluminous type of volcanic activity. At present submarine acoustic networks provide the only comprehensive set of data on deep-ocean volcanism.

Sixty-two volcanic eruptions were reported in 1991, and 165 volcanoes erupted during the decade 1980 to 1989. Some eruptions continue for a long time. Stromboli in the Aeolian islands, north of Sicily, has been in continuous eruption for the past two and a half millennia. During the past decade fifteen volcanoes have erupted frequently. They are Stromboli, Erta Ale in Ethiopia, Manam, Langila and Bagana in Papua New Guinea, Yasour in Vanuatu, Semeru and Dukono in Indonesia, Suwanose-Jima and Sakura-Jima in Japan, Santa Maria and Pacaya in Guatemala, Arenal in Costa Rica, Sangay in Ecuador, and Erebus in Antarctica. Volcanic eruptions are usually of short duration. Some 10 per cent of recorded eruptions last for less than one day. The spectacular June 10, 1886, eruption of Tarawera, near Rotorua, in New Zealand took place between 0100 and 0600 hours local time, but associated craters formed on the site of Lake Rotomahana continued to be active for another 26 days. Most eruptions end in less than a hundred days. Few last longer than three years. The median duration of recent eruptions is seven weeks.

4.2   The magnitude of volcanic eruptions

When assessing the volcanic risk posed by a volcano it is customary to evaluate its past behaviour as an aid to predicting likely future activity. If one is informed that there has been an eruption in any part of the world one can use the Smithsonian Institution's publication *Volcanoes of the World* (Simkin *et al.*, 1981) to ascertain its past behaviour. This publication lists 1300 volcanoes that have been active during the past 10 000 years. Unfortunately the data on many of these volcanoes are incomplete because they have not erupted in historic times. Some volcanoes, particularly the most explosive types, remain dormant for hundreds, or even thousands, of years. When such volcanoes are reactivated after a long period of dormancy, the composition of the material extruded, and the style of eruption, may have changed. For example, the cataclysmic eruption of Tambora Volcano in Indonesia in April 1815 took place after 5000 years of dormancy. This eruption was probably the largest ever produced by this volcano, and the materials extruded seem to be its most evolved products (Self *et al.*, 1984, p. 659).

Explosive eruptions result from one of two processes. In magmatic eruptions the volatiles that propel the explosion are mainly derived from within the magma. This is because in the near-surface environment magma decompresses and this results in the explosive exsolution of volatiles that entrain and expel pyroclastic materials. The second type of explosive behaviour is called hydromagmatic. It happens when hot magma interacts explosively with an external supply of water such as a crater lake or the sea.

Volcanic explosivity index

A fundamental problem in volcanology is to find objective criteria for measuring the size of explosive volcanic eruptions, particularly if these eruptions occurred before 1960. Simkin *et al.* (1981) use a simple volcanic explosivity index (VEI) developed by

Newhall and Self and published in 1982. This index uses volume of ejecta, height of eruptive column, duration of continuous blast, and eyewitness descriptions. It is primarily a measure of the destructive potential of a volcanic eruption. Voluminous, quiet effusions of basaltic lava, such as the historic eruptions of Mauna Loa on Hawaii, are given a volcanic explosivity index of naught, whereas the eruptive event that generated the huge Huckleberry Ridge Tuff in the Yellowstone Volcanic Centre is given a tentative rating of eight. The most explosive event associated with the 1815 eruption of Tambora Volcano in Indonesia has been assigned a volcanic explosivity index of six.

In 1987 Legrand and Delmas proposed a glaciological volcanic index (GVI) based on measurements of ice-cores from Greenland and Antarctica. This index must be used with care because it shows a distinct bias towards volcanic eruptions that took place at high latitudes and those eruptions characterised by high sulphate emissions. Ice-core records are of most value when historical records enable local and global volcanic events to be separated.

Magmatic central eruptions

The following is a brief description of the principal types of magmatic central eruptions. The gentle effusive eruptions with volcanic explosivity indices of naught or one are called Hawaiian-type eruptions. Strombolian-type eruptions have frequent, small to moderate discharges of incandescent scoria and bombs. Their volcanic explosivity indices range from one to two. A Vulcanian-type of eruption ejects moderately large volumes of pyroclastic materials, and the height of its eruption column is usually between 3 and 15 km. Its volcanic explosivity index ranges between two and four. Plinian eruptions are exceptionally powerful, continuous gas blasts. They eject between 0.1 and 10 km^3 of pyroclastic materials and this usually includes huge amounts of pumice. The eruptive columns of Plinian eruptions are normally higher than 15 km and their volcanic explosivity indices range from four to six. The term Plinian is used to describe large eruptions equivalent in magnitude to the AD 0079 eruption of Somma-Vesuvio (see Section 1.2).

Ultraplinian eruptions are colossally large. They eject more than 10 km^3 of pyroclastic materials. As their name suggests these eruptions are normally

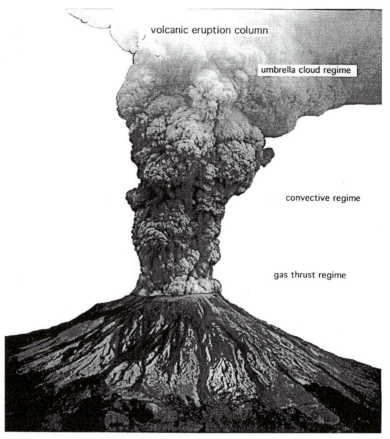

Fig. 4.1 Physical regimes that operate in a volcanic eruption column.

considerably larger than the AD 0079 eruption of Somma-Vesuvio. Their eruptive columns are over 25 km in height and they usually inject significant volumes of pyroclastic materials into the stratosphere. Eruptions of this explosivity are generally associated with volcano-tectonic depressions. These negative landforms develop because colossal explosions are more likely to produce craters than construct positive landforms. The dispersal of pyroclastic products during this type of eruption is normally so widespread that little is deposited close to the vent, and the materials that are so deposited are not thick enough to balance the subsidence produced by the rapid expulsion of magma from beneath the volcano. Many hazardous volcanoes of this type spend long periods of dormancy in the benign guise of tranquil lakes. Walker (1980) called these sunken ultraplinian volcanic centres inverse volcanoes.

Volcanic eruption columns

The coarse debris ejected from volcanoes follows ballistic trajectories, but most pyroclastic materials of smaller size and weight become entrained in volcanic eruption columns. These columns are gas-solid systems that can be considered to belong to three physical regimes (see Fig. 4.1). The gas thrust system operates at the base of the eruption column where it occupies only a small fraction of its total height. Exit velocities generally vary between 100 and 600 m/s. Upon leaving the vent the density of the erupting column is significantly greater than the air it displaces. As the larger pyroclasts fall out of the column air mixes into the column and is heated, producing a rapid decrease in bulk density. When the bulk density of the gas thrust system becomes less than that of the surrounding atmosphere, a convective regime develops. The pyroclasts are supported by the turbulence and thrust produced by the convective system. If the pyroclasts are small and hot, heat exchange with the air is rapid and the convective system grows rapidly.

In the density stratified atmosphere a convecting column soon reaches a level where its bulk density equals that of the surrounding atmosphere. At this level momentum enables some materials to ascend further and spread out radially to form an umbrella-shaped cloud. Winds often displace the eruption column laterally. Changes in the abundance, or nature, of the materials being ejected, or changes in the size of the vent, may trigger the collapse of the eruption column. When this happens most of the pyroclastic materials and the trapped gases and air plunge to the ground where they may become part of a fluidised system that travels rapidly outwards. Such pyroclastic flows are typical of silicic plinian and ultraplinian eruptions.

The composition of the materials erupted largely determines the violence of an eruption. Basaltic lavas have low silica contents and low viscosities. They are likely to erupt at higher temperatures and have a lower volatile content than silicic materials. As a result of these properties basaltic materials have low explosivity indices and often display fire fountaining and the quiet extrusion of freely flowing lava. In silicic magmas the combination of high viscosity and high volatile content often results in powerful explosive eruptions, with the lava being violently fragmented during the eruption process.

| 4.3 | Lavas, pyroclastic fall deposits and pyroclastic flow deposits |

The classification of pyroclastic rocks was considered in Section 2.13. Terms such as air-fall deposit (or pyroclastic fall deposit), pyroclastic flow deposit, lahar, and pyroclastic surge deposit were defined. This Section contains a more detailed examination of the processes responsible for producing these deposits. As the name implies air-fall deposits fall through the air from an eruption column and accumulate on the surface. Their distribution pattern is normally unrelated to topography. The following properties are frequently used to describe these deposits: (1) proportion of crystals; (2) proportion of lithic clasts; (3) mean grain size of clasts; (4) local maximum bed thickness; (5) local maximum

clast size; (6) size sorting of clasts; and (7) vesicularity of pyroclasts. Once such information has been collected from various sampling areas, isopleth maps are constructed for the various parameters described above. Such maps enable volcanologists to identify the type of eruptive activity responsible for producing particular air-fall deposit. For example Pyle (1989) uses measurements of maximum clast size and deposit thickness to figure out the type of eruptive activity suited to generating particular air-fall deposits.

Hawaiian-type eruptions

Most air-fall deposits produced by Hawaiian-type eruptions are small in volume and aerial distribution. In these eruptions the magma is normally a low-viscosity basalt that erupts with an easy separation of the gas and liquid phases. Liquid lava is sprayed into the air in the fashion of a fire fountain. The narrow jets of incandescent liquid soon break up into droplets and larger volcanic bombs. When the rate of ejection is high the lava droplets may remain liquid. On landing they are likely to coalesce at the base of the fountain to form a clastogenic lava (or clastolava). When the rate of eruption is lower the droplets and bombs are likely to weld together to form steep-sided spatter cones. Although fire fountaining is a spectacular feature of Hawaiian-style eruptions such eruptions usually produce insignificant volumes of pyroclastic materials (see Fig. 4.2).

Strombolian-type eruptions

Strombolian-type eruptions are frequent, small to moderate discharges of incandescent scoria and bombs. Stromboli Volcano is a cone that rises about 3000 m above the floor of the Southern Tyrrhenian Sea (see Figs 4.3 and 4.4). It is the northernmost island in the Aeolian Archipelago to the north of Sicily. The volcano's summit crater usually contains five or six small (1–10 m in diameter) active vents. Normally there are about six explosive eruptions per hour, and the duration of each event is less than a few minutes. In a single explosion the mass of lava ejected usually varies between 10 and 1000 kg (Gilberti *et al.*, 1992, p. 536). Infrequently there are more violent explosive events or eruptions of lavas. The last large explosive eruption took place on

Fig. 4.2 Basaltic lava from the East Rift of Kilauea Volcano flowing quietly down a road to the east of Kamoamoa Campground on the big island of Hawaii, July 1989.

Stromboli in 1930 and produced about 70 000 m³ of tephra. Central eruptions of Strombolian-type are typical of many volcanoes. For example, this type of activity is the most common style of explosive volcanism exhibited by Etna in Sicily. On Etna this type of eruption usually consists of a series of discrete explosions that throw clasts to heights of between a few metres and many hundreds of metres (Chester *et al.*, 1985, p. 126; McClelland *et al.*, 1989, pp. 52–76).

During a Strombolian-type eruption the lava is normally more viscous and richer in silica than the material discharged during a Hawaiian-type eruption. In a typical Strombolian eruption there is a high volume of escaping gas that fragments the lava and produces highly vesicular clasts or scoria. Angular clasts are usually thrown only a few hundred metres into the air and on returning to the ground they usually land near their parent vent. The chemical composition of these air-fall deposits is variable, but they are usually basaltic andesites or basaltic trachyandesites. Stromboli has erupted materials that range in composition from basalt to latite. These

materials include potassic trachybasalts, shoshonites, basaltic andesites and andesites (Rosi, 1980, pp. 16–17). The older volcanic rocks are mainly in the basaltic andesite to andesite range, whereas the younger rocks are mainly shoshonites and latites (see Table 4.1). Etna has extruded various basaltic rocks and a suite of sodic rocks that range from hawaiites to benmoreites (Chester *et al.*, 1985, p. 240).

Vulcanian-type eruptions

In typical Vulcanian-type eruptions between a million and a hundred million cubic metres of pyroclastic materials are ejected explosively. These pyroclasts are propelled upwards in an eruption column that varies between 1 and 15 km in height. Vulcanian eruptions are often the first phase in a protracted eruptive event and during such eruptions part of the volcanic edifice may be destroyed. The initial explosions of the 1980 eruption event at Mount

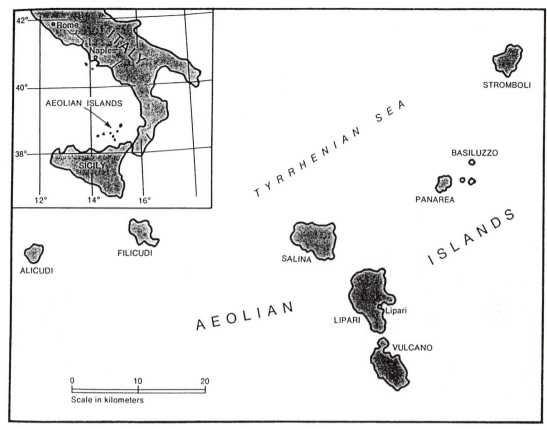

Fig. 4.3 Sketch map of the Aeolian Archipelago in the South Tyrrhenian Sea, Italy. After Bullard, 1977, p. 225. Locations of all the active volcanoes of Italy are shown in Figure 11.6.

Saint Helens in the western USA were of the Vulcanian-type (volcanic explosivity index = three), but the later cataclysmic eruption of 18 May was of the Plinian-type (volcanic explosivity index = five). Immediately prior to a Vulcanian-type eruption the magma usually has a high yield strength and viscosity (see Tables 4.2 and 4.4). The compositions of these magmas are usually in the andesite or trachyandesite range.

This type of eruption is named after Vulcano, the southernmost island in the Aeolian Archipelago (see Figs 4.3, 4.5 and 4.6). Since its last eruption between 1888 to 1890 it has only produced fumarolic activity. Most of this activity is confined to the deep crater (Gran Cratere), inside the largest cone on the island. This cone is known as La Fossa and it is mainly constructed of silicic pyroclastic rocks. La Fossa lies within a roughly circular depression that formed between 11 000 and 15 000 years ago. The rocks that

have erupted on Vulcano have exceptionally diverse compositions (Keller, 1980), and range from leucite-bearing phonotephrites to rhyolites. La Fossa is mainly composed of foid-bearing latites, latites, quartz trachytes and rhyolites (see Table 4.2).

Plinian-type eruptions

Plinian-type eruptions are exceptionally powerful explosive events that eject between a hundred million cubic metres and $10 \, km^3$ of pyroclastic materials. Section 1.2 contains an evaluation of the historical significance of the 0079 AD eruption of Somma-Vesuvio which was admirably described by Pliny the Younger (see Fig. 4.7). The air-fall phase of this prototype Plinian eruption lasted from about 1300 hours (local time) on August 24 to 0800 hours on August

Table 4.1 Major element composition of rocks of the latest volcanic cycle from near the summit, Stromboli, Italy. After Rosi, 1980, p. 17.

Number	B35	C14	B24	B38	C12	C13	Mean	Recalculated
SiO_2	52.83	53.61	54.19	54.25	56.78	59.18	55.14	55.71
TiO_2	0.92	1.02	0.97	0.92	0.82	0.73	0.9	0.91
Al_2O_3	18.12	18.9	19.16	18.55	17.5	17.43	18.28	18.47
Fe_2O_3	3.55	2.79	1.94	2.73	2.62	2.05	2.61	2.64
FeO	4.22	4.78	5.08	4.54	3.94	3.58	4.35	4.4
MnO	0.16	0.15	0.15	0.15	0.16	0.15	0.15	0.15
MgO	4.63	2.16	2.86	3.92	2.87	1.89	3.05	3.08
CaO	8.7	8.4	8.14	8.1	6.73	5.47	7.59	7.67
Na_2O	2.92	3.47	3.1	3.01	3.63	4	3.36	3.39
K_2O	2.46	2.91	2.81	2.63	3.56	3.66	3.01	3.04
P_2O_5	0.41	0.75	0.58	0.51	0.47	0.43	0.53	0.54
L.O.I.	1.08	1.06	1.02	0.69	0.92	1.43	1.03	—
Total	100	100	100	100	100	100	100	100

Fig. 4.4 Weak explosive eruption of incandescent lava from vents in Stromboli Volcano, Aeolian Archipelago, Italy.

25. During the initial Plinian stage the white pumice was extruded and the eruption column reached a height of about 27 km (Lirer *et al.*, 1993; Mues-Schumacher, 1994). Later, at an early stage in the explosive phase that ejected the grey pumice, the eruption column rose to its maximum height of 33 km. The eruption produced about 1 km^3 of dense-rock equivalent of white pumice, and 2.6 km^3 of dense-rock equivalent of grey pumice.

Most of this air-fall pumice deposit was dispersed to the southeast in the direction of Pompeii. Other pyroclastic materials occur interbedded with the pumice deposits. The grey pumice, for example, contains at least five pyroclastic surge deposits

Fig. 4.5 View of the main crater (Gran Cratere) of Vulcano, with Lipari Island in the distance. Lipari and Vulcano are the southern islands of the Aeolian Archipelago, Italy.

Fig. 4.6 Ngauruhoe Volcano in central North Island, New Zealand. Its volcanic explosivity index is usually in the range 2–3.

Table 4.2 Major element composition of volcanic rocks from the Fossa di Vulcano, Italy. After Keller, 1980, p. 48.

Number	V154	V227	V40	V185	V205	V228
SiO_2	55.4	58.3	58.6	59	73	68.8
TiO_2	0.7	0.7	0.6	0.5	0.1	0.1
Al_2O_3	18.3	17.5	16.3	16.2	13.6	13.8
Fe_2O_3	3.85	2.3	3.75	3.95	1.15	1.75
FeO	2.75	2.7	2.45	2.2	1.05	1.4
MnO	0.13	0.07	0.12	0.13	0.08	0.09
MgO	2.5	1.3	2.6	2.65	0.15	0.6
CaO	4.9	2.7	4.9	4.75	1	1.7
Na_2O	4.2	4.1	4.1	3.7	4.5	3.5
K_2O	5.8	6.8	5.6	5.55	5.15	4.9
P_2O_5	0.52	0.3	0.34	0.15	0.05	0.08
H_2O^+	0.9	2.4	0.4	0.6	0.6	4.1
Total	99.95	99.17	99.76	99.38	100.43	100.82

Fig. 4.7 Present crater of Somma-Vesuvio Volcano in southern Italy.

(Sigurdsson *et al.*, 1985). A 5–10-cm-thick transitional layer lies between the white and grey pumice layers. It contains three different types of pumice clasts. These clasts include the common white and grey types, plus a light grey component that Mues-Schumacher (1994, p. 391) calls the boundary pumice. Table 4.3 reveals that the boundary pumice is intermediate in chemical composition between the other types of pumice. The white pumice and the boundary pumice are phonolites whereas the grey pumice is a tephriphonolite. These rocks have distinct compositions with higher than normal potassium/sodium ratios and this is reflected in their mineralogy as they contain crystals of biotite, leucite and sanidine.

Earlier in this Section it was recorded that the cataclysmic eruption of Mount Saint Helens on 18 May

Table 4.3 Mean major element composition of the white, grey and boundary pumice from the Boscoreale section of the AD 79 deposits of Somma-Vesuvio, Italy. After Mues-Schumacher, 1994, p. 390.

	Mean white pumice	Mean boundary pumice	Mean grey pumice
SiO_2	56.13	54.47	54.22
TiO_2	0.25	0.36	0.49
Al_2O_3	22.47	20.48	19.68
Fe_2O_3	1.6	3.41	2.49
FeO	1.07	n.a.	2.07
MnO	0.12	0.12	0.12
MgO	0.33	0.93	2.09
CaO	2.85	6.25	5.66
Na_2O	5.37	4.66	4.29
K_2O	9.76	9.2	8.67
P_2O_5	0.05	0.12	0.22
Total	100	100	100

Table 4.4 Major element composition for each eruptive episode of Mount Saint Helens, 18 May to 7 August 1980. After McClelland *et al.*, 1989, p. 362.

Number	1	2	3	4	5	Mean	Recalculated
SiO_2	64.13	64.19	63.72	63.49	63.28	63.76	63.92
TiO_2	0.58	0.6	0.64	0.57	0.64	0.61	0.61
Al_2O_3	17.61	17.92	18.04	17.87	17.51	17.79	17.83
FeO total	4.04	3.99	4.24	4.44	4.39	4.22	4.23
MgO	1.88	1.91	1.99	2.2	2.17	2.03	2.04
CaO	4.9	5.06	5.16	5.22	5.3	5.13	5.14
Na_2O	4.63	4.83	4.7	4.97	4.89	4.8	4.82
K_2O	1.26	1.29	1.25	1.26	1.23	1.26	1.26
P_2O_5	0.15	0.15	0.15	0.15	0.16	0.15	0.15
Total	99.18	99.94	99.89	100.17	99.57	99.75	100

1980 was a Plinian event. The dacitic materials it released are more typical of Plinian-type eruptions than the unusual leucite phonolite pumice generated by the AD 79 eruption of Somma-Vesuvio. The events of 18 May started with a strong earthquake that caused slumping on the northern flank of the volcano (Foxworthy and Hill, 1982). This resulted in the collapse of a huge mass of rock and ice, removing about 1 km of overburden from an intruding cryptodome. The sudden removal of pressure on the cryptodome and its associated hydrothermal system caused a series of explosions. They blasted a mixture of hot gas and rock upwards and to the north. A ground-hugging surge of pyroclastic materials blasted out of the volcano at a velocity of over 600 km/h (Hoblitt and Waitt, 1989). Within a 10-km zone around the northern sector of the volcano all life was extinguished. Trees were stripped of their branches, snapped off near the ground, or uprooted. An area of about 600 km^2 was devastated. As the pyroclastic surge moved outwards its velocity decreased and it shed its load. While the outward motion of the surge was slowing down a huge convective system developed above the various layers of hot materials.

The main Plinian-type eruption lasted for about nine hours, with the eruption column reaching an altitude of at least 24 km. Most of the pyroclastic materials entrained in this system were carried east by the prevailing winds. The chemical composition of the materials erupted from Mount Saint Helens between 18 May and 7 August 1980 are given in Table 4.4. All this material is dacitic in composition. The high viscosity of dacitic magma had a profound influence on the style of eruption. Lava of this composition is likely to produce volcanic domes, coulee and cryptodomes. Volcanic domes are steep-sided, rounded extrusions of lava that form bulbous masses above and around volcanic vents. Short, stubby flows of viscous lava are called coulee. They are essentially transitional between normal lava flows and low lava

Fig. 4.8 Crater Lake in Oregon looking southeast to Kerr Notch. This lake occupies a nearly circular caldera that is about 10 km in diameter. The ancestral cone, called Mount Mazama, collapsed during an ultraplinian explosive phase about 7000 years ago. In the Pleistocene Mount Mazama was extensively glaciated and its old U-shaped glacial valleys now appear as notches in the caldera rim.

domes. Cryptodomes, such as the one that was emplaced over a period of two years prior to the cataclysmic eruption of Mount Saint Helens in May 1980, are plugs of magma, or lava (if degassed), that intrude volcanic cones and deform the surface topography without breaking through to the surface.

Ultraplinian-type eruptions

This type of eruption ejects more than 10 and sometimes more than 1000 km^3 of pyroclastic materials. If Plinian eruptions are considered to be the type that makes history (Francis, 1993, p. 127), then ultraplinian eruptions may be categorised as the type that change geography. The eruptive columns developed during these colossal explosive events may exceed 45 km. Eruptions of this magnitude are extraordinary because they can change global weather patterns for more than a year.

The last two ultraplinian eruptions took place in April 1815. They were part of a more extensive eruptive episode that emanated from Tambora Volcano on the island of Sumbawa, in the Lesser Sunda Islands of Indonesia (Self *et al.*, 1984; Sigurdsson and Carey, 1989; Sigurdsson and Carey, 1992; Stothers, 1984). Prior to this catastrophic eruptive episode Tambora was about 4000 m high. Its elevation is currently only 2650 m and what had once been the central summit area is now a caldera that is about 6.5 km in diameter and over 1 km deep. Caldera collapse probably occurred towards the end of the climactic eruption. This caldera is one of the largest to have formed in the Holocene. It is similar in size to Crater Lake Caldera (Mount Mazama) in Oregon that developed during an ultraplinian eruption about seven thousand years ago (see Fig. 4.8).

On the Sanggar Peninsula, where Tambora Volcano is located, the extensive pumice fall deposits from the April 1815 eruption are overlain by a thick incipiently welded, grey ignimbrite. During

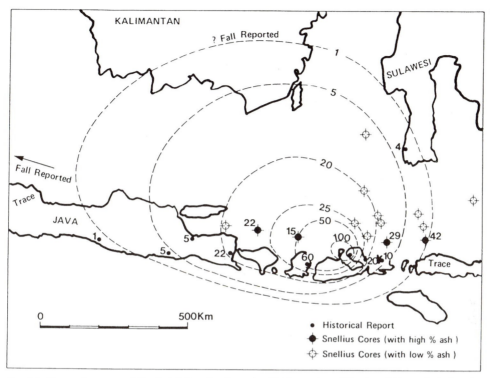

Fig. 4.9 Dispersal pattern of distal air-fall tephra from the 1815 eruption of Tambora Volcano in Indonesia. After Self *et al.*, 1984, p. 661.

the eruptive episode tephra buried most of the cultivated land on the islands of Sumbawa and Lombok. Fig. 4.9 shows the dispersal pattern of distal air-fall tephra from the 1815 eruption. It shows that tephra was deposited on Java, Kalimantan and Sulawesi. About 92 000 people died directly or indirectly from the eruption and the tsunami it generated. During the two ultraplinian-type eruptions the heights of the eruptive columns were 33 and 43 km respectively. Huge amounts of fine-grained tephra, gases and aerosols were injected into the stratosphere. According to Sigurdsson and Carey (1992, p. 16) the sulphur released during these eruptions formed a stratospheric aerosol with a mass equivalent to 1.75×10^{11} kg of sulphuric acid. This estimate is supported by data from ice-cores. When aerosols are injected into the stratosphere they reflect sunlight and absorb some solar radiation. This results in a lowering of temperature at the Earth's surface.

Most of the pyroclasts extruded from Tambora Volcano during the ultraplinian eruptive phase were latites. If one substitutes appropriate values for Fe_2O_3 and FeO in major element data from these rocks (Table 4.5), and calculates their CIPW norms one discovers that they are nepheline normative latites (nepheline = 3.4 per cent). These latites are the most evolved rocks that ever erupted from Tambora Volcano. They may represent a transition in magma chemistry and style of eruption. The total mass of pyroclastic materials erupted during April 1815 was about 50 km^3 of dense-rock equivalent.

The Tambora eruptive event produced what has been poetically called 'the year without summer'. Dust and aerosols screened out and absorbed substantial amounts of the sunshine. This caused the great central European famine of 1816–1817 (Harington, 1992). The bleak European summer of 1816 inspired Lord Byron to write his gloomy poem on darkness that described how 'morn came and went – and came, and brought no day'.

The Toba Caldera Complex in northern Sumatra, Indonesia, is the largest Quaternary caldera discovered on Earth (Chesner and Rose, 1991). It is an elongate basin-shaped feature, that is about 100 km long, and covers an area of 2270 km^2. Lake Toba occupies much of the caldera complex. A resurgent

Table 4.5 Major element composition of volcanic rocks from the 1815 eruption of Tambora Volcano, Indonesia. After Sigurdsson and Carey 1992, p. 23 (sample 1) and Self *et al.*, 1984, p. 662 (samples 2 and 3).

Number	1	2	3	Mean
SiO_2	58.31	56.33	57.91	57.53
TiO_2	0.66	0.63	0.58	0.62
Al_2O_3	20.25	19.9	19.83	19.99
FeO	4.85	5.23	4.82	4.97
MnO	0.25	1.8	1.34	1.13
MgO	1.38	0.18	0.21	0.59
CaO	2.88	4.58	3.94	3.8
Na_2O	5.57	5.04	5.03	5.21
K_2O	5.69	5.86	6.01	5.85
P_2O_5	0.16	0.45	0.33	0.31
Total	100	100	100	100

dome that is bisected by a graben lies within the caldera complex. Four huge eruptions are known to have issued from the caldera complex. The 2800 km³, Youngest Toba Tuff, erupted about 74 000 years ago. Since this cataclysmic event a series of lava domes have been extruded on Samosir Island and along the western ring fracture of the caldera complex. The other major eruptions have been dated at 1.2 million years ago, 0.84 million years ago and 0.50 million years ago. This indicates a mean repose period between principal eruptions is about 0.38 million years. Chesner and Rose (1991, p. 353) estimate that about 3400 km³ of volcanic materials have erupted from this volcanic complex during the last 1.2 million years. If one estimates that shallow magma chambers usually extrude about 10 per cent of their volume, then it is likely that a volume of about 34 000 km³ of subvolcanic rock and magma now lies beneath the Toba Volcanic Complex. A glance at a map of southeast Asia reveals how threateningly close this huge volcanic complex is to the major cities of Kuala Lumpur, Singapore and Padang.

Taupo Volcanic Centre

The term ultraplinian-type eruption was first used to describe pyroclastic deposits from the AD 0186 eruption of the Taupo Volcanic Centre of North Island, New Zealand. During the last 10 000 years this centre has produced eight plinian to ultraplinian eruptions. The AD 186 event included an ultraplinian eruption that triggered significant changes to the Earth's weather patterns. These changes include spectacular coloured sunsets. They are described in both Roman and Chinese official records. For example, it was recorded that on several occasions during the reign of Emperor Ling Ti (AD 0168–0189) the sun rose looking as red as blood and lacking in heat.

All the recent activity in the Taupo Volcanic Centre has originated in the depression now occupied by Lake Taupo. The lake is currently surrounded by low plateaus of pyroclastic materials. During the AD 186 eruptive event about 100 km³ of pyroclastic materials were extruded. The principal products were: (1) early air-fall pyroclastic rocks; (2) an ultraplinian air-fall deposit (24 km³); and (3) an ignimbrite deposit (30 km³) and its co-ignimbrite air-fall deposit (20 km³).

During the early air-fall stages of the eruption the Initial Ash, Hatepe Plinian Pumice, Hatepe Ash and Rotongaio Ash were deposited. The Initial Ash is a low-volume, bedded, mainly fine-grained ash and lapilli, air-fall deposit that formed from a vent-clearing phreatomagmatic explosion. It was probably triggered by contact between magma and near-surface water. The Hatepe Plinian Pumice was the first open-vent phase. It is a well sorted, coarse-grained, bedded, air-fall deposit. The Hatepe Ash lies conformably over the pumice unit and consists of poorly sorted materials that grade from lapilli to fine ash. It was probably generated by a phreatoplinian eruption of a vesiculating magma. The Rotongaio Ash is a grey, obsidian-rich, pumice-poor, fine ash generated by a phreatoplinian eruption of non-vesiculating magma.

The later Taupo Ultraplinian Pumice is a pumice lapilli, airfall deposit that covered approximately 30 000 km² to a depth of more than 10 cm. This material has a particularly low density. About 80 per cent of it was swept out to sea by strong westerly winds.

According to Walker (1980, p. 69) the eruption column exceeded 50 km in height. This is about twice the height of the eruption column responsible for the AD 0079 white pumice deposits of Somma-Vesuvio. The great height of the Taupo eruption column is attributed partly to the high eruption rate of approximately 1 000 000 m^3/s, and to the physical characteristics of the materials discharged. Many of the pyroclasts were small and hot and able to rapidly release their heat thus optimising the convection processes within the eruption column. One consequence of the great size of this eruption column is that the pumice it deposited attained its greatest thickness about 20 km down-wind of the vent. During this ultraplinian eruption the upper part of the magma chamber beneath the Taupo Volcanic Centre was drained. This resulted in the collapse of its roof and a sudden increase in the rate of discharge through the greatly enlarged volcanic vent. The rate of discharge was far too large to persist for longer than a few minutes. Then the huge eruption column collapsed catastrophically and produced the hot, fast moving, fluidised phases that flowed out to form the Taupo Ignimbrite.

This high-energy pyroclastic flow deposit has a near circular distribution although it had to surmount a variety of topographic obstacles that lay in its path. The highly erosive flow-head moved at a velocity of about 300 m/s and travelled up to 90 km from the vent. For example, some ignimbrite was deposited in Red Crater in the Tongariro National Park that is about 400 m above and at a distance of 45 km from the vent (Horomatangi Reefs). An area of over 20 000 km^2 was devastated by the Taupo ignimbrite flow. It has been estimated that all this material was erupted in approximately six minutes. All six stages of the AD 186 eruption took place within a period of less than a month and possibly less than two days.

According to Wilson *et al.* (1995) the Taupo Volcanic Zone contains about 15 000 km^3 of rhyolitic materials that erupted from its central chain of calderas. During the last 340 000 years it has been the most frequently active and productive rhyolitic volcanic system on Earth. Hochstein (1995) has estimated that the long-term extrusion rate of this zone is about 400 kg/s/100 km.

Hydromagmatic central eruptions

Most volcanoes are located near or beneath oceans or lakes and interactions are likely to take place between these external sources of water and magma or degassed lava (Self and Sparks, 1978). The 1883 eruption of Krakatau in Indonesia prompted much of the early interest in hydromagmatic eruptions. This was because Verbeek (1884) in his excellent description of the various eruptive events suggested that the cataclysmic event was probably caused by seawater rushing into the collapsing caldera and interacting explosively with hot magma. More recent studies (Self and Rampino, 1981) have shown that it is unlikely that this devastating eruption was triggered by magma–seawater interaction. They suggest that the major tsunami associated with the eruption were generated when several km^3 of pyroclastic flow materials plunged into the sea following the collapse of a huge eruption column. During the 1883 eruption of Krakatau both near sea-waves (tsunami) and distant sea-waves were produced. The latter are now usually interpreted as being the result of a strong coupling between the sea and air-waves generated by the exploding volcano.

The principal types of hydromagmatic eruptions are the Surtseyan, and the phreatoplinian. Surtseyan-type eruptions are the hydromagmatic equivalents of Strombolian- or weaker Vulcanian-type magmatic eruptions. Pyroclastic deposits extruded by Surtseyan-type eruptions are characterised by: (1) poor size sorting of fragment; (2) an average grain size in the fine range and; (3) an overall thickness that decreases rapidly away from the vent. Many of these deposits are finely laminated showing that they were erupted in repeated explosions of short duration. Other deposits produced by hydromagmatic activity contain accretionary lapilli or small spheres of fine ash with a hard outer shell and a more friable core.

Surtsey is a young volcano that lies southwest of the small island of Geirfuglasker in the Vestmannaeyjar (Westman Islands; eyjar = islands), off the southern coast of Iceland. It is located at the southern end of Iceland's extremely active eastern volcanic zone. Evidence of a submarine eruption in this area was first recorded on November 14, 1963. It started at a depth of about 130 m below sea-level, and during the first four months it deposited about 0.4 km^3 of volcanic materials. On April 4 1964 one of the vents developed a water-tight lining. Lava fountains formed and later lava flowed over the tephra protecting it from further erosion by the sea. This effusive phase continued intermittently until May 1965 when the new island covered an area of 2.8 km^2. All the volcanic materials were of basaltic composition (see

Table 4.6: Major element composition of volcanic rocks from Surtsey Volcano (1), after Tilley *et al.*, 1967.

SiO_2	46.56
TiO_2	2.02
Al_2O_3	15.93
Fe_2O_3	1.61
FeO	10.32
MnO	0.2
MgO	9
CaO	10.51
Na_2O	3.21
K_2O	0.51
P_2O_5	0.26
H_2O^+	0.02
Total	100.15

Table 4.6). After it had erupted the fresh basaltic glass was rapidly transformed into a hydrated material called palagonite.

The remote Kavachi submarine volcano (9.02°S, 157.95°E) in the Solomon Sea to the west of Honiara has recently produced many Surtseyan-type eruptions (McClelland *et al.*, 1989, pp. 192–195). It has been the site of at least eight island-forming eruptions since 1939. Once they have emerged from the sea these islands are soon demolished by marine erosion. For example, in early September 1976 a pile of pyroclastic materials emerged from the sea. Later a cone formed and on September 7 lava was seen to be pouring out of the central vent. Small explosive eruptions occurred at about one-minute intervals. Pyroclastic activity was maintained until at least October 10. Since then the island has once again been destroyed by the sea.

Phreatoplinian eruptions are the most explosive type of hydromagmatic central eruption. The term was introduced by Self and Sparks (1978) who used it to describe large hydromagmatic-type eruptions that produced widespread pyroclastic deposits of silicic composition. The explosivity of these eruptions is equivalent to the most explosive Vulcanian-, or a Plinian-type magmatic eruption. Self and Sparks (1978) and other volcanologists have described phreatoplinian eruptive episodes that took place during the AD 186 eruption of Taupo in central North Island, New Zealand and also during the 1875 eruption of Askja in central Iceland. These eruptions all took place beneath lakes situated within preexisting calderas. The volumes and dispersal patterns of the phreatoplinian deposits are usually similar to those produced by Plinian-type magmatic eruptions. It is thus likely that both types of deposit are discharged from broadly similar eruption columns. The distinctive characteristic of phreatoplinian deposits is their extremely fine grain-size. This fine-grained material is even typical of areas close to their parental vents.

Auckland Volcanic Field

The city of Auckland in North Island, New Zealand, is an ideal site to study both magmatic and hydromagmatic eruptions (see Fig. 4.10). It contains almost 50 young volcanoes (Searle, 1981; Kermode, 1991). They first became active about 100 000 years ago. The youngest and largest is Rangitoto Volcano. It was active only a few hundred years ago. During the development of the present Auckland Volcanic Field about 3 km^3 of lava, scoria, lapilli tuff and ash was deposited over an area of about 100 km^2. Most of the activity was in the area adjoining the Auckland isthmus. A typical eruption usually begins with the explosive excavation of a crater up to 800 m in diameter and 100 m deep. Such explosion craters are usually surrounded by tuff rings containing deposits up to 30 m thick. The deposits are usually composed of comminuted material from the underlying sedimentary units, together with layers of black vesicular lapilli. Twenty-three of the Auckland explosion craters subsequently produced scoria cones. The Auckland Volcanic Field also contains 15 small lava fields comprising flows that are up to 25 m thick. Some smaller flows form aprons at the bases of their parental volcanoes or remain confined within their craters. The total area covered by lava is about 75 km^2.

Rangitoto Volcano occurs as an island at the entrance of Waitemata Harbour. Its symmetrical cone dominates the skyline to the northeast of the city. The lower slopes of the volcano are lava covered with a gentle gradient of about 12°, whereas the upper part of the edifice has steeper slopes. Scoria is the characteristic material of the steeper slopes and it surrounds a summit crater that is 60 m deep and 150 m in diameter (see Fig. 4.11). A smaller crater occurs to the east of the main crater and a complex series of mounds, ridges and depressions occur to the north. On its flanks one can readily observe a well developed system of lava tubes linked by lava trenches. Small steep-sided spatter cones, or hornitos, have developed above some lava tubes. The evolution of Rangitoto Volcano probably began in a similar way to Surtsey, with an extended series of hydro-

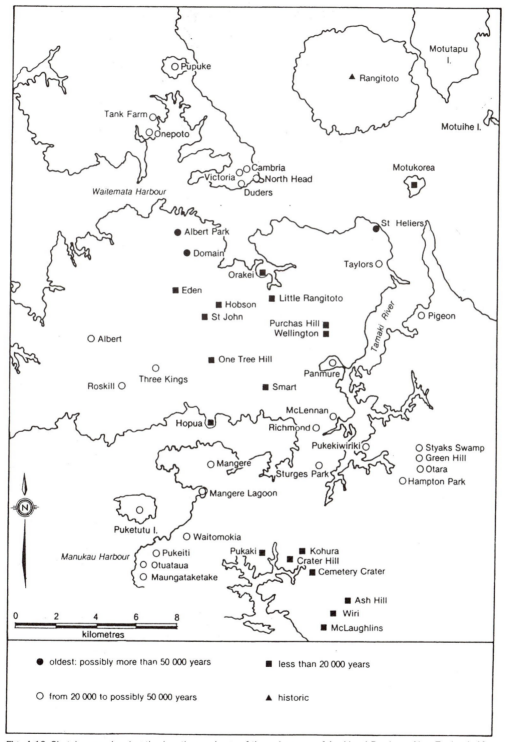

Fig. 4.10 Sketch map showing the location and age of the volcanoes of Auckland Province, New Zealand. After Searle, 1981, p. 46.

Fig. 4.11 Rangitoto Volcano has produced an island at the entrance of Waitemata Harbour in Auckland, New Zealand. Its symmetrical silhouette dominates the skyline to the east of the city.

magmatic eruptions. Most of the volcano was extruded about 500 years ago when sea-level was at approximately its present elevation.

The rocks of the Auckland Volcanic Field are alkali basalts, basanites and phonotephrites. They usually have fine-grained, porphyritic to vitrophyric textures and contain euhedral to subhedral phenocrysts of olivine (Fo_{90-80}), set in a groundmass of clinopyroxene (diopside-augite), plagioclase (labradorite-andesine), olivine and iron-titanium oxides. Most of these rocks are strongly olivine and nepheline normative. Nepheline is usually an occult mineral, but it occurs in blocks ejected from the Domain Volcano in central Auckland.

Submarine volcanic activity

Submarine volcanic eruptions differ greatly depending on the composition of the magma and the depth below sea-level at which the eruption occurs. It is revealing to begin this discussion by investigating the flow of basaltic lava into the sea. In recent times this activity has been observed on Hawaii on many occasions. Such lava is usually encased in a lava tube and it flows smoothly into the sea. If the waves break open the lava tube a thin film of steam develops and insulates the hot lava from the much cooler water preventing violent interaction between the two. Intermittently small explosions occur and throw fragments of partly quenched lava into the air. Clouds of steam and water vapour rise into the air as heat is exchanged between the lava and the seawater. If the lava tubes transport the lava below the level of active wave action, the process is likely to be a placid, thus enabling scuba divers to approach and film in safety. Submarine eruptions of this type often produce pillow lavas (Moore, 1975). Individual pillows are frequently about a metre across and they may occur in extensive piles. Films have been made that show the process of pillow formation. In such films one sees a steep underwater flow front comprised of numerous pillows. As one looks one notices a reddish glow and then the rapid extrusion of a rounded blob of red lava. The new pillow-shaped blob cools quickly and on losing its bright colour it comes to look like all the other pillows that surround it. This process of pillow formation is then repeated at another location on the flow front. Most pillow lavas are basaltic in composition, but similar pillow structures occur in lavas of diverse composition and of all ages including the Archean.

Water is usually the main volatile component in the materials that erupt under the sea. At a pressure equivalent to a depth of about 3 km sea-water is at its critical point and highly compressed steam behaves in much the same way as liquid water. This means that below 3 km, or at a lesser depth called the maximum volatile fragmentation depth, explosive volcanic activity stops. Cashman and Fiske (1991) have recorded the presence of pyroclastic rocks at depths of about 1.5 km at several sites in the western Pacific. Fisher and Schmincke (1984, p. 274) argue that silicic volcanoes erupting at depths down to 500 m below sea-level may give rise to floating pumice, ejecta and steam that rises to considerable heights above the surface of the sea.

Below the maximum volatile fragmentation depth a new class of pyroclastic rocks is likely to form. These materials include hyaloclastites and pillow breccias. The former are volcaniclastic materials produced by the explosive shattering of volcanic glass when it is quenched on coming into contact with water, ice, or water-saturated material such as pelagic sediment. Pillow lavas are often fragmented around their margins.

Mid-oceanic ridge volcanism is the most abundant type of volcanism on Earth yet it is hidden from view. If it could be seen it would probably be unspectacular. The remarkable feature of this type of volcanism is that it is essentially continuous for the life of a particular ocean basin. It may be active for over a hundred million years. Strange as it may seem the fastest spreading ridges are usually the least impressive topographically. This is because on the slower spreading ridges the lithosphere is able to cool, thicken and subside closer to the ridge. This results in a more imposing topographic feature.

Both the fast and slow spreading ridges are segmented. The fast-spreading ridges are characterised by elongate axial shield volcanoes with most of the eruptive activity focussed along their crest. On the slow-spreading ridges the active zone consists of a long, narrow rift valley that contains innumerable small volcanoes that form a discontinuous central ridge. Pillow and sheet lavas have been described from the flanks of all these mid-oceanic ridge volca-

noes. Pillow basaltic lavas are probably the most common volcanic rocks on Earth.

Subglacial eruptions and jökulhlaups

Subglacial eruptions present a special type of interaction between water and hot volcanic materials. Such eruptions are difficult to study, but in Iceland the sudden release of water from such an eruption may produce huge floods. Such floods are called jökulhlaups that means glacier bursts in Icelandic (jökull is the word for glacier or ice sheet). Grímsvötn Volcano beneath the Vatnajökull, Iceland's largest ice cap, and Katla Volcano beneath the smaller Myrdalsjökull ice cap in the south, have produced many jökulhlaups in historic times. According to McClelland et al. (1989, p. 556) Grímsvötn caldera has produced more confirmed subglacial eruptions than any other volcano in the world. Its jökulhlaups usually last for about three weeks and they happen about once or twice a decade.

At present most jökulhlaups occur in high latitudes. An exception is Antisana Volcano in Ecuador. It is only 55 km from the equator, nevertheless it has produced subglacial eruptions and jökulhlaups. At various times in the past when the surface of the Earth was covered with large sheets of ice, this type of eruption was probably more common. Subglacial eruptions are likely to be the normal type of eruption on the ice covered planets and satellites of the outer Solar System.

4.4 Volcanic hazards

The present decade has been designated the international decade for natural hazard reduction. Scientists associated with the United States Academy of Sciences have suggested the following programme for reducing volcanic hazards: (1) identification and mapping of all active or potentially active volcanoes; (2) assessment of the hazards posed by these volcanoes through the study of their deposits and eruptive history; (3) quantitative assignment of the intensity and magnitude of all their historic eruptions; (4) set up baseline geophysical and geochemical databases on volcanoes, particularly those in densely populated areas; (5) set up training and education programmes in volcanology in all countries where volcanoes are a potential threat; (6) form expert international volcanic crisis assistance teams that can respond rapidly to volcanic emergencies; (7) develop and coordinate emergency warning, evacuation and response methods and techniques; and (8) evaluate the environmental impact of eruptions on the Earth's weather and climate.

The environmental impact of volcanic activity is usually called volcanic risk. Risk maps ought to be used in planning the development of areas menaced by volcanic activity. For example, Scandone *et al.* (1993) have evaluated the volcanic risk associated with the towns surrounding Somma-Vesuvio in Italy. There is a large measure of irony in their document because it is only concerned with hazards emanating from Somma-Vesuvio, when about 25 km to its west lies the huge Campi Flegrei volcanic complex that has spread ash over much of the eastern Mediterranean Sea (see Section 1.2).

Volcanic activity has claimed more than 260 000 lives in the past 400 years (Tazieff and Sabroux, 1983; Blong, 1984; Wright and Pierson, 1992). Examples of particularly destructive events include: (1) the death of 20 000 in the Mount Etna area of Italy in 1669 (most of these deaths are directly attributable to a swarm of volcanogenic earthquakes); (2) the 1792 collapse of the cone of Unzen Volcano in Japan that killed about 14 300 people; (3) the 1815 eruption of Tambora Volcano that killed 92 000 people; (4) the 1883 eruption of Krakatau in the Sunda Strait that killed 36 000 people (most were drowned by tsunami); (5) the 1902 eruption of Mount Pelee on the island of Martinique in the West Indies where a series of pyroclastic surges killed 29 000 people; (6) in 1919 Kelut Volcano in Indonesia produced a lahar that killed 5110 people on its flanks; and (7) the 1985 eruption of Nevado del Ruiz volcano in Colombia killed 22 000 when a series of lahars swept down the slopes of the volcano. Volcanic gas is another hazard. In February 1979 on the Dieng Plateau on the Indonesian island of Java a small predawn explosive eruption released carbon dioxide that suffocated 149 people. Similarly in August 1986, 1700 of the people who had been living around Lake Nyos, a crater lake, in Cameroon, West Africa, were suffocated by a dense cloud of carbon dioxide.

Volcanic eruptions often produce serious repercussions on people, their airports and aeroplanes, agriculture, buildings, economic activity, electricity supply systems, fresh and waste water systems, machinery, motor vehicles, railway systems, roads, shipping, social activities and telecommunications.

The main volcanic hazards are lava flows, tephra falls and volcanic ballistic projectiles, pyroclastic flows and debris avalanches, lahars and jökulhlaups, volcanogenic seismic activity, volcanogenic tsunami, atmospheric effects, release of gases and aerosols, and diseases and starvation that develop as a result of volcanic activity.

Lava usually occurs in lava lakes, or it pours out of vents and forms streams that flow at velocities that range from a few metres per hour to more than 60 kph. Fast flowing lava is characteristic of steep slopes and narrow channels. The temperature of mobile lava is usually in the range between 880°C and 1130°C. It can easily set fire to trees and wooden structures. In the past major eruptions of flood lava have been highly disruptive as they have covered large areas and released vast quantities of volatiles into the atmosphere.

Projectiles of various sizes are ejected during most eruptions and may be most destructive. In September 1979 a 30-second explosive eruption on Mount Etna in Sicily showered several hot blocks of up to 25 cm in diameter on a group of about 150 tourists. Nine were killed and twenty-three injured. Motor vehicles provided little protection against these hot and heavy projectiles.

Pyroclasts carried upwards in a volcanic eruption column eventually fall on to the surface. Such air-fall deposits are usually dangerous only in areas close to the erupting vent. Thick deposits of tephra may cause roofs to collapse. In May 1973 the eruption of Heimaey Volcano (Eldfell Vent) in the Vestmannaeyjar, south of Iceland, resulted in about 65 houses being buried beneath air-fall deposits. Tephra from the huge 1815 eruption of Tambora buried much of the cultivated land on several Indonesian islands. The resulting starvation and disease claimed many thousands of lives.

In recent years encounters between in-flight commercial and military aircraft and high-rising eruption columns have caused problems in western North America, Southeast Asia and the southwestern Pacific. When volcanic ash is ingested by an aircraft engine, it usually melts in the high temperature area within the engine. Subsequently the melt is deposited as a glassy layer on the cooler surfaces of the turbine vanes. This produces a loss in thrust. At night it is difficult to detect clouds of volcanic ash from aircraft using normal onboard sensors. Infrared detectors are more likely to discriminate between volcanic cloud and meteorologic cloud.

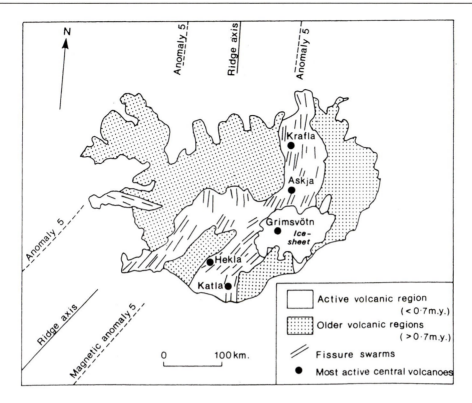

Fig. 4.12 Sketch map showing that Iceland is where the mid-Atlantic oceanic ridge emerges from the sea. Iceland's neovolcanic zone is linked to the mid-oceanic ridge system.

Volcanic ash produced by the 1989–90 eruptions of Redoubt Volcano in Alaska affected both commercial and military air operations in the Anchorage area. Problems included damage to aircraft and losses resulting from rerouting and cancellation of flights. Five commercial aircraft were damaged. A Boeing 747-400 was seriously damaged when it encountered an ash cloud as it was descending to land. Repairs to its avionics, engines and structure cost in excess of US $80 million. The loss of revenue from curtailed operations at Anchorage International Airport during the several months following the eruption is estimated to have totalled about US $2.6 million.

Pyroclastic flows and surges are a major threat to life and property. Both may be powerful enough to destroy most standing structures, knock down forests and kill animal life by asphyxiation and extreme heat. In 1902 a series of surges swept through the city of Saint Pierre in Martinique. They set the city and its large supplies of rum ablaze. This resulted in the death of about 29 000 people. These surges from the volcano, Mount Pelee, travelled across the waters of Saint Pierre harbour setting fire to ships and killing or severely injuring those on board. During the eruption of Krakatau surges travelled some 20 km over a sea covered in floating pumice to blast the coast of Sumatra. In the notorious AD 0079 eruption of Somma-Vesuvio many of the residents of Pompeii and Herculaneum were overcome by pyroclastic surges and flows that swept down from the volcano. Herculaneum was eventually buried beneath more than 20 m of pyroclastic deposits.

Volcanoes and climate

Large volcanic eruptions can produce changes in climate on a local, regional, or even a global scale. Interest in the question of whether eruptions are able to influence global weather patterns goes back to at least the late eighteenth century when the huge Lakagigar flood eruption ravaged southeast Iceland (see Figs 4.12 and 4.13). Benjamin Franklin (1784),

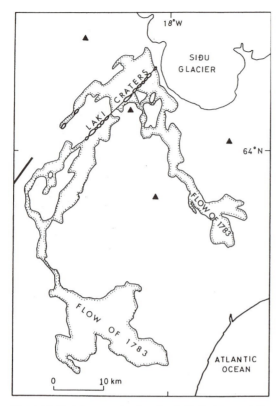

Fig. 4.13 Sketch map showing the Laki Craters and the extent of the 1783 Lakagigar flood eruption in southern Iceland. Adapted from Bullard, 1977, p. 308 and others.

summer in 1783. In Europe, parts of Asia and North America the warm summer was followed by one of the coldest winters on record. Exceptionally low temperatures prevailed throughout the whole of 1784 (Wood, 1992).

Large explosive eruptions often produce veils of fine materials that remain in the stratosphere for several years (Lamb, 1992). Veils diminish the brightness of the Sun and brighten the background sky. In the mid- and low-latitudes the speed at which the fine particulate materials that form a veil can be transported around the planet varies from two to six weeks. Volcanic materials injected into the lower stratosphere at low latitudes usually spread globally, whereas materials injected at high latitudes are entrained within a particular wind system and spread over the latitudinal zones affected by the wind system. These materials are thus likely to remain in the hemisphere where the volcanic eruption occurred.

The oceans and their circulation have a profound effect on climate. Over time submarine volcanism must influence the topographic and thermal characteristics of extensive areas of the ocean floor. It thus appears self-evident that volcanism is important in the gradual modification of the patterns followed by ocean currents. Recently there has been some speculation about whether large submarine eruptions are able to change the temperatures of enough water to trigger local, or even global, changes in weather patterns.

the US Ambassador to France, observed and described the volcanic ash and haze over Europe. He speculated on a possible link between the abnormal temperature fluctuations of that time and the huge eruption taking place in Iceland. The Lakagigar flood eruption produced the worst famine in Icelandic history. About a quarter of the human population of Iceland died (White and Humphreys, 1994, p. 181), yet the eruption was not directly responsible for any loss of life.

The huge volume of erupting lava ($12.3\,km^3$) moved slowly enough to enable people to move out of its way. Most of the deaths were caused by the release of about a hundred million tonnes of sulphur dioxide and fluorine. Together they produced a blue haze that collected over most of Iceland. For a while all the rain that fell was acidic. It killed or stunted the growth of the vegetation. This ushered in the horrifying 'haze famine'. The blue haze was then transported across Europe. It caused a remarkably warm

Mount Pinatubo

To conclude this section we will investigate the sequence of volcanic events that occur when a huge volcanic eruption takes place in a populated area. Mount Pinatubo Volcano is located in the southern part of the island of Luzon, about 100 km northwest of Manila, in the Philippines. On April 2, 1991 a series of small steam-induced explosions ended more than 600 years of quiescence. A monitoring and alert network was immediately set up. About 58 000 people were moved prior to the major eruption of June 15. Most of the 320 people who died during the eruption were killed when the weight of accumulating ash collapsed the roofs of their houses. During April and May a volcanic hazard map was prepared. Until the end of May almost all volcanogenic earthquakes occurred in a cluster roughly 5 km north-northwest of the summit, at depths of between 2 and 6 km.

Sulphur dioxide measurements registered a tenfold increase between May 13 and May 28. This was interpreted as evidence of juvenile magma rising to the surface. During the latter part of May and the first week in June the seismic source shifted towards a point beneath the steaming vents. Several episodes of harmonic tremor occurred. This type of seismicity is usually produced by magma movement. Simultaneously the sulphur dioxide flux decreased. Between June 6 and 7 an intense swarm of shallow earthquakes detonated beneath the vent area and a significant inflationary tilt was recorded on the upper east flank of the volcano. Continuous low-level emissions of ash began in early June and increased in volume when a dome appeared on June 7. After the dome had been extruded earthquake activity stopped in the northwest sector. It became concentrated beneath the dome and it continued to grow.

At 0851 hours on June 12 the first of a series of cataclysmic eruptions lifted a volcanic eruption column to over 19 000 m above sea level. Further cataclysmic explosive eruptions occurred at 2252 hours on June 12 when the eruption column reached 24 000 m, and at 0841 hours on June 13 when the eruption column once again reached 24 000 m. These explosions destroyed most of the new dome and excavated a 250 m diameter crater immediately southeast of the former dome. Renewed eruptions began at 1309 hours on June 14 and became increasingly frequent during the night. Shortly after 2320 hours on June 14 an infrared video camera recorded the rapid opening of a fissure vent on the eastern flank of the volcano.

A major lateral blast occurred at 0555 hours on June 15. At 1342 hours the signals from all local seismometers became saturated. Satellite images reveal an eruption column reaching up to over 35 000 m at 1500 hours. This eruption column rose above the tropopause and then expanded in all directions. Pyroclastic materials cascaded out of the column and covered an area of about 100 km². In so doing they buried all prehistoric pyroclastic flow deposits. The hastily prepared, worst-case hazard map forecast the full extent of these actual pyroclastic flows related to the volcano.

On June 15 typhoon Yunya passed within 50 km of Mount Pinatubo. The combined effect of a huge eruption column and cyclonic winds scattered ash over a huge area. Rain from the typhoon wetted the ash and increased its weight resulting in the col-lapse of many domestic structures. Rain also led to the remobilisation of pyroclastic materials and lahars caused extensive damage on all sides of the volcano. When the ash and cloud cleared a new caldera about 2 km in diameter was discovered. Ash continued to vent until the end of July. The number of devastating lahars increased during the following rainy season.

4.5 Extraterrestrial volcanism and planetary development

The surfaces of all the terrestrial planets and some of the satellites of the Jovian planets show volcanic features that developed under diverse physical conditions (Beatty and Chaikin, 1990). According to Carr (1987, p. 137) as one travels outwards in the Solar System one encounters 'stranger and stranger forms of volcanism'. Prior to the Mariner 10 encounters (1973–75) little was known about the surface of Mercury because it is particularly difficult to observe from Earth. Even now much still remains obscure because both the resolution and coverage of the images obtained by Mariner 10 are poor. Superficially Mercury looks like the Moon, yet much is different (Strom, 1987). It has an unusually large core and thermal models predict that it experienced a period of protracted heating and melting during its early history. This probably produced the extensive flood lavas. They cover most of its intercrater plains and partly fill some of its old impact craters. As this small planet cooled, its lithosphere thickened, eruptions became less frequent and the planet began to contract, triggering major faulting and the growth of extensive fault scarps.

Venus

Venus and Earth have much in common. They have similar radii, masses and densities. Their surfaces show evidence of having been shaped by meteorite bombardment, volcanism and tectonic activity (Head *et al.*, 1992; Cattermole, 1994). At present the surface temperature on Venus is remarkably high, as is its atmospheric pressures (see Section 1.5). This harsh surface environment would be expected to affect the style of volcanic eruptions, the character of subvolcanic processes and the shape of volcanic landforms. Volcanic eruptions on

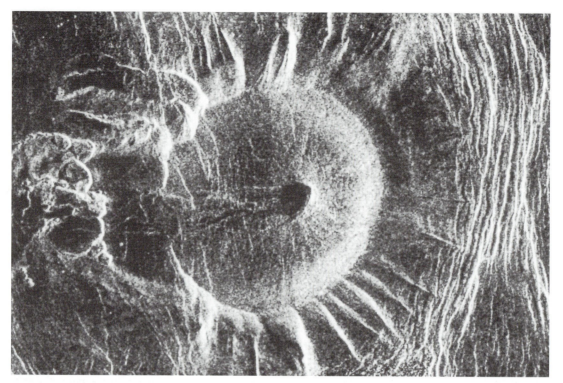

Fig. 4.14 Large volcanic dome with scalloped margin from northeast of Alpha Regio on Venus. The summit pit is about 5 km in diameter. Magellan image F-MIDR 20s003.

Venus take place at pressures similar to those found at a depth of about 1 km beneath the Earth's oceans. Materials ejected into the dense and layered Venusian atmosphere are likely to be strongly affected in a negative way by atmospheric drag and in a positive way by convective thrusting. It is anticipated that most Venusian eruptions would consist of the relatively fluid emission of lava. The mobility of the lava, particularly if it was basaltic in composition, would be greatly enhanced by the high surface temperature and pressure. Both of these parameters reduce the effective viscosity of lava. High pressure would also inhibit degassing. Images of Venus show that its various volcanic landforms are distributed globally and not restricted to linear belts as on Earth (Saunders *et al.*, 1992).

According to Newcott (1993, p. 42) the surface of Venus is 'rent by rift valleys, scarred by comets and asteroids, and blackened by seas of hardened lava'. More than 85 per cent of it is covered by flood lavas and other volcanic products. On its surface there are 156 volcanoes that exceed 100 km in diameter, 274 volcanoes of between 20 and 100 km in diameter,

and tens of thousands of smaller volcanic edifices. Most of the small volcanoes occur in clusters that average about 150 km in diameter. Each cluster is interpreted as being the surface manifestations of a mantle plume. Long lava channels are another remarkable feature of the surface of Venus. One of these channels is nearly 2 km wide and 6800 km long. It occurs to the north of the highland area called Aphrodite Terra.

The surface of Venus is embellished by some bizarre volcanic landforms. For example, the arachnoids are raised circular features surrounded by fractures that resemble spider-webs. Arachnoids probably formed when magma rose to just below the surface, causing doming and fracturing. Later cooling, assisted by magma withdrawal, resulted in collapse and the creation of a complex pattern of radial and concentric fractures. Another volcanic feature that is typical of Venus is the steep-sided volcanic dome. They are sometimes called pancakes (see Fig. 4.14). A line of seven of these pancake-like domes has been discovered in Alpha Regio. Their mean diameter is about 25 km. They are heavily frac-

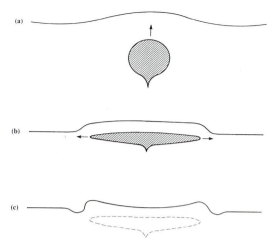

Fig. 4.15 Sketch showing how coronae are likely to form on the surface of Venus. (**a**) A rising diapir domes the surface rocks. (**b**) On impinging on the base of the uppermost rigid layer the diapir spreads radially and flattens. (**c**) The surface rocks relax as the diapir cools and contracts. Adapted from Cattermole, 1994, p. 161.

tured, with steep margins and a small central collapse pit. It is suggested that they were produced by a viscose, silicic lava oozing onto the surface and cooling as it spread out radially. Individual domes may range from 15–90 km in diameter, but they are usually only a few hundred metres high.

A unique Venusian volcano-tectonic feature is the coronae (see Fig. 4.15). They are large, hundreds of kilometres in diameter, ovoid-shaped, caldera-like features surrounded by rings of ridges. Most of them are concentrated in groups or chains. It has been suggested that they were generated by small mantle plumes (Stofan and Saunders, 1990). When viewed from above, the plate tectonics process on Earth appears to produce horizontal movements between plates and fragments of plates. On Venus surface movements are largely vertical. Various volcanoes, volcano-magmatic landforms and the highland regions appear to have developed above sites of mantle upwelling. Complementary areas of sinking appear to be represented by areas of lithospheric thickening and compression.

The Moon

The year 1994 was an important one, possibly the most important, in the exploration of the Moon. It saw the return of over a million new images collected by the Clementine One Mission. This lightweight spacecraft digitally imaged the whole lunar surface under first-rate lighting conditions. These new images cover the range of wavelength between 0.3 and 9.5 μm (ultraviolet/visible to near-infrared). They are at present inspiring new interpretations of the volcanological evolution of our close neighbour in space (Bruning, 1994; Gillis, 1994).

The Moon is uniquely important in our quest to understand the origin and evolution of the Earth. Its origin and that of the Earth are closely linked (Hartmann *et al.*, 1986). It also contains a wealth of information about processes that operated in the Hadean Era. In the pre-Nectarian Period between 4.55 and 3.95 giga-years ago the singular anorthositic crust of the Moon formed (Ashwal, 1993), and nine groups of impact basins were blasted out of its surface. The Nectarian Period spans the interval between the two huge impacts that created the Nectaris Basin (circa 3.92 giga-years) and the Imbrium Basin (circa 3.85 giga-years). During the Early Imbrium Epoch the surface of large areas of the Moon was covered by ejecta from the Imbrium and Orientale impacts. Ejecta was transported for distances of up to 1000 km from the crater rims. Material excavated during these huge impacts also triggered the formation of secondary craters with their own ejecta sheets. The Late Imbrium Epoch was characterised by widespread eruptions of mare basalts.

The mare lavas occupy about 17 per cent of the total surface area of the Moon. Most of this material was erupted between 3.9 and 3.2 giga-years ago. The flow patterns of these lavas reveal that they were usually more fluid, and capable of flowing longer distances, than the continental flood basalts of Earth. Individual flows of mare basalt are up to 600 km long. They flowed down slopes of only one in a hundred. The landforms produced during these eruptions show that they were normally quiet and lacking in significant explosive activity. Volcanic craters and domes are rare, but sinuous rilles are common. As these rilles appear to have been diverted by high ground, and observed to follow preexisting depressions, they are usually interpreted as open lava channels.

The lunar highlands are densely covered by impact craters that abut and overlap one another to form a continuously cratered surface. Approximately 25 per cent of the samples collected from the highlands have textures resembling normal basaltic lavas. Detailed examination is however likely to reveal that these rocks were generated by impact melting. As was noted in Section 2.12 three components are normally found in the surface rocks of the highland areas (Heiken *et al.*, 1991). They are ANT, KREEP and a meteoritic component. The anorthosites of the ANT, or anorthosite-norite-troctolite suite, are responsible for the light colour of the lunar highlands as viewed from Earth (see Section 6.13).

Soon after fragments of anorthosite had been recovered by the Apollo 11 Mission the concept of a magma-lava ocean was introduced to account for their origin. It was postulated that in the pre-Nectarian Period the outer layer of the Moon was partly or completely molten and an anorthositic crust formed as a flotation cumulate in a high-temperature silicate ocean. The complementary ferromagnesian phases sank into the magma-lava ocean where they later became source materials for the younger mare basalts. Perhaps the most compelling evidence in support of a magma-lava ocean is the global distribution of anorthosite on the lunar surface. The great depth of this ocean can be gauged from the observation that the complementary ferromagnesian cumulates are not revealed even at the depths excavated by the large impacts. The pre-Nectarian age of the anorthosites, and the geochemical uniformity of the KREEPy materials also support the concept of a magma-lava ocean. Major uncertainties remain as to the depth and initial composition of the hypothetical magma-lava ocean.

Longhi and Ashwal (1985) have proposed a two-stage petrogenetic model for the origin of the lunar anorthosites. In stage one a lunar crust forms that is composed of a stack of overlapping layered intrusions. During the second stage these rocks are reheated and the less dense anorthositic layers are mobilised and move upwards to form the anorthositic layer.

Most petrologists were surprised by the discovery of KREEP because it contains high concentrations of elements characteristic of highly differentiated, residual magmas. KREEP is now usually considered to represent the final 2 per cent residual liquid that evolved during the initial cooling of the Moon's crust. The highland crust is probably about 60 km thick. Its rocks contain some 25 per cent alumina, whereas the Moon as a whole probably contains less than 6 per cent of this oxide. Potassium, uranium and thorium are all concentrated near the lunar surface by two orders of magnitude in excess of their total lunar abundances (Taylor, 1982, p. 243). Europium is concentrated in the highland feldspathic crust to about the same degree as aluminium.

The following hypotheses have been proposed for the origin of the Moon: (1) it formed independently of the Earth but was later captured by that planet; (2) it formed simultaneously with the Earth; (3) it formed from the mantle of a small planet that impacted into the Earth; and (4) it formed mainly from materials removed from the Earth's mantle by a series of small impact events. At present, most planetologists favour the large impact theory. This is because it explains the rotational speed of the Earth and the Moon and the orientation of the Moon's orbit around the Earth (Taylor, 1982). Trace element abundance data, particularly for the elements vanadium, chromium and manganese, support the idea that a significant proportion of the Moon was derived from the Earth's mantle. It is likely that the Moon has had a complex evolutionary history. Part of it being derived from the Earth's early lithosphere, and the rest from fragments of the mantles of asteroids that collided with the Earth and were then blasted into orbit around the Earth. According to some planetologists the metallic cores of these asteroids that once impacted with the Earth are now part of the Earth's core.

Mars

Interest in Mars is currently building up as the date of the landing of the Mars Pathfinder and the Sojourner micro-rover approaches. There is also renewed interest in the possibility that meteorites from Mars contain traces of primordial life. In 1971 when the Mariner Nine spacecraft orbited Mars for the first time its instruments recorded vast clouds of rust-coloured dust whipped up into its rarefied atmosphere by strong winds. The only permanent features that were clearly visible from the orbiting spacecraft were four dark spots that rose above the cloaking dust-cloud. When the dust settled the spots emerged as four huge volcanoes. Many of the volcanic landforms on Mars are between ten and a hundred times larger than their equivalents on Earth. It is likely that the huge younger volcanoes on Mars owe their great size to: (1) high eruption rates; (2) a lack of horizontal lithospheric plate movement; and (3) long eruptive histories (Carr, 1987, p. 131).

Another spectacular feature imaged by the Mariner Nine spacecraft was a vast rift valley that was over 4000 km long. This vast gash in the planet's surface has since been named after the spacecraft, the

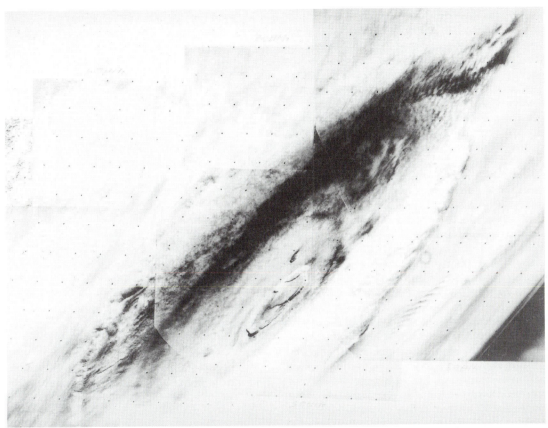

Fig. 4.16 Image of the summit of enormous Olympus Mons shield volcano in the Tharsis Province of Mars. Viking image 17 444.

Valles Marineris. In its central section it is over 7 km deep and in the area where three parallel canyons merge the chasm is over 600 km wide (Mutch *et al.*, 1976). The Martian landscape also contains thousands of smaller valleys that appear to have been carved by water.

The volcanic landforms on Mars range in size from several hundred kilometres across down to the resolution limit of available images (Basaltic Volcanism Study Project, 1981, p. 772). In a few provinces spectacular volcanoes rise above the volcanic flows and wind-blown deposits of the northern plains. As recorded in Section 1.5 the younger volcanism is concentrated in a few areas, such as the Tharsis and the Elysium volcanic provinces. Both of these large provinces have developed on huge elongate topographic bulges or domes. The Tharsis Dome is cut by an extensive system of fractures that are connected to the Valles Marineris. Three huge shield volcanoes called Ascraeus Mons, Pavonis Mons and Arsia

Mons (see Fig. 4.16) are supported by the dome. Enclosed within the summit caldera of Arsia Mons is a fissure vent surrounded by spatter ramparts.

The 24 km high shield volcano Olympus Mons is situated on the northwestern edge of the dome. Olympus Mons is surrounded by a roughly circular outward facing cliff 550 km in diameter and in places over 6 km high. Lava flows can be traced down the flanks of the volcano, over the cliff and across the plains beyond. Another huge volcano, called Alba Patera, lies to the north of the Tharsis Dome. Paterae are found on both Mars and Io. The term is applied to large, low relief volcanic landforms. Alba Patera is a well preserved volcanic edifice that rises about 7 km above the surrounding plains. Its most remarkable features are that it is over 1500 km in diameter, and its central caldera complex is about 150 km in diameter. Cattermole (1989, p. 107) has described it as a very low shield volcano whose profile is largely the result of the unimpeded effusion of large volumes of

Fig. 4.17 Image of Ceranius Tholus in the northern part of the Tharsis Province of Mars. It is about 100 km in diameter and lava flows, lava channels and collapsed lava tubes radiate out from its central caldera. Viking image 516 A 24.

very fluid lavas. Some paterae, such as Tyrrhenna Patera, appear to be largely composed of ash and are surrounded by poorly preserved, deeply-eroded, bedded deposits.

Another Martian landform that appears to have been produced by a central eruptive mechanism is the tholus. This landform is smaller and contains steeper slopes than the shield volcanoes. For example, Ceranius Tholus in the northern part of the Tharsis Province is about 100 km in diameter. Lava flows, lava channels and collapsed lava tubes radiate out from its central caldera (see Fig. 4.17).

A significant characteristic of the terrestrial planets is the asymmetry of their major surface features. On Earth the Pacific occupies a complete hemisphere, whereas on the Moon the maria are mainly concentrated on the nearside, with the farside standing about 5 km higher than the nearside. The heavily cratered and older southern hemisphere of Mars is about 3.5 km higher than the younger somewhat smooth volcanic plains of the northern hemisphere. Sleep (1994) has proposed that a type of plate tectonics was active on Mars for some two hundred million years during its early development. Spreading took place along a single 8000 km long equatorial belt. This activity led to the development of the younger thinner crust of the northern lowlands. This tecto-magmatic event was also probably an important factor in the cooling of the interior of this small planet. The huge volcanoes of the Tharsis Volcanic Province now lie astride the equator and directly over the postulated subduction zone. Possibly the Tharsis Dome and the volcanic and magmatic activity associated with it developed in response to continued magma generation at the principal site of subduction after crustal spreading had ceased.

Io

Io is the innermost of the four large, or Galilean, satellites of Jupiter. Ganymede and Callisto are larger than Mercury; whereas Io and Europa are much the same size as the Moon. During the Voyager One encounter with Jupiter in March 1979 Io was studied for six and a half days and nine separate eruption columns were observed. Four months later, during the Voyager Two encounter, eight of these volcanoes were still erupting. In the Voyager One images of Io the eruptive centres and lava lakes appear dark, whereas the sulphur-rich flows appear in various hues of red, orange and white (see Plate 4.1). Most of the volcanoes display little topographic relief. This is probably because the lavas are highly fluid, and Io's crust is unable to support the weight of lofty volcanic landforms.

Many of the vents are surrounded by halo-like features that are likely to be deposits of volcanic ejecta. The heights of the eruption columns, or plumes, vary between 60 and 300 km. Most of the characteristics of these features can be matched by ballistic ejections of materials at vent velocities of between 0.5–1 km/s. On Earth vent velocities rarely exceed 0.1 km/s. The shapes of Io's eruption plumes have been likened to the plumes released by geysers. It is probable that if a large geyser similar in size to Old Faithful in Yellowstone Park, Wyoming, was to erupt on Io, with its lower surface gravity and lack of atmosphere, the geyser would eject a plume to an altitude of more than 35 km.

Io contains many large paterae. The largest is Pele and it is about 200 km in diameter. It is rimless, with steep inner walls, and a flat floor. Flows have been traced down the flanks of about half of the calderas on Io. Some of these flows, as for example those

issuing from Ra Patera, are several hundred kilometres long and tens of kilometres wide (see Plate 4.1).

Sulphur covers large areas of the surface of Io and many of the young, red, orange and white lava flows appear to be essentially composed of this element. Both sulphur and sulphur dioxide seem to play important roles in the present explosive volcanism on the satellite. The normal daytime surface temperature on Io is $-148°C$, but it has numerous warm spots where the mean temperature is about 27°C. In areas of active volcanism the temperature rises to between 325° and 430°C. At these temperatures sulphur is molten, but silicate lavas remain solid. The mean global heat-flow on Io is about 2000 milliwatts/m^2. This is over 30 times greater than the global heat-flow value for the Earth. Most of the heat released on Io's surface is derived from tidal stresses that are constantly building up and being released. The make-up of these tidal forces is constantly changing in response to changes in the relative positions of Jupiter's other satellites. Europa the satellite second closest to Jupiter also shows signs of volcanism, but on this ice-covered planet the process is cryovolcanism.

Geyser-like plumes have also been described from Triton, Neptune's largest satellite. The surface of this satellite has been described as the coldest known place in the Solar System, yet it displays a great variety of landforms that have been interpreted as indicating a complex history of cryovolcanic and possibly normal volcanic activity (Smith, 1990, pp. 127–129). Like Io, much of Triton's heat is derived from internal tidal stresses. Such stress and the heat it generates was probably enormous in the past before Triton developed its present circular orbit.

References

Ashwal, L.D., 1993. *Anorthosites*. Springer-Verlag, Berlin, 422 pp.

Basaltic Volcanism Study Project (BVSP), 1981. *Basaltic Volcanism on the Terrestrial Planets*. Pergamon Press Inc., New York, 1286 pp.

Beatty, J.K. and Chaikin, A. (eds) 1990. *The New Solar System*. Third Edition, Cambridge University Press, Cambridge, 326 pp.

Blong, R.J. 1984. *Volcanic Hazards*. Academic Press, Orlando, Florida, 424 pp.

Bruning, D., 1994. Clementine maps the Moon. *Astronomy*, 22 (7), pp. 36–39.

Bullard, F.M., 1977. *Volcanoes of the Earth*. Revised Edition, University of Queensland Press, Saint Lucia, Australia, 579 pp.

Carr, M.H., 1987. Volcanic processes in the Solar System. *Earthquakes and Volcanoes*, 19 (4), pp. 128–137.

Cashman, K.V. and Fiske, R.S., 1991. Fallout of pyroclastic debris from submarine volcanic eruptions. *Science*, 253, pp. 275–281.

Cattermole, P., 1989. *Planetary Volcanism: A Study of Volcanic Activity in the Solar System*. Ellis Horwood Limited, Chichester, UK, 443 pp.

Cattermole, P., 1994. *Venus: The Geological Story*. UCL Press, London, 250 pp.

Chesner, C.A. and Rose, W.I., 1991. Stratigraphy of the Toba Tuffs and the evolution of Toba Caldera Complex, Sumatra, Indonesia. *Bulletin of Volcanology*, 53, pp. 343–356.

Chester, D.K., Duncan, A.M., Guest, J.E. and Kilburn, C.R.J., 1985. *Mount Etna: The Anatomy of a Volcano*. Chapman and Hall, London, 404 pp.

Fisher, R.V. and Schmincke, H.-U., 1984. *Pyroclastic Rocks*. Springer-Verlag, Berlin, 472 pp.

Foxworthy, B.L. and Hill, M., 1982. *Volcanic Eruptions of 1980 at Mount St. Helens: The First 100 days*. Geological Survey Professional Paper 1249, US Printing Office, Washington, D.C., 125 pp.

Francis, P., 1993. *Volcanoes: A Planetary Perspective*. Clarendon Press, Oxford University Press, Oxford, 443 pp.

Franklin, B., 1784. Meteorological Imaginations and Conjectures. *Manchester Literary and Philosophical Society Memoirs and Proceedings*, 2, p. 122.

Gilberti, G., Jaupart, C. and Sartoris, G., 1992. Steady-state operation of Stromboli Volcano, Italy: constraints on the feeding system. *Bulletin of Volcanology*, 54, pp. 535–541.

Gillis, J., 1994. Inside the Batcave: The Clementine 1 Mission. *Lunar and Planetary Bulletin*, 71, pp. 2–5.

Harington, C.R. (ed.), 1992. The year without summer? World Climate in 1816. *Canadian Museum of Nature*, Ottawa, 576 pp.

Hartmann, W.K., Phillips, R.J. and Taylor, G.J. (editors) 1986. *Origin of the Moon*. Lunar and Planetary Institute, Houston, 781 pp.

Head, J.W., Crumpler, L.S., Aubele, J.C., Guest, J.E. and Saunders, R.S., 1992. Venus volcanism: classification of volcanic features and structures, associations, and global distribution from Magellan data. *Journal of Geophysical Research*, 97 (E8), pp. 13 153–13 197.

Heiken, G.H., Vaniman, D.T. and French, B.M., 1991. *Lunar Sourcebook: A User's Guide to the Moon*. Cambridge University Press, New York, 736 pp.

Hoblitt, R.P. and Waitt, R.B., 1989. *General information: Recent volcaniclastic deposits and processes at Mount St. Helens Volcano, Washington*. New Mexico Bureau of Mines and Mineral Resources Memoir 47, pp. 51–54.

Hochstein, M.P., 1995. Crustal heat transfer in the Taupo Volcanic Zone (New Zealand): comparison with volcanic arcs and explanatory heat source models. *Journal of Volcanology and Geothermal Research*, 68 (1–3), pp. 117–151.

Keller, J., 1980. The island of Vulcano. *Rendiconti Societa Italiana di mineralogia e petrologia*, 36 (1), pp. 369–414.

Kermode, L.O., 1991. Whangaparaoa-Auckland, Infomap 290, Sheet R 10/11, 1:100 000, Department of Survey and Land Information, Wellington, New Zealand.

Lamb, H.H., 1992. *Climatic History and the Modern World*, Methuen, London, 387 pp.

Legrand, M. and Delmas, R.J., 1987. A 220 year continuous record of volcanic H_2SO_4 in the Antarctic ice sheet. *Nature*, 327, pp. 671–676.

Lirer, L., Munno, R., Petrosino, P. and Vinci, A., 1993. Tephrostratigraphy of the AD 79 pyroclastic deposits in perivolcanic areas of Mt. Vesuvio (Italy). *Journal of Volcanology and Geothermal Research*, 58, pp. 133–149.

Longhi, J. and Ashwal, L.D., 1985. Two-stage models for lunar and terrestrial anorthosites: Petrogenesis without magma ocean. Proceedings of the fifteenth Lunar and Planetary Science Conference, Part 2. *Journal of Geophysical Research*, 90 (Supplement), pp. 571–584.

McClelland, L., Simkin, T., Summers, M., Nielsen, E. and Stein, T.C., 1989. *Global Volcanism 1975–1985*, Prentice Hall–Englewood Cliffs, New Jersey, 655 pp.

Moore, J.G., 1975. Mechanism of formation of pillow lava. *American Scientist*, 63, pp. 269–277.

Mues-Schumacher, U., 1994. Chemical variations of the AD 79 pumice deposits of Vesuvius. *European Journal of Mineralogy*, 6, pp. 387–395.

Mutch, T.A., Arvidson, R.E., Head, J.W., Jones, K.L. and Saunders, R.S., 1976. *The Geology of Mars*, Princeton University Press, Princeton, N.J., 400 pp.

Newcott, W., 1993. Venus Revealed. *National Geographic*, 183 (2), pp. 36–59.

Newhall, C.G. and Self, S., 1982. The Volcanic Explosivity Index (VEI): An estimate of explosive magnitude for historical volcanism. *Journal of Geophysical Research*, 87, pp. 1231–1238.

Pyle, D.M., 1989. The thickness, volume and grain size of tephra fall deposits. *Bulletin of Volcanology*, 51, pp. 1–15.

Rosi, M., 1980. The Island of Stromboli, *Rendiconti Societa Italiana di mineralogia e petrologia*, 36 (1), pp. 345–368.

Saunders, R.S. and the Magellan Mission Team, 1992. Special Magellan Issue. *Journal of Geophysical Research*, 97 (E8), pp. 13 063–13 689.

Scandone, R., Arganese, G. and Galdi, F., 1993. The evaluation of volcanic risks in the Vesuvian area. *Journal of Volcanology and Geothermal Research*, 58, 263–271.

Searle, E.J., 1981. *City of Volcanoes: A Geology of Auckland*. Second edition, Longman Paul Limited, Auckland, 195 pp.

Self, S. and Rampino, M.R., 1981. The 1883 eruption of Krakatau. *Nature*, 292, pp. 699–704.

Self, S., Rampino, M.R., Newton, M.S. and Wolff, J.A., 1984. Volcanological study of the great Tambora eruption of 1815. *Geology*, 12, pp. 659–663.

Self, S. and Sparks, R.S.J., 1978. Characteristics of pyroclastic deposits formed by the interaction of silicic magma and water. *Bulletin Volcanologique*, 41, pp. 196–212.

Sigurdsson, H. and Carey, S., 1989. Plinian and co-ignimbrite tephra fall from the 1815 eruption of Tambora volcano. *Bulletin of Volcanology*, 51, pp. 243–270.

Sigurdsson, H. and Carey, S., 1992. The eruption of Tambora in 1815: environmental effects and eruption dynamics. In Harington, C.R. (editor), *The Year Without Summer? World Climate in 1816*, pp. 16–45. Canadian Museum of Nature, Ottawa, Canada, 576 pp.

Sigurdsson, H., Carey, S., Cornell, W. and Pescatore, T., 1985. The eruption of Vesuvius in AD 79. *National Geographic Research*, 1, pp. 332–387.

Simkin, T., 1993. Terrestrial volcanism in space and time. *Annual Review of Earth and Planetary Sciences*, 21, pp. 427–452.

Simkin, T., Siebert, L., McClelland, L., Bridge, D., Newhall, C. and Latter, J.H., 1981. *Volcanoes of the World*. Smithsonian Institution, Hutchinson Ross, Stroudsburg, Pennsylvania, 232 pp.

Sleep, N.H., 1994. Martian Plate Tectonics. *Journal of Geophysical Research*, 99 (E3), pp. 5639–5655.

Smith, B.A., 1990. The Voyager Encounters. In Beatty, J.K. and Chaikin, A. (editors), *The New Solar System*, pp. 107–130, Third Edition, Cambridge University Press, Cambridge, 326 pp.

Stofan, E.R. and Saunders, R.S., 1990. Geologic evidence of hotspot activity on Venus: Predictions for Magellan. *Geophysical Research Letters*, 17, pp. 1377–1380.

Stothers, R.B., 1984. The great Tambora eruption in 1815 and its aftermath. *Science*, 224, pp. 1191–1198.

Strom, R.G., 1987. *Mercury: The Elusive Planet*, Cambridge University Press, Cambridge, 197 pp.

Taylor, S.R., 1982. *Planetary Science: A Lunar Perspective*, Lunar and Planetary Institute, Houston, 481 pp.

Taylor, S.R., 1992. *Solar System Evolution: A New Perspective*, Cambridge University Press, 307 pp.

Tazieff, H. and Sabroux, J.-C. (editors) 1983. *Forecasting Volcanic Hazards*, Elsevier, Amsterdam, 635 pp.

Thorarinsson, S., 1970. *Hekla: A Notorious Volcano*. Almenna Bokafelagid, Reykjavik, Iceland, 55 pp.

Tilley C.E., Yoder, H.S. and Schairer, J.F., 1967. Melting Relations of Volcanic Rock Series. *Carnegie Institution of Washington Yearbook*, 65, pp. 260–269.

Verbeek, R.D.M., 1884. The Krakatoa eruption. *Nature*, 30, pp. 10–15.

Walker, G.P.L., 1980. The Taupo pumice: product of the most powerful known (ultraplinian) eruption? *Journal of Volcanology and Geothermal Research*, 8, pp. 69–94.

White, R.S. and Humphreys, C.J., 1994. Famines and cataclysmic volcanism. *Geology Today*, September–October 1994, pp. 181–185.

Wilson, C.J.N., Houghton, B.F., McWilliams, M.O., Lanphere, M.A., Weaver, S.D. and Briggs, R.M., 1995. Volcanic and structural evolution of Taupo Volcanic Zone, New Zealand: a review. *Journal of Volcanology and Geothermal Research*, 68 (1–3), pp. 1–28.

Wood, C.A., 1992. Climatic effects of the 1783 Laki eruption. In Harington, C.R. (editor), *The Year Without Summer? World Climate in 1816*, pp. 58–77. Canadian Museum of Nature, Ottawa, Canada, 576 pp.

Wright, T.L. and Pierson, T.C., 1992. Living with Volcanoes. *U.S. Geological Survey Circular*, 1073, 57 pp.

CHAPTER 5

Partial melting and movement of magma

Magma reservoirs may be viewed as great mechanical capacitors that, during magma influxes, slowly accumulate potential energy.

Ryan, 1994a, p. xx.

5.1 Magma generation

Magma is the most important concept in igneous petrology. The origin, evolution and movement of magma has a crucial influence on planetary development. Magmas are generated by diverse processes and at different times during the evolution of a planet one of these processes is likely to become dominant. Immediately after accretion, the surface heat-flow on the terrestrial planets was higher than at present. This produced widespread melting and it may have produced lava-magma oceans. During the past 4.5 giga-years heat production on these planets has been in steady decline. Takahashi (1986, p. 9379) has suggested that the reason why olivine tholeiitic lavas are currently dominant on Earth is because the maximum temperature now available in the source region is between 1300–1400°C. In the Archean when komatiitic lavas were more common the maximum temperature in the source region was probably 200–400°C higher. His experiments on the high-pressure melting of peridotite also show that komatiitic magmas are likely to be generated by partial melting of mantle peridotite in the pressure range five to eight gigapascals.

On Earth most melting currently takes place in the outer core and contiguous lowermost mantle and in various zones within the upper mantle and crust. In the upper mantle partial melting is usually the result of: (1) upwelling of the asthenosphere beneath oceanic spreading centres; (2) the release of volatiles into upper mantle materials from a subducted slab; or (3) the upward movement of mantle plumes and their impingement on the lithosphere. Most partial melting in the crust results from the intrusion, or underplating, of picrobasaltic or basaltic magma.

Such magma provides heat, and it is also likely to interject some chemical components into the new anatectic magma. Another important factor in generating anatectic magma is the presence in the crustal rocks of hydrous or other readily fusible phases.

There are three principal ways of generating magma: (1) adding heat at constant pressure; (2) lowering the pressure at constant temperature; or (3) lowering the initial melting temperature of the source rock by adding volatile components. At present most magmatic activity is confined to a few relatively narrow belts that girdle the Earth. The bulk of this activity takes place on, or beneath, the mid-oceanic ridges where about 4 km^3 of volcanic materials and 9.5 km^3 of intrusive materials are inserted each year (see Section 7.2). Subduction systems provide the second most important site for magmatic activity. About 0.9 km^3 of volcanic rocks and 1.2 km^3 of intrusive materials are erupted or emplaced at convergent plate boundaries each year. Most of this magma is generated in the mantle-wedge where upwelling mantle materials react with volatiles released from the descending plate of oceanic lithosphere. The volatiles reduce the melting temperature of the upwelling mantle materials and as this magma moves upwards its volume is augmented by decompression melting (see Section 8.3).

Decompression melting typically occurs during the ascent of diapirs and mantle plumes, the up-arching of the lithosphere, or when faulting or fracturing causes a decrease in pressure. Olson (1994) has investigated the dynamics of magma generation in mantle plumes and in particular the large plumes that generate flood basalts. According to him decompression melting is likely to occur in a lens-shaped zone at a depth of about 100 km, and the initial melting pulse from a single large plume can produce $1\,000\,000 \text{ km}^3$

of magma in a few million years beneath normal continental lithosphere, and up to ten times more magma beneath thin oceanic lithosphere. After the initial melting event the plume constricts to a much narrower body and the rate of magma production decreases by about two orders of magnitude.

5.2   Melting processes

In order to explain why large batches of relatively homogeneous magma are constantly being produced in the upper mantle, we will survey the simplified mantle-rock system, forsterite (Fo)–diopside (Di)–pyrope garnet (Py) at four gigapascals (see Fig. 5.1). This ternary system is assumed to be an ideal system in which M represents the mean composition of the upper mantle. If the temperature of an assemblage of these minerals of composition M is raised to 1670°C melting will begin. The lowest melting point of any ratio of components in the system is at E which is a eutectic, or pseudo-eutectic. All the early melt is of E composition and as melting proceeds the bulk composition of the solid phases moves towards R. In doing this their bulk composition passes through successive compositions on the line MR. When about 30 per cent of the system is liquid, the bulk composition of the solid phases reaches R. Now only two solid phases, forsterite and pyrope, remain.

The melt produced isothermally at 1670°C may be removed in batches or continuously. At one extreme, called fractional melting, each minuscule amount of liquid is removed from the source minerals as soon as it forms. Once removed it is unable to react further with the crystalline residue. When the composition of the residue reaches R melting stops, only to resume when the temperature is raised to 1770°C. Partial melting now produces a liquid of composition D. No liquid of intermediate composition is produced. This means that fractional melting is able to produce two or more magmas of conspicuously different composition from the same source region. If all the initial liquid of E composition remains in contact with the residual solid phases, forsterite and pyrope, during the partial melting process, as the temperature is raised the composition of the liquid gradually changes from that of E and goes towards D. At C all the pyrope is melted. If the temperature is now increased the composition of the liquid moves directly towards forsterite as it is the only remaining

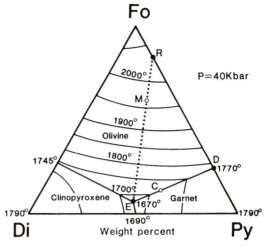

Fig. 5.1 The simplified mantle-rock system, forsterite (Fo)–diopside (Di)–pyrope garnet (Py) at four GPa. E is the piercing point composition. M is the postulated chemical composition of the upper mantle. Adapted from Davis and Schairer, 1965, p. 124 and Yoder 1976, p. 108.

solid phase. At M the whole system is liquid. A succession of batches of liquids with compositions that vary from E via C to M may thus be produced if the residual solid phases remain in equilibrium with the evolving liquid. Our brief review of this simplified mantle-rock system clearly shows how large volumes of magma of constant major element composition can be produced by the partial melting of the major phases found in the upper mantle. For example, the partial melting of about 30 per cent of the mass of this system is likely to yield a melt with a composition equivalent to E.

Takahashi (1986) determined the melting phase relations of a dry fertile lherzolite in the pressure range 0.001 to 14 GPa (see Table 5.1). His starting material contained about 58 per cent olivine, 25 per cent orthopyroxene, 15 per cent clinopyroxene and 2 per cent spinel (by weight). Olivine is the liquidus phase across the whole range of pressures used in these experiments. The mineral that crystallises second changes with increasing pressure, with chromian spinel crystallising at atmospheric pressure, calcium-poor orthopyroxene up to 3 GPa, pigeonitic pyroxene up to 7 GPa, and a pyrope garnet crystallises above 7 GPa. As can be seen in Table 5.1 the compositions of the partial melts become progressively more magnesium-rich at elevated pressures. At 3 GPa they are picrites but at higher pressure they become komatiitic in composition.

Table 5.1: Chemical composition of partial melts of peridotite that formed near the dry solidus at various pressure from 0.001 to 14 GPa. After Takahashi, 1986, p. 9377.

Pressure, GPa	0.001	1	3	5	8	10.5	14
Temperature, °C	1200	1325	1550	1700	1800	1950	1900
SiO_2	54.2	49.2	46.9	47.2	46.6	45.8	45.2
TiO_2	0.7	0.6	0.9	0.2	0.2	0.2	0.2
Al_2O_3	15.1	17.7	11	5.1	4.6	4.2	4.2
FeO	5.6	6.7	7.8	9.3	8.8	9.5	7.9
MnO	0.1	0.1	0.2	0.1	0.2	0.2	0.2
MgO	8.4	9.5	19.2	31.3	34.9	35.9	37.8
CaO	12.3	11.4	12.2	4.8	3.9	3.4	3.7
Na_2O	2.1	2.9	1.2	0.5	0.3	0.3	0.3
Cr_2O_3	0.1	0.1	0.4	0.5	0.4	0.5	0.5
Total	98.6	98.2	99.8	99	99.9	100	100

5.3   Zone melting

When considering the melting process it is instructive to examine the extreme process of zone melting. It illustrates how major and minor element abundances can become decoupled during melting. The term zone melting was originally used to describe a metal purification process, where a zone of melting was passed along a metal bar. Impurities, or incompatible elements, were preferentially partitioned into the melt and swept along in the zone of melting as it moves from one end of the bar to the other. Within those zones of the mantle where incipient melting is taking place a body of magma may rise by a process called solution stoping. In the roof-rocks above the magma the solid phases go into solution because they are at a lower pressure than the main body of magma. As the roof-rocks partially melt and become pliable other phases crystallise at the base of the rising magma body. This results in all those elements, or their ions, (1) that are too large or too small, or differ too much in valency and bond type to replace the ions in the minerals being deposited, and (2) are not abundant enough to form minerals of their own, being partitioned into the magma (Harris, 1957, p. 199). These incompatible elements include the elements phosphorus, potassium, rubidium, zirconium, niobium, caesium, barium, tantalum, lead, thorium and uranium (see Section 2.14). A unique feature of zone melting is that it concentrates the incompatible elements from a large volume of rock into a relatively small volume of magma. Wall-rock reaction is a similar process. It takes place when a magma in transit reacts with, or partially melts, the phases it encounters in the wall-rocks. Disequilibrium melting is a rare process. It may happen when magma is rapidly extracted from its source rocks. Partial melting is usually complex with the degree of melting varying from a maximum at the centre of the zone of melting to zero at its margins. Many magmas that move into the lithosphere are essentially a mixture of a series of magmas generated in slightly different physical environments. In Section 3.5 on the characteristics of the Icelandic plume it was recorded that mantle plumes are likely to entrain materials from their surroundings as they advance towards the surface.

5.4   Buoyancy and the movement and storage of magma

Magmas of diverse composition move at different velocities. Rhyolitic magmas are normally viscous and move relatively slowly. Some volatile-charged magmas, such as the lamproites and kimberlites, travel at up to 15 m/s. The main forces that act within a planet and cause magma to rise are buoyancy and magmatic hydrostatic pressure (Yoder, 1976, pp. 186–9).

Buoyancy is the upward thrust exerted on a body of any shape or density when it is immersed in a fluid. The buoyant force is equal to the sum of all the forces that the fluid exerts on the body. It is always upwards because the pressure underneath the body is greater than the pressure above it, whereas the pressures on the sides essentially cancel each other out. If the buoyant force is greater than the weight of the immersed body, the body will tend to move upwards. Informally such bodies are sometimes said to have positive buoyancy. When the buoyant force is approximately equal to the weight of the immersed body, the body tends to remain stationary. This is informally called neutral buoyancy. If the buoyant force is less than the weight of the body, the body tends to sink. This is sometimes called negative buoyancy (Ryan, 1987, 1994b; Walker, 1989).

Weight is the gravitational pull of a planet on a body. It differs with position on, or within, a planet. The surface gravity on the Moon is about one-sixth that on the surface of the Earth. In the Earth at the present time the distribution of mass is such that the gravitational force is practically constant within the crust and mantle, but decreases steadily at deeper levels (Anderson and Heart, 1976).

On Earth basalt appears to erupt in all geological settings. This means that magma of this composition is able to rise through a range of different types of continental and oceanic lithosphere. The average density of the crust is about $2750 \, kg/m^3$, whereas the densities of most basaltic magmas are lower than this. This helps to explain the high abundance and wide distribution of basaltic volcanoes. Mafic intrusions have a similar high abundance and wide distribution. Some of these intrusions depict the pathways followed by basaltic magmas on their way to the surface. Others appear to record intrusive events unrelated to volcanism, or the evolution of subvolcanic magma chambers. Basaltic magmas have considerable freedom to move through the lithosphere because of their low viscosity and relatively low density. As one would expect they normally come to rest in a zone where their density and that of the surrounding rocks is equal. This is the level of equivalent density or, as it is sometimes called, the level of neutral buoyancy. Below this level the rocks are denser than the magma and above it they are less dense. The level of equivalent density for a basaltic magma may occur at the surface, at shallow depths below a variety of rocks, at the base of a silicic body of magma, or at deeper levels within the continental crust. Picrobasaltic magmas are of greater density. They are likely to reach their level of equivalent density deep within the continental crust or even at its base. This often results in the underplating of the lower crust. The ponding of magmas at their level of equivalent density provides opportunities for high-pressure fractional crystallisation, assimilation of crustal rocks and the anatectic generation of silicic magmas.

One might ask how magmas are able to break out of magma chambers established at their level of equivalent density. If such a magma is located deep within the crust it is likely to cool, differentiate and produce less dense evolved magmas. These magmas will probably rise whereas the dense cumulates that remain may sink. In seemingly gravitationally stable magma chambers beneath volcanoes magma is often forced to move laterally when new batches of magma arrive. The magma chamber expands; if this results in a release of pressure some of the magma may vesiculate, become less dense, and erupt. Vesiculation is triggered by various processes that include the rupture of a magma chamber, the local concentration of gases towards the top of a body of magma, or the influx of a hot or volatile-rich batch of new magma. Once a vesicular magma begins to rise a runaway situation often develops because the magma is continually moving into a lower pressure environment where decompression encourages gas to exsolve and bubbles to expand and decrease the bulk density of the magma.

5.5 Diapirs

Magmas commonly move through the overlying rocks as diapirs (Greek: *dia* = through; *peiro* = piercing). It may be difficult to visualise how a diapir that is 10 km in diameter can rise and shoulder aside solid-looking country rocks. In south-east Papua New Guinea there are several surface-breaching, granitic gneiss domes, such as the Goodenough Dome, that appear to be doing just this. In sedimentary sequences low-density salt domes are often observed to have thrust their way through beds of more dense rocks. In order to understand how diapirs are emplaced one has to appreciate that the over-all strength of a body decreases with increasing size. Galileo Galilei, in his famous dialogues concerning two new sciences, illustrated this idea very well when he stated that 'if one wished to maintain in a great giant the same proportion of limbs as found in ordinary man one must either find a harder and stronger material for making the bones, or one must admit a diminution of strength in comparison with men of medium stature; for if his height be increased inordinately he will fall and be crushed by his own weight'.

At the level of the individual phases in the source rocks melting begins along grain boundaries. In the upper mantle one would expect it to occur at three- or four-phase junctions. With further partial melting the source rock develops an interconnected network with some large local pockets of melt (Kohlstedt and

Chopra, 1994). This means that the initial partial melting of typical upper mantle rock is likely to produce a material with interconnected pores that can facilitate the movement of small volumes of magma. Once partial melting has produced a picrobasaltic magma in the upper mantle both the magma and the residual phases are likely to be less dense than the original source rock. This happens because: (1) the magma occupies a larger volume than the solid phases it replaces; (2) the residual mineral assemblage is depleted in dense minerals, such as garnet; and (3) the partial melting process lowers the Fe/Mg ratios of the residual olivines and pyroxenes and this lowers their densities. The density of forsterite is $3200 \, kg/m^3$, but fayalite is $4390 \, kg/m^3$. As a picrobasaltic magma and the residual solid phases are all less dense than the original undepleted mantle they tend to rise together in a buoyant diapir. If the diapir retains most of its heat as it rises the decreasing ambient pressure will encourage further partial melting.

5.6   Magma chambers

Geochemical, geophysical and petrologic models all require some volcanoes to have high-level magma chambers. They develop in various shapes and sizes and range from simple chambers to complex networks of interconnecting dykes and sills. Geophysical studies have, for example, revealed the complex shape of the magma chamber beneath Kilauea Volcano in Hawaii (see Fig. 5.2). Other well-known magma chambers lie beneath the Long Valley–Mono basin of eastern California (see Fig. 5.3), and the Yellowstone caldera of Wyoming.

Recent seismic investigations of the eastern shoulder of the East African Rift System have revealed a prominent body of low-velocity materials beneath the volcanoes of the Chyulu Hills in southern Kenya (Ritter *et al.*, 1995, p. 273). In this area the lithosphere is about 115 km thick. The anomalous materials occur in a 50 km wide zone beneath the young volcanoes in the southern part of the Chyulu Hills. In this area the seismic P-wave velocity of the crust and upper mantle is reduced by about 4 per cent relative to the surrounding region. The zone of reduced velocity is interpreted as delineating a series of small magma chambers surrounded by a more extensive thermal aureole. Magma chambers may be very large. It has been proposed that the huge

East African lithospheric dome that extends for about 1500 km from lakes Turkano and Albert in the north to Lake Malawi in the south, and even the much larger Tharsis dome on Mars (see Fig. 1.12), are surface expressions of huge magma chambers, or areas of crustal underplating.

The two most abundant types of basaltic rocks, the mid-oceanic ridge basalts and flood basalts are likely to have evolved in enormous magma chambers. The mid-oceanic ridge basalts evolve in large, elongate magma chambers that developed at shallow depths beneath the crests of mid-oceanic ridges. Continental flood basalts are likely to have evolved from primary picrobasaltic magmas that become trapped by their high density at the mantle–crust boundary. The narrowness of the zone of magma emplacement at oceanic spreading centres shows that the magma that collects beneath these areas is focused towards the spreading axis (Sparks and Parmentier, 1994). This focusing of magma results from the up-slope flow of magma along sloping decompacting boundary layers that develop at the top of the zone of partial melting and magma accumulation. Many magma chambers are thermally and compositionally zoned and once a low density layer develops at the top of a magma chamber it will tend to act as a density filter that tends to prevent denser magmas from reaching the surface. A large zoned magma chamber that has been exhumed by erosion can be inspected in the south-western part of the island of Saint Helena in the Atlantic Ocean.

5.7   Emplacement of plutonic and subvolcanic rocks

The way in which plutonic and subvolcanic rocks are emplaced into the crust depends on their viscosity and the manner in which the country rocks respond to being intruded. Field data show that the country rocks may either flow plastically, or pseudoplastically, away from a magma, or they may deform by fracturing. Brittle fracture is common in the upper crust, but it is found at much deeper levels in cratonic areas. In contrast to this in areas of high heat flow plastic deformation is likely to occur at relatively shallow depths. The granitic rocks, for example, may be emplaced by forcibly compacting and radially distending their host rocks, or by piecemeal stoping in brittle rocks.

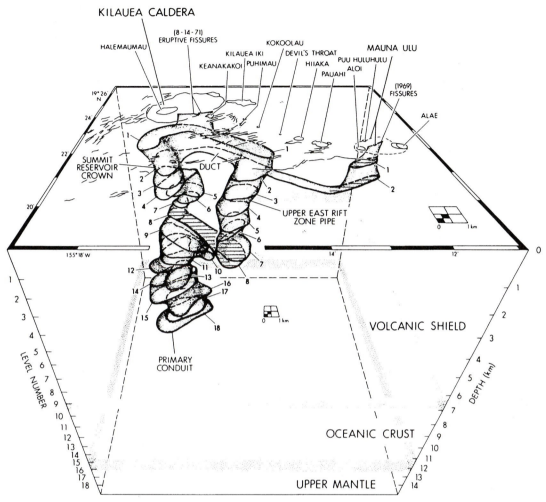

Fig. 5.2 Perspective view northward and downward into the interior of Kilauea volcano, Hawaii, showing current magma conduits. Modified from Ryan *et al.*, 1981, and Decker *et al.*, 1987, p. 1005.

When considering the emplacement of plutonic rocks it is convenient to divide the crust into three intensity zones, each being defined by a characteristic range of temperatures and pressures, and having a distinctive style of deformation (Buddington, 1959; Pitcher, 1993, pp. 154–181). The epizone is the upper intensity zone. It is characterised by brittle country rocks. The mesozone is the intermediate zone characterised by more ductile country rocks, whereas the catazone is the deepest intensity zone where the country rocks are highly ductile and usually deform plastically or pseudoplastically. Plutonic rocks emplaced into the catazone zone usually have concordant contacts.

Ductility is a measure of the amount of deformation a rock can sustain before fracturing or faulting. The outer shape of an igneous intrusion is usually governed by differences in ductility and density between the magma and the surrounding country rocks. The following are some of the intrusive mechanisms used by magmas emplaced into the epizone: (1) lateral magmatic wedging; (2) doming of the roof rocks; (3) piecemeal stoping; (4) ring-fracture stoping, or cauldron subsidence; and (5) fluidisation-assisted intrusion. Daly (1933, p. 267) used the term stoping to describe the subsidence, or ascent, of blocks of country rocks enclosed within a magma that was either higher or lower in density than the

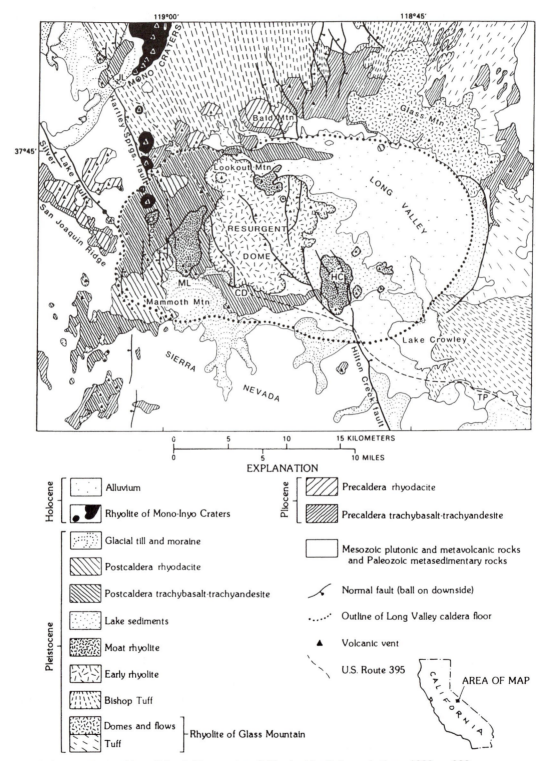

Fig. 5.3 Geological map of Long Valley Caldera, eastern California. After Bailey and others, 1989, p. 229.

engulfed blocks. He claimed that by continued fracturing of wall and roof rocks and continued immersion and subsidence of the fragments a magma chamber can be enlarged both upwards and sideways.

Plutonic rocks that were emplaced into the mesozone are typically surrounded by zones of contact metamorphism/metasomatism and local deformation. The size and nature of the contact aureole depends on the size and temperature of the intrusion and the composition, permeability and initial temperature of the country rocks. In the catazone silicic materials usually move upwards as diapirs and assume a streamlined shape, similar to that of an inverted drop of water. As such bodies rise they heat, deform, compact and displace the surrounding rocks. On entering the mesozone the head of the diapir slows down relative to the tail and the diapir flattens out and becomes mushroom-shaped (Ramberg, 1970 and 1981).

Batholiths are large, generally discordant, bodies of plutonic rocks that have an outcrop area greater than $100 \, km^2$. This term is derived from the Greek word *bathos* which means depth. Careful mapping of many large batholiths has revealed that they often consist of clusters of smaller intrusive bodies. Smaller plutonic bodies with outcrop areas of less than $100 \, km^2$ are called stocks. Lopoliths are large generally concordant bodies of plutonic rocks that have plano-convex or lenticular shapes. This term is derived from the Greek word *lopas* meaning a basin. Huge batholiths, such as the Coastal Batholith of Peru, usually consist of extended clusters of smaller intrusions that coalesced after being channelled into the same zone of crustal weakness (see Fig. 5.4). Such intrusive complexes often contain basic rocks that may have supplied some of the heat that facilitated the emplacement of the more viscous granitic rocks.

References

Anderson, D.L. and Heart, R.S., 1976. An Earth model based on free oscillations and body waves. *Journal of Geophysical Research*, 81, pp. 1461–1475.

Bailey, R.A., Miller, C.D. and Sieh, K., 1989. Long Valley caldera and Mono-Inyo Craters volcanic chain, eastern California. *New Mexico Bureau of Mines and Mineral Resources* Memoir 47, pp. 227–254.

Buddington, A.F., 1959. Granite emplacement with special reference to North America. *Bulletin of the Geological Society of America*, 70, pp. 671–747.

Daly, R.A., 1933. *Igneous Rocks and the Depths of the Earth*. Second Edition, McGraw-Hill Book Company, New York, 598 pp.

Davis, B.T.C. and Schairer, J.F., 1965. Melting relations in the join diopside–forsterite–pyrope at 40 kilobars and one atmosphere. *Carnegie Institution of Washington, Yearbook*, 64, pp. 123–126.

Decker, R.W., Wright, T.L. and Stauffer, P.H., 1987. *Volcanism in Hawaii*. United States Geological Survey, Professional Paper 1350, Washington, 1667 pp.

Harris, P.G., 1957. Zone refining and the origin of potassic basalts. *Geochimica et Cosmochimica Acta*, 12, pp. 195–208.

Kohlstedt, D.L. and Chopra, P.N., 1994. Influences of basaltic melt on the creep of polycrystalline olivine under hydrous conditions. In Ryan, M.P. (editor), *Magma Systems*. Chapter 3, pp. 37–54, Academic Press, San Diego, 398 pp.

Olson, P., 1994. Mechanics of flood basalt magmatism. In Ryan, M.P. (editor), *Magma Systems*. Chapter 1, pp. 1–18, Academic Press, San Diego, 398 pp.

Pitcher, W.S., 1978. The anatomy of a batholith. *Journal of the Geological Society*, London, 135, pp. 157–182.

Pitcher, W.S., 1993. *The Nature and Origin of Granite*. Blackie Academic and Professional, Glasgow, 321 pp.

Ramberg, H., 1970. Model studies in relation to intrusion of plutonic bodies. In Newall, G. and Rast, N. (editors), *Mechanisms of Igneous Intrusion*, pp. 261–286, Geological Journal, Special Issue Number 2, 380 pp.

Ramberg, H., 1981. *Gravity, Deformation and the Earth's Crust*. Second edition, Academic Press, London, 452 pp.

Ritter, J.R.R., Fuchs, K., Kaspar, T., Lange, F.E.I., Nyambok, I.O. and Strangl, R.L., 1995. Seismic images illustrate the deep roots of the Chyulu Hills Volcanic Area, Kenya. *EOS, Transactions, American Geophysical Union*, 76, 28, pp. 273 and 278.

Ryan, M.P., 1987. Neutral buoyancy and the mechanical evolution of magmatic systems. In Mysen, B.O. (editor), *Magmatic processes: physicochemical principles*, pp. 259–287, The Geochemical Society, Special Publication Number 1, 500 pp.

Ryan, M.P. (editor), 1994a. *Magma Systems*. Academic Press, San Diego, 398 pp.

Ryan, M.P., 1994b. Neutral-buoyancy controlled magma transport and storage in mid-ocean ridge magma reservoirs and their sheeted-dyke complex: a summary of basic relationships. In Ryan, M.P. (editor), *Magma*

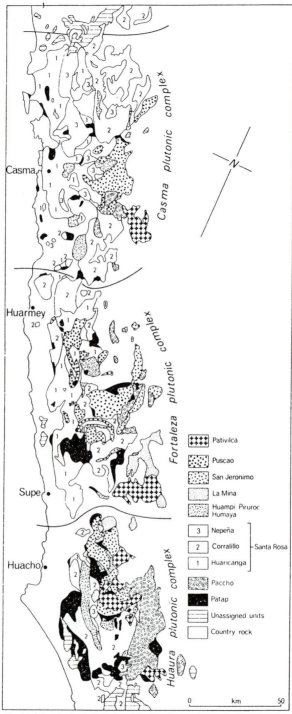

Fig. 5.4 Geological map showing the size and outline of a complex batholith. This is the Lima segment of the Coastal Batholith of Peru. After Pitcher, 1978, p. 162.

Systems. Chapter 6, pp. 97–138, Academic Press, San Diego, 398 pp.

Ryan, M.P., Koyanagi, R.Y. and Fiske, R.S., 1981. Modelling the three-dimensional structure of macroscopic magma transport systems: application to Kilauea Volcano, Hawaii. *Journal of Geophysical Research*, 86, pp. 7111–7129.

Sparks, D.W. and Parmentier, E.M., 1994. The generation and migration of partial melt beneath oceanic spreading centres. In Ryan, M.P. (editor), *Magma Systems*. Chapter 4, pp. 55–76, Academic Press, San Diego, 398 pp.

Takahashi, E., 1986. Melting of a dry peridotite KLB-1 up to 14 GPa: implications on the origin of peridotitic upper mantle. *Journal of Geophysical Research*, 91 (B9), pp. 9367–9382.

Walker, G.P.L., 1989. Gravitational (density) controls on volcanism, magma chambers and intrusions. *Australian Journal of Earth Sciences*, 36, pp. 149–165.

Yoder, H.S., 1976. *Generation of Basaltic Magma*. US National Academy of Sciences, Washington, D.C., 265 pp.

CHAPTER 6

Evolution of diverse rock associations

It is difficult to see how so simple and natural a process of solidification as fractional crystallisation can fail to be carried out in at least some rocks.

Becker, 1897, p. 260.

6.1 Diversity and magmatic differentiation

At the beginning of the twentieth century the enormous diversity of igneous rocks was recognised as an important petrological enigma (Harker, 1909). Current petrological literature contains over 1580 separate igneous rock names (Le Maitre *et al.*, 1989, p. 35). Fortunately, when the abundance and distribution of these rocks is examined, one soon discovered that less than 50 names are in common use (Chayes, 1975, p. 545). One also finds that most igneous rocks do not crop out alone. They belong to rock associations that keep recurring in space and time. It is normal for individual members of a rock association to be part of a comagmatic suite. The various processes responsible for the evolution of these suites are frequently grouped together and called the magmatic differentiation processes (Wilson, 1993, pp. 611–624).

In 1825 Scrope introduced the idea of a dominant primary magma. He suggested that various differentiation processes can produce a range of igneous rocks from a single parental magma. Later Bowen (1928, p. 3) evaluated this idea and furnished compelling corroborative experimental evidence. Layered intrusions and the comagmatic suites of rocks that issue from volcanic complexes provide field evidence that many magmatic differentiation processes operate during the cooling and crystallisation of magmas. Layered intrusive bodies include the well-exposed and studied Skaergaard Intrusion of East Greenland and the huge ($66\,000\,km^2$) Bushveld Complex of southern Africa. Intrusions of this type characteristically contain systematic sequences of magmatic rocks that usually grade from ultramafic cumulate rocks at their base, via basic and intermediate rocks, to a lesser amount of silicic rocks at the top (Wager and Brown, 1986). They display the sequence of rocks that is likely to evolve when a body of subalkalic basaltic magma is subjected to a long period of magmatic differentiation.

The formation of the core was probably the first major differentiation process that took place within the Earth. It consisted of the separation of the metal sulphide core from silicate phases. This process has been likened to the extraction of metals from rocks that are mainly composed of silicate phases. When examined in detail the evolution of the layered Earth is found to be complex because some planetesimals and planetoids that accreted to form the planet had undergone differentiation prior to merging with the Earth (see Section 13.4).

6.2 Liquid immiscibility

Liquid immiscibility is a process by which an initially homogeneous liquid, or magma, separates into two or more compositionally distinct liquid phases (see Fig. 6.1). When this happens the droplets of the minor phase or phases become suspended in the major phase. The liquids usually differ significantly in density, thus they normally separate from one another (Roedder, 1979, pp. 34–35). Anyone who has ever observed a bottle containing salad dressing will be familiar with the idea of liquid immiscibility, because the oil and vinegar in the dressing are unlikely to mix unless shaken vigorously. Many subalkaline basalts contain two co-existing glass phases. These glasses usually differ from one another in colour and composition. For example, liquid immiscibility is observed in the tholeiitic basalts from Kohala Volcano in the northern part of the big island of Hawaii. The dark

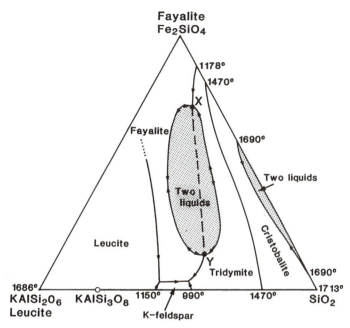

Fig. 6.1 A phase diagram of the system Fe$_2$SiO$_4$–KAlSi$_2$O$_6$–SiO$_2$ at 0.1 MPa (one atmosphere), showing two stable fields where liquid immiscibility occurs. Modified after Roedder, 1951.

glass contains 40.8 per cent SiO$_2$, 35.6 per cent FeO, 6.3 per cent TiO$_2$, 3.1 per cent Al$_2$O$_3$, 8.0 per cent CaO, 1.3 per cent Na$_2$O and 4.1 per cent P$_2$O$_5$, whereas the lighter glass contains 77.0 per cent SiO$_2$, 3.2 per cent FeO, 0.3 per cent TiO$_2$, 11.5 per cent Al$_2$O$_3$, 3.1 per cent CaO, 3.4 per cent Na$_2$O and 0.3 per cent P$_2$O$_5$.

Similar late-stage, liquid immiscibility is observed in the mesostasis of the mare basalts, and various continental flood basalts. Experimental studies by McBirney and Nakamura (1974, p. 352) have shown that the granophyres (silica-rich rocks) and iron-rich gabbros of the Skaergaard Layered Intrusion probably evolved as complementary immiscible liquids. There is also credible laboratory evidence of liquid immiscibility between alkalic silicate liquids and carbonate rich fluids (Kjarsgaard and Hamilton, 1988). Liquid immiscibility is believed to hold the key to understanding the evolution of the carbonatite–ijolite–nephelinite rock association (see Section 12.6).

6.3   Soret effect

Diffusion is expected to take place in magmas in response to compositional gradients. It also occurs in homogeneous, non-convecting magmas that are subjected to thermal gradients. This type of diffusion is called the Soret effect and the diffusing species move up or down the thermal gradient to equalise the distribution of internal energy. The characteristics of a compositional gradient depend on the magnitude of the Soret coefficient. It usually varies both in sign and amplitude from component to component and it also depends on the nature of the temperature gradient within a magma. Components with positive Soret coefficients accumulate at the colder ends of the gradient, whereas those with negative coefficients concentrate at the hotter ends. The Soret effect is usually of only local importance in the differentiation of the common silicate magmas.

Walker and De Long (1982) investigated the Soret effect in melts of mid-oceanic ridge basalt and discovered that the liquid at the hotter end of their experimental vessel was andesitic in composition, whereas the liquid at the cold end had an iron-rich picrobasaltic composition. This is unlike the normal trend in fractional crystallisation, where the low temperature products of differentiation are silica-rich. Large temperature differences are likely to persist in the stagnant magma that usually develops in the border zone of magma chambers. Care must be taken when examining the rocks in the border zone of a plutonic body. Materials that may appear at first sight to be a chilled margin may have a complex gradational composition produced by the Soret effect.

6.4   Gaseous transfer

Prior to eruption, magmas usually contain volatile components that are held in solution by the confining pressure. When the pressure is lowered fluid is likely to separate from the magma and form bubbles. Such vesicular magmas have lower bulk densities and are .more buoyant than the initial magma. Once the bubbles have expanded to occupy about 75 per cent of the volume of the rising magma their continued growth becomes difficult because the liquid that has to be displaced for new bubbles to grow only occurs in thin sheets between the bubbles. Continued exsolution of gas causes the pressure within the bubbles to increase above the ambient pressure and the larger bubbles burst. Strombolian type volcanic eruptions result from the bursting of bubbles and this process of bubble formation may continue for many years (see Section 4.3).

When a new batch of magma is introduced into a magma chamber the fluids released by the reaction between the new and old magmas are likely to surge upwards and selectively scavenge and move those components that are preferentially distributed into the fluid phase. This magmatic differentiation process is called gaseous transfer. A magma in an isolated magma chamber may convect because compositional or thermal variations produce differences in density. This type of convection is called free convection. If the magma chamber is connected to a feeder channel the introduction of new batches of magma may set up a system of forced convection. This process is particularly important in blending new and residual magmas in magma chambers that are repeatedly replenished.

Normally when magma contacts the near vertical walls of a magma chamber a temperature gradient develops with heat being conducted into the country rocks. On cooling the magma in the border zone contracts, becomes denser and sinks. Meanwhile the hotter magma towards the centre of the reservoir rises activating free convection. This implies that magmas that cool against a vertical contact are gravitationally unstable. When the contact is horizontal the cooling pattern is more complex, because the

physical conditions required for the onset of convection in a sheet-shaped body of liquid is given by the Rayleigh ratio. This ratio considers specific heat, thermal expansion, thermal conductivity, viscosity, acceleration due to gravity, density, height of sheet and difference in temperature between the bottom and top of the layer of magma.

6.5   Convection currents in magma chambers

For the whole of the present century the role of convection currents in magmatic differentiation has been a topic of lively discussion (cf. Becker, 1897). Such currents are usually transient phenomena because they are self-inhibiting. Once they have developed they increase the rate of cooling and this limits and eventually inhibits flow. Their life-span essentially depends on the amount of heat that can be lost through the roof and walls of the magma chamber and on the amount of new hot magma injected into the chamber. Convection currents of variable velocity, depth and direction probably play a prominent role in the cooling and differentiation of most magmas that occupy magma chambers (Sparks et al., 1993). Convective motions in high-level magma reservoirs are normally sufficiently vigorous to keep crystals in suspension. Each mineral probably has its own critical concentration that must be exceeded before it can accumulate on the floor of the magma chamber. When crystallisation takes place in the border zone of a magma chamber the residual liquid is likely to be removed by the convecting magma, thus the main body of magma is likely to display the effects of differentiation without having cooled sufficiently to have directly participated in the crystallisation process. Nielsen (1990) has shown that the liquid line of descent may be strongly influenced by the amount of crystallisation that takes place in the boundary layer and the composition of the phases incorporated into the boundary facies.

Different types of currents may exist simultaneously in a cooling body of magma. For example it may contain both a slow (about 0.06 m/h), widespread, continuous, convective circulation and a faster (about 125 m/h) intermittent density current of somewhat lower volume. The upward component in these currents provides a winnowing mechanism for separating sinking minerals of different size and density. Magmatic density currents probably produce some size-graded layers and scour-channels (trough banding structures), observed in layered igneous intrusions (see Section 6.12).

Hildreth (1979, p. 43) proposed that the hydrous silicic magma that is now evolving in the upper part of the magma chamber beneath the Long Valley Caldera in California was produced by the combined effects of convective circulation, internal diffusion, complexation (polymerisation) and wall-rock exchange. He called this process convection-driven thermo-gravitational diffusion. It is essentially a differentiation process that produces and preserves compositionally zoned magma reservoirs. Hildreth (1981, p. 10 153) asserts that most large eruptions of silicic magma tap magma bodies that are thermally and compositionally zoned.

Once a density stratified magma body has been established the pattern of convection is likely to be complex (Turner and Gustafson, 1978). This can be illustrated by considering what happens when a batch of hot saline water is released at the bottom of a body of cooler and less saline water. The properties of salinity and temperature have opposing effects on the density of the introduced liquid. High salinity denotes increased density, whereas higher temperature causes thermal expansion and lower density. When the new hot, saline liquid is inserted into the main body of liquid it produces a double-diffusive (or multi-diffusive) effect. This is because heat and salt diffuse into the surrounding liquid at different rates. The new batch of liquid rises until its density matches that of the surrounding liquid. It then spreads out horizontally and heat is exchanged rapidly upwards. The new batch of liquid contracts, becomes more dense and moves downwards thus executing a convective cycle. By analogy it is proposed that double-diffusive (or multi-diffusive) convection may operate in a densely stratified magma where it would produce a series of roughly horizontal layers, each containing a battery of small, irregular convection cells. Each of these separately convecting layers is likely to become uniform in temperature and composition, but the whole magma reservoir would contain a series of step-like gradients in temperature and composition.

Irvine (1980, p. 367) used the process of double-diffusive convection to explain the evolution of the Muskox Layered Intrusion in the Coppermine River area of northern Canada. He also suggested that in some magma reservoirs convection in the bottommost layer may become coupled with the upward flow of intercumulus liquid squeezed out of the pile

of cumulus minerals collecting at the base of the magma. The process of self-imposed filter pressing is likely to encourage reaction between the unconsolidated crystals and the rising liquid. This process has been called infiltration metasomatism (Irvine, 1980, p. 341).

6.6 Magma blending

After examining the volcanic rocks of Iceland, Bunsen (1851) concluded that the intermediate rocks were produced by mixing of basic and silicic magmas. Harker (1909) referred to this commingling of magmas as reverse differentiation. His expression implies that one has first to generate a highly differentiated rhyolitic magma and then reverse the differentiation process by mixing it with a more primitive basaltic magma. Bunsen's ideas are still being debated. This is because some of the rhyolitic rocks found in areas like Iceland (see Section 3.5) seem to have been produced by the remelting of small pockets of silicic rocks within thick sequences of essentially basaltic rocks (McBirney, 1993). This means that in such areas there is always the possibility of mixing between mantle-derived picrobasaltic-basaltic magmas and rhyolitic magmas produced by local magmatic differentiation.

In areas of subduction zone magmatism such as the continental margin arcs the partial melting of crustal rocks is common and there is always the possibility of the commingling of magmas with different compositions (see Figs 6.2 and 8.3). Many large layered igneous complexes such as the Bushveld Igneous Complex of southern Africa reveal evidence of the commingling, or blending, of magmas (Gruenewaldt *et al.*, 1985, p. 807). Such blending is common in magma reservoirs where there is continuous magmatic differentiation, periodically interrupted by the introduction of new batches of magma.

When intermediate rocks contain petrographic or chemical evidence of magma mixing it is often difficult to decide whether the magmas that commingled were autonomous or comagmatic. Many rocks that contain evidence of magma mixing seem to have evolved in compositionally zoned magma chambers (McBirney, 1980). With regard to the chemical evidence of commingling one would expect magmas generated by magma mixing to plot as straight lines on normal oxide versus oxide variation diagrams, whereas magmas generated by fractional

Fig. 6.2 Commingling of lava at the edge of Glass Creek Dome, Inyo Volcanic Chain, Long Valley–Mono Basin area, eastern California. Black obsidian commingling with an intruding light grey rhyolite to produce a rock with a marbled cake appearance.

crystallisation would tend to plot as curved lines (see Section 2.8).

6.7 Fractional crystallisation

If lava cools rapidly it may produce a single glass phase. It is more common for lava or intruded magma to contain some minerals. When minerals react continually with a magma the congelation process is called equilibrium crystallisation. If some or all of the minerals are prevented from equilibrating with the magma the process is called fractional crystallisation. This style of crystallisation was first outlined by Becker in 1897 and it is now regarded as the most important magmatic differentiation process. It operates whenever minerals separate from a residual magma. This may be done by the mantling of early-formed minerals or the mechanical removal of minerals from a magma. Bowen (1928, pp. 63–91) in his book on the evolution of igneous rocks devoted a

whole chapter to the fractional crystallisation of basaltic magmas.

Evidence that the process of fractional crystallisation has operated during the evolution of many rocks is readily uncovered in the compositions of mesostases of most porphyritic basic and intermediate volcanic rocks. These mesostases are normally significantly more differentiated than the bulk compositions of the rocks. They are usually depleted in magnesium and calcium, and enriched in silicon, sodium and potassium. For example, the glassy mesostases in many andesitic lavas are dacitic in composition.

6.8 Reaction series

To understand completely how magmas crystallise one has to examine the reaction principle. This critical idea was as we have already noted independently expounded by Bowen (1922) and Goldschmidt (1922) (see Section 1.2 and Figs 1.7 and 1.8). Bowen identified two discrete types of reaction series. The plagioclase feldspars exemplify his continuous reaction series. Calcium-rich plagioclase forms at high temperatures and with continuous reaction a more sodium-rich plagioclase crystallises at lower temperatures. If complete reaction or chemical equilibrium is maintained among the phases in a continuous reaction series all the minerals will have identical compositions at any particular temperature and pressure. Complete reaction is seldom achieved during the cooling of silicate magmas. This is because: (1) diffusion in solids is usually slow compared to the rate of mineral growth, and (2) minerals are often physically separated from the main body of magma. The common ferromagnesian minerals (olivine → pyroxene → amphibole → biotite) form a discontinuous reaction series.

To provide a concrete illustration of the reaction principle in petrogenesis Bowen (1928, p. 60) devised an idealised flow-sheet depicting the phases that form during the cooling and fractional crystallisation of a subalkaline basaltic magma (see Fig. 1.7). At almost the same time Goldschmidt (1922, p. 6) produced a broadly similar diagram (see Fig. 1.8). His diagram also named the types of magma that are likely to form at progressively lower temperatures. In 1979 Osborne reviewed the reaction principle in the light of new experimental studies and introduced two new reaction series. The first was the reaction series that operates in a closed system, whereas the second operated under oxygen-buffered conditions. Fractional crystallisation in a closed system results in the removal of the early formed olivines, Mg-rich pyroxenes and Ca-rich plagioclases, and produces a differentiated residual magma impoverished in magnesium and calcium but enriched in silicon, iron, sodium and potassium. Under oxygen-buffered conditions, magnetite forms early, and the residual magma is impoverished in magnesium, calcium and iron, and enriched in silicon, sodium and potassium.

6.9 Lava lakes

Lava lakes have been likened to laboratories dedicated to the study of the cooling, crystallisation and differentiation of large volumes of natural silicate liquid at low pressure. In recent times Kilauea Volcano in Hawaii has provided several readily accessible lava lakes (see Plate 6.1 and Fig. 6.3). The liquidus temperature of the subalkaline basaltic lava in these lakes is about 1200°C whereas the solidus temperature is about 980°C. Olivine (Fo_{80}) is normally the first mineral to crystallise at a temperature of 1190°C. It is soon followed by augite ($Wo_{40}En_{49}Fs_{11}$) at 1183°C, and plagioclase (An_{70}) at 1168°C. During the crystallisation of these Hawaiian lava lakes the plagioclases changed in composition from An_{70} to An_{59}. A more sodic plagioclase and a small amount of alkali feldspar usually crystallises in the mesostasis. Olivine ceases to crystallise at 1100°C when its composition is about Fo_{65}. The congealed Hawaiian lava lakes also may contain coarse-grained segregation veins of more differentiated rock. These veins are essentially similar in composition to the liquid that oozes into drill holes when they penetrate the outer solid shell of partially solid lava lakes. Such segregation veins and oozes are usually rhyolitic in composition.

Magmatic differentiation is likely to occur while a magma is moving through the lithosphere. The differentiation processes that operate in this milieu include: (1) the crystallisation and removal of early-formed minerals; (2) the nucleation of early-formed minerals onto feeder walls and the incorporation of the interprecipitate liquid into the magma; (3) reaction between the magma and the wall-rocks of the feeder; and (4) the escape of volatile components that exsolve as the magma rises to higher levels and lower pressures.

6.10 Flowage differentiation

A unique process that operates during the movement of a magma is called flowage differentiation. It is based on the observation that when solid particles, irrespective of their shape and size, flow through a conduit they tend to concentrate in the central part of the flow. Flowage differentiation may, for example, concentrate early-formed crystals towards the centre of a sill. Such a process is considered to have operated during the intrusion of the extensive Whin Sill in the north of England. According to Dunham and Strasser-King (1981, pp. 27–28) flowage differentiation concentrated the early-formed crystals towards the centre of the sill. Crystal settling was then inhibited by the high viscosity of the partly crystallised magma.

Fig. 6.3 Halemaumau crater at the summit of Kilauea Volcano has contained several active lava lakes in the past two hundred years. Since 1924 there has been little prolonged lava-lake activity.

6.11   Assimilation and anatexis

The process of assimilation occurs when a magma incorporates and digests foreign material (Fig. 6.4). This process is not important in the evolution of most magmas because there is an enormous discrepancy between the specific heat capacities of silicate magmas and the heats of solution of silicate country rocks. Bowen (1928, p. 187) showed that magmas usually react with and convert included fragments of wall rock into a material that contains phases that are in equilibrium with the magma. He also pointed out that a magma saturated in one member of a reaction series is effectively supersaturated in all higher temperature members of that reaction series. In brief, this means that rhyolitic magma is unlikely to assimilate much basalt but a picrobasaltic, or basaltic, magma may assimilate granite, or the phases typically found in granite, even when they are present in sedimentary or metamorphic rocks. Anatexis is the name given to the process by which magmas, particularly rhyolitic magmas, are generated, or their volume is augmented, by the selective assimilation of crustal rocks. In the past decade Sr-Nd-Pb isotopic studies of comagmatic suites have corroborated the idea that assimilation coupled with fractional crystallisation (AFC) is often important in the generation of magmas in continental settings.

Sorensen and Wilson (1995) conducted a detailed neodymium and strontium isotopic study of the rocks of the Fongen-Hyllingen layered intrusion. The intrusion crops out about 60 km southeast of Trondheim in Norway and it contains rocks that range from dunites and troctolites to quartz-bearing ferrosyenites. Isotope geochemistry was used to trace the complex cooling and differentiation history of the intrusion. The initial isotopic variations were produced by the assimilation of crustal wall and roof rocks and repeated mixing between new primitive, uncontaminated magma and evolved, contaminated residual magma. Most blending of isotopes took place in the roof zone of the magma chamber. Here buoyant contaminated magmas mixed with buoyant differentiated magmas and produced a complex hybrid magma. This magma floated on top of the

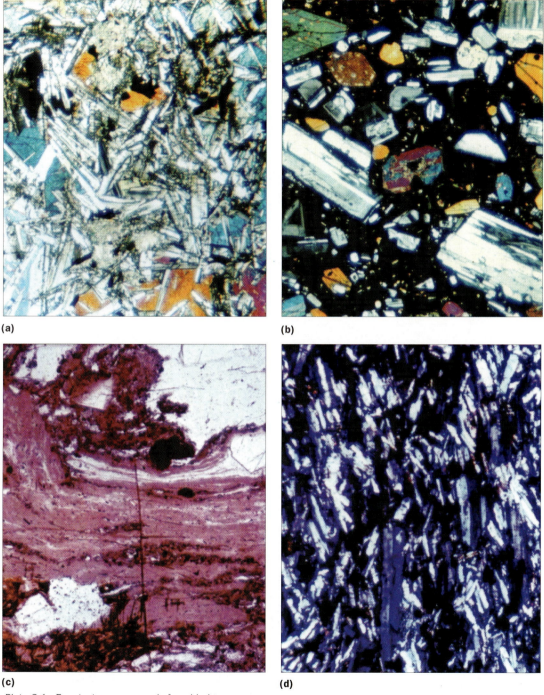

Plate 2.1 Four textures commonly found in igneous rocks: (a) a microgabbro or dolerite with an ophitic texture (crossed polars); (b) a phorphyritic andesite with large crystals of zoned plagioclase, hornblende and augite set in a glassy groundmass (crossed polars); (c) a glassy dacite that shows flow banding; and (d) a trachyte displaying a trachytic texture (crossed polars).

Plate 3.1 The viscosity of the basaltic lava from the East Rift of Kilauea Volcano, Hawaii is revealed by the geologist as he collects a sample of the flow.

Plate 4.1 Image of the surface of Io showing flows issuing from Ra Patera. These flows are several hundred kilometres long and tens of kilometres wide. Note the eruptive centres and lava lakes appear dark, whereas the sulphur-rich flows appear in various hues of red, orange and white. Voyager 1 image P-21277C

Plate 6.1 Kupaianaha a lava lake that was active in July 1989 on the big island of Hawaii

Plate 6.2 Chromitite and anorthosite layers from the Dwars River area of the Bushveld Layered Complex, northern South Africa

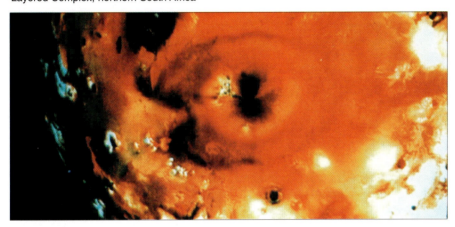

Plate 6.3 A Voyager One view of Io, the innermost of Jupiter's major satellites, showing its extremely active surface covered by volcanic vents and sulphur flows. Voyager 1 image P-21226C.

Plate 7.1 An illustration of the great thicknesses of Karoo flood lavas exposed in the spectacular Cathedral Peak area of Drakensberg Mountains on the border between South Africa and the Kingdom of Lesotho.

Plate 8.1   Young essentially andesitic volcanoes from Tongariro National Park, central North Island New Zealand. Mount Ruapehu is in the background, the symmetrical cone is Mount Ngauruhoe, and Red Crater is in the foreground.

Plate 9.1   Champagne Pool in the Waiotapu geothermal area where bubbles of carbon dioxide constantly rise and break the surface. The orange coloured precipitates observed near the edge of the pool contain amorphous arsenic, and antimony-sulphur compounds. These compounds contain up to 80 parts per million gold, 175 parts per million silver, 320 parts per million thallium and 170 parts per million mercury.

Plate 11.1   Recent dark phonolitic lavas spread out radially from the summit vent of Teide Volcano on Tenerife in the Canary Islands (Pico de Teide, 3718 metres).

Fig. 6.4 Rafts of metasomatised siltstone caught up in the Cape Granite, Llandudno, Cape Peninsula, South Africa. Note the feldspar megacrysts in both the granite and the metasiltstone.

main body of more dense, less contaminated and less differentiated magma.

Initially the degree of contamination decreased downwards and produced an isotopic gradient. Later complex convection currents seem to have dragged some contaminated magma down to lower levels. The final stage of magma injection produced forced convection and mixing between the new magma and the resident magma. This resulted in the generation of a new layered body, with its own isotopic characteristics. During this final stage in the evolution of the complex, crystallisation took place in a closed system.

6.12   Layered igneous rocks

Igneous layering is a conspicuous feature of many large basic, ultrabasic, and less commonly felsic, intrusions (Parsons, 1987; Wager and Brown, 1968, p. 3). This type of layering is well displayed in the central part of the Tertiary age, Cuillin gabbro com-

plex on the Isle of Skye in northwestern Scotland. When Geike (1897, volume 2, p. 327) first saw these layered rocks he thought they were banded Precambrian gneisses. Later he and Teal found that the layers contained essentially the same minerals, but in each layer the minerals were present in different proportions (see Table 6.1 and Fig. 6.5). Individual layers within some large intrusions such as the Bushveld Layered Complex of northern South Africa may persist for distances of up to 150 km (see Fig. 6.6).

Repetitive layering is a prominent feature that usually separates major layered intrusions from simple sills. The sills are essentially equivalent to a single macro-unit in a major layered intrusion. In the layered series of major intrusions the cumulus minerals change in composition from base to top. Plagioclases become more sodic, whereas the olivines and pyroxenes become less magnesium-rich and more iron-rich. This gradual change in composition produces a subtle layering called cryptic layering (see Table 6.1). A spectrum of rocks that includes dunite, peridotite, pyroxenite and chromitite usually occur in the lowermost units of major layered intrusions (see

Table 6.1 Essential minerals present in the various rocks of the Layered Series, Skaergaard Intrusion. After Wager and Brown, 1968, p. 26–27.

Height in metres	Zone	Plagioclase An content	Clinopyroxene Ca:Mg:Fe	Pigeonite Ca:Mg:Fe	Olivine Fo content
2500	Upper	30	43:0:57		0
2400	Upper	33			2
			40:1:59		
2300	Upper				
		35.5	41:11:48		11
2200	Upper				
		37			17
2100	Upper		40:21:39		24
		38			26
2000	Upper				
		38.5	38:27:35		
1900	Upper				
1800	Upper	39			31
		43	35:33:32		34
1700	Upper				
			35:35:30		
1600	Upper	44.5			40
	Middle	46	35:37:28	9:45:46	
1500	Middle				
1400	Middle				
1300	Middle				
		49	36:38:26	9:48:43	
1200	Middle				
1100	Middle				
1000	Middle	51	37:39:24		
900	Middle				
800	Middle	51			53
	Lower	54			55
700	Lower				
600	Lower	56	38:41:21	9:56:35	57
		57			59
500	Lower				
400					
300	Lower	62			63
200	Lower				
100	Lower	66			67
0	Lower				

Fig. 6.5 Igneous layering in the Cuillin Gabbro Complex on the Isle of Skye in the Inner Hebrides, Scotland.

Plate 6.2). Many of these intrusions are ellipse-shaped lopoliths.

Most major layered intrusions are either late Archean or early Proterozoic in age. The Stillwater Complex of Montana is 2.7 giga-years old, the Great Dyke of Zimbabwe and the Jimberlana complex of Western Australia are 2.45 giga-years old, whereas the Bushveld and Molopo Farms Complex of southern Africa is 2.05 giga-years old. These early Precambrian intrusions are noted for their extensive economic deposits of chromium and platinum group elements (Hatton and Gruenewaldt, 1990, pp. 56–82). Their immense size and the large size of their ore deposits may reflect high heat production during this stage in the evolution of the Earth.

An examination of their chilled zones and their trends in fractional crystallisation reveals that most major layered intrusions evolved from parental basaltic, picrobasaltic or komatiitic magmas. Crystallisation appears to have mostly taken place from the floor upwards to form gently dipping layers of cumulates. Towards the sidewalls the dips of the various layers are usually steeper. During the initial stages in the crystallisation of these large intrusions there was

usually repeated mixing of new batches of magma. This process was in part responsible for the extensive development of rocks such as dunites, peridotites, pyroxenites and chromitites in the lower zones of the intrusions. Various processes have been proposed to account for the layering. They include gravity settling, winnowing of crystals in convection currents, concentration of crystals in rapidly moving magmatic density currents (Irvine, 1980, p. 368), double-diffusive (or multi-diffusive) convection (Turner and Chen, 1974), magma mixing, gaseous transfer, infiltration metasomatism and adcumulus growth (Irvine, 1980, pp. 370–373).

As we have already noted cumulates usually contain clusters of minerals that have settled out of a magma. As these aggregates accumulate the individual cumulus crystals are partially enclosed by an intercumulus liquid that has approximately the same chemical composition as the main circumjacent magma. Some cumulate rocks are special because they are monomineralic, yet they appear to be wholly composed of cumulus phases. Some anorthosites are monomineralic rocks that only contain bytownite (An_{84-88}). As the intercumulus liquid

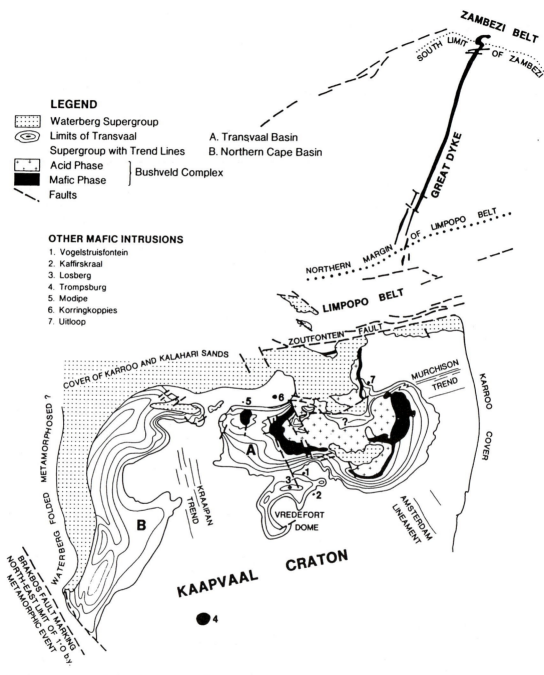

Fig. 6.6 Sketch map showing the location of the Bushveld Layered Complex of South Africa, the layered complexes of the Great Dyke of Zimbabwe and seven other smaller mafic intrusive bodies. After Hunter, 1975, p. 2.

could have provided only some of the chemical components required for the growth of these crystals, Hess (1939, p. 431) and others have proposed that the growth of these minerals into the interstices between the authentic cumulus parts of the crystals was encouraged by diffusion operating through the intercumulus liquid. Wager, Brown and Wadsworth (1960) called this process adcumulus growth, and they used the term adcumulate to describe the rocks generated by this process. The economically important titanomaghemite (19.5 per cent TiO_2) layers in the Upper Zone of the eastern part of the Bushveld Layered Complex were produced by adcumulus growth (Wager and Brown, 1968, p. 378). Rocks chiefly composed of cumulate minerals, and interstitial minerals that crystallised from intercumulus liquid are called orthocumulates. Cumulate rocks that contain a small amount of intercumulus material and are intermediate between orthocumulates and adcumulates are sometimes called mesocumulates.

According to the modelling carried out by Irvine (1980, p. 370) adcumulus growth is likely to take place deep in a cumulate pile. As the pile compacts under its own weight the cumulus minerals recrystallise because of continuous reaction with the migrating intercumulus liquid. Adcumulus growth is not confined to layered complexes. Irvine (1980, p. 373) states that some of the largest coherent bodies of adcumulates are the dunite cores of Alaskan-type zoned complexes, such as the one at Union Bay in Alaska. Dunites of this type may have initially formed as the result of flowage differentiation in the feeder pipe of a volcano (Murray, 1972, p. 331). Later the olivine content of the rock is likely to have been increased by the passage of fluids that promoted adcumulus growth.

McBirney has made several detailed field and experimental studies of the mechanisms responsible for the differentiation of the Skaergaard Layered Intrusion of Kangerlugssuaq in East Greenland. In a review paper McBirney (1995) suggested that after being emplaced the magma crystallised concurrently on the floor, along the walls, and beneath the roof of the magma chamber. The Layered Series crystallised on the floor, the Upper Border Series formed under the roof and the Marginal Border Series crystallised along the walls. Eventually the upper and lower crystallisation fronts converged resulting in the evolution of the Sandwich Horizon. While this horizon was crystallising its interstitial liquid continued to evolve and migrate upwards. These processes account for

the high concentration of incompatible elements in the upper part of the Sandwich Horizon.

During the development of the Layered Series the crystals accumulating on the floor had an average density that was greater than their parental magma. As each batch of crystals settled additional interstitial liquid was expelled. If permeability in the crystal mush was high the interstitial liquid was removed by porous flow, if the crystal mush was less permeable the liquid was forced to escape along slender vertical channels. The discovery of pipe-like bodies of coarser grained rocks that are more differentiated than the rocks they abut is interpreted as evidence of the earlier presence of such channels. The addition of interstitial liquid to the main magma reservoir produces a more evolved magma. This process has been called compaction-driven fractionation (McBirney, 1995, p. 426). The broad compositional differences between the Layered Series and the Upper Border Series are probably best explained by the downward migration of dense, iron-rich liquids during the main stages of differentiation; followed by the upward movement of more buoyant residual magma during the later stages of magmatic differentiation.

6.13 Anorthosite saga

Anorthosites are leucocratic plutonic rocks that consist essentially of plagioclase. They may contain up to 10 per cent mafic minerals, such as pyroxene, olivine and iron-titanium oxides (see Fig. 2.5). They frequently occur as distinctive light-coloured units within the layered sequences of major igneous complexes (Wager and Brown, 1968), but they also occur in several other tectonic and thermal settings (Ashwal, 1993; Isachsen, 1969). Originally the rock and the mineral (plagioclase), were called labradorstein, after the island of Labrador in eastern Canada where it was first described in about 1770 (de Waard, 1968, p. 1). In 1862 Hunt produced a descriptive catalogue of crystalline rocks of Canada, and on page 65 he described a crystalline rock that consisted of almost pure 'anorthic feldspar' as anorthosite. At this time the feldspars were divided into two varieties, the 'orthose' or monoclinic and the 'anorthose' or triclinic type (Ashwal, 1993, p. 1).

The anorthosites are a special group of rocks because: (1) each type of anorthosite has distinctive petrographic features; (2) each type usually belongs to a distinct rock association; and (3) anorthosites are crucially important in the evolution of the lunar crust (see Section 4.5). In his recent review of the anorthosites Ashwal (1993, p. 3) divided the anorthosites into six principal types. They are: (1) Archean calcic type; (2) Proterozoic massif type; (3) layered complex type; (4) oceanic type; (5) xenolithic type; and (6) extra-terrestrial type. The major element compositions of selected examples of the main types of anorthosites are given in Table 6.2. As one would anticipate, all types have relatively high aluminium and calcium abundances. The lunar and Archean types have par-ticularly high calcium and low alkali contents when compared with the common massif type. Anortho-sites from layered intrusions and the oceanic basins are much more variable in composition. This shows that while most of the anorthosites from layered complexes are superficially similar they often evolve from parental magmas with different chemical compositions.

Archean calcic-type anorthosites

The distribution of Archean anorthosites is extremely wide. Many crop out in greenstone belts where they are associated with basaltic and picrobasaltic rocks. Archean anorthosites often contain distinctive equi-dimensional calcic plagioclase megacrysts that give the rock a porphyritic or spotted appearance. These megacrysts are typically unzoned, with compositions of An_{80} or greater. They range in diameter from 0.5 to 30 cm. The megacrystic texture is particularly com-mon in some complexes, such as the Bad Vermilion Lake Complex of Ontario in Canada. In other com-plexes it is either sporadically developed or poorly preserved. Rocks of this texture are especially abun-dant in the Superior Province of the Canadian Shield and in southwestern Greenland (Ashwal, 1993, p. 14). Basaltic dykes, sills and flows that contain simi-lar plagioclase megacrysts crop out in many Archean terranes.

Examples of well-known Archean calcic anortho-site complexes include the Fiskenaesset Complex of West Greenland (2.87 giga-years), the Messina layered intrusion (3.15 giga-years) of the Limpopo mobile belt in southern Africa, the Sakeny anortho-sites of Madagascar and the rocks of the Sittampundi area of India. The Fiskenaesset Complex lies between the coastal village of Fiskenaesset and the Greenland icesheet about 75 km inland from the coast. It is an elongate sheet-like layered intrusions that is at least 200 km long (Windley, 1969). Magmatic structures, such as mineral-layering, size-graded layering, trough banding and internal slump structures are well pre-served. Individual layers contain rocks such as anorthosites, leuco-gabbros and gabbros, and chro-mitites. In the Majorqap qava outcrop the main anorthosite layer has an average thickness of 200 m (Bridgwater et al., 1976, p. 45–49).

Ashwal (1993, pp. 74–81) has reviewed ideas about the origin of the Archean calcic anorthosites and in so doing has highlighted the finding that most of these anorthosites were originally emplaced at shal-low depths (p. 74). Some Archean anorthosite com-plexes share features with the younger, major layered intrusions, but they are characteristically more calcic and less likely to display cryptic variation in compo-sition. The megacrystic texture that is so typical of Archean anorthosites is normally absent in the younger, major layered intrusions. Isotopic and stra-tigraphic dating has shown that the Archean anorthosites are not analogous to the calcic lunar anorthosites that formed shortly after the Moon accreted.

Many attempts have been made to discover the chemical composition of the parental magma respon-sible for the Archean anorthosites. Such research has generally concluded that these anorthosites are cumulates or mushes of cumulate crystals that evolved from a picrobasaltic to basaltic parental magma. It is usually difficult to find unambiguous evidence that relates to the tectonic setting of Archean anorthosites. According to Ashwal (1993, p. 80) it is likely that they evolved within the oceanic crust. This idea is supported by the common associa-tion of Archean anorthosites and pillow lavas.

Proterozoic massif-type anorthosites

Most of the Earth's anorthosites are Proterozoic in age and of the massif type (Leelanandam, 1987). The following are some of the distinctive features found in Proterozoic massif-type anorthosites (Ashwal, 1993, p. 83). They are composite intrusions that fre-quently crop out over areas of thousands of square kilometres, and usually consist of several discrete plutonic bodies. The dominant rock types in these complexes are typically anorthosite, leuco-norite, leuco-gabbro and leuco-troctolite. Lesser amounts of norite, gabbro, troctolite, and rocks enriched in iron-titanium oxides, are also usually present.

Table 6.2 Major element chemistry of the main types of anorthosite. After Ashwal, 1993.

	Archean anorthosite p.34, No. 1 Fiskenaesset W. Greenland	Archean anorthosite p.34, No. 2 Sittampundi India	Massif type anorthosite p. 146, No 3 Harp Lake Labrador	Massif type anorthosite p. 146, No 7 Kunene Angola/Nambia	Layered intrusion anorthosite p. 245, No 2 Bushveld South Africa	Layered intrusion anorthosite p. 245, No 5 Rhum Scotland	Ocean basin anorthosite p. 264, No 1 Vema F.Z. Indian Ocean	Ocean basin anorthosite p. 264, No 9 Mid-Cayman Rise, Caribbean	Lunar, 'ferroan anorthosite' p. 318, No 1 Apennine Front Apollo 15, Moon
SiO_2	44.85	46.65	53.02	53.72	48.68	43.86	55.94	49.24	44.08
TiO_2	0.04	0.07	0.27	0.34	0.08	0.19	0.58	0.05	0.02
Al_2O_3	33.14	32.24	28.07	27.13	30.65	32.86	25.81	26.55	35.49
Fe_2O_3	0.07	0.23	0.42	0.66	*	0.96	0.52	*	*
FeO	0.78	1.02	1.38	0.66	1.7	0.98	0.78	3.02	0.23
MnO	0.02	0.01	0.02	0.01	0.02	0.01	0.03	0.06	0
MgO	0.64	0.5	0.98	0.97	1.44	2.71	1.01	6.26	0.09
CaO	17.83	15.96	11.02	10.1	14.79	16.58	8.28	12.01	19.68
Na_2O	0.99	2.13	4.29	4.99	2.23	1.43	6.18	2.63	0.34
K_2O	0.07	0.05	0.44	0.59	0.21	0.1	0.3	0.06	0
P_2O_5	nd	0.01	0.06	0.04	0.03	0.02	0.01	0	nd
LOI	nd	0.39	nd	0.66	nd	0.31	0.41	nd	<0.01
Total	98.43	99.26	99.97	99.87	99.83	100.01	99.85	99.88	99.93

*Total iron as FeO.

Troctolite is a gabbroic rock that contains calcic plagioclase and olivine, but little or no pyroxene (see Fig. 2.5). Monzonitic, quartz monzonitic and granitic rocks of the mangerite-charnockite suite are typically also present. Little cumulate peridotite, pyroxenite or dunite is normally found in close association with this type of anorthosite. These intrusive complexes contain some large iron-titanium oxide economic mineral deposits such as the Sanford Hill deposit that occurs in association with the Marcy Anorthosite of the Adirondack Mountains of northern New York State.

Most massif-type anorthosites are coarse-grained and their plagioclase is usually in the range An_{40-60}. Minor and accessory minerals include low-calcium pyroxene (enstatite and inverted pigeonite), olivine, ilmenite, magnetite, clinopyroxene and apatite. Some rocks contain high-aluminium orthopyroxene megacrysts that may have originally crystallised at high pressure. Remarkable as it may seem, most massif-type anorthosites were emplaced during the Mesoproterozoic Era (1.6 to 1.0 giga-years). This discovery prompted much speculation, including the conjecture that a single momentous 'anorthosite event' generated all the massif-type anorthosites. The largest outcrops of this type of anorthosite crop out in the Lac Saint-Jean Complex in Quebec, Canada (17 000 km²), the Kunene Complex, that lies astride the Namibia–Angola border and then extends north into Angola (15 000 km²) and the Harp Lake Complex of Labrador in Canada (10 000 km²). Other large bodies of massif-type anorthosite are reported from the Kola Peninsula of northwestern Russia, and from eastern Siberia.

According to Ashwal (1993, p. 99) many massif type anorthosites are emplaced into Proterozoic granitic gneisses. Some bodies appear to have been emplaced at or near unconformities between gneiss and overlying sequences of younger supracrustal rocks. Such an intrusion has, for example, been reported from a coastal section of Labrador, where the Nain anorthosites are observed to intrude between old gneiss and an overlying sequence of younger, layered metamorphic rocks. Ashwal (1993, p. 217) has proposed that the massif anorthosites evolved in the following manner:

1 A basaltic or picrobasaltic magma became ponded in a large pancake-shaped magma chamber in the lowermost crust.

2 Cooling and fractional crystallisation led to the sinking of the early formed mafic minerals, such as olivine and pyroxene, to the base of the magma chamber.

3 Plagioclase was buoyant at the prevailing pressures. It floated upwards and accumulated as an anorthositic mush at the top of the magma chamber.

4 The crystal mush was less dense than the surrounding lower crustal rocks and part of it became a component in the series of diapirs of evolved magma that ascend from the magma chamber.

5 At higher levels in the crust the massif anorthosite complexes form as a result of the emplacement and coalescence of a sequence of these diapirs with plagioclase-mush tops.

6 At all stages during their ascent the diapirs are likely to incorporate fragments (inclusions) of the crustal rocks they pass through.

7 Back in the deep-seated magma chamber, heat from the crystallising magma is likely to induce anatexis thus producing silicic magmas.

8 The anatectic magmas would crystallise to form the monzonitic, quartz monzonitic and granitic rocks that normally accompany the massif-type anorthosites.

Anorthosites from layered complexes

Plagioclase-rich rocks crop out in most mafic layered complexes, but their abundance varies greatly. The 4500 m thick Mineral Lake Intrusion from near Mellen in northwestern Wisconsin is reported by Olmsted (1969, p. 149) to contain only 1 per cent pyroxenitic and peridotitic rocks, but 73 per cent plagioclase-rich rocks (anorthosite and olivine leuco-gabbronorite). When considered as one unit the rocks of this intrusion contain 69.1 per cent modal plagioclase. In stark contrast some layered igneous intrusions, such as the Skaergaard Intrusion, contain only minor volumes of plagioclase-rich rocks. The huge Bushveld Layered Complex of northern South Africa contains only 6.5 per cent anorthosite. Anorthosite-bearing layered complexes are not concentrated in one geological era but appear to have been emplaced throughout most of geological time. The Stillwater Layered Complex that occupies the northern margin of the Beartooth Mountains in Montana is Archean in age (2.701 giga-years), whereas the huge Dufek Layered Complex in the Pensacola Mountains of Antarctica is Mid Jurassic in age (172 mega-years).

The textures of the anorthosites in these complexes are variable, but most rocks are interpreted as either orthocumulates or adcumulates. Anorthosite layers typically occur above the ultramafic zones that occupy the base of these complexes. The Stillwater complex contains about 18 per cent anorthosite, and most of it is concentrated in the Middle Banded Zone where it is the dominant rock. A distinctive feature of many layered complexes, including this complex, is the presence of thick layers of anorthosite. Both at the bottom and at the top of the Middle Banded Zone there are such layers. They are 350 m and 570 m thick, respectively.

The upper anorthosite zone in the middle to upper Proterozoic Kalka Layered Complex in the South Australian segment of the Giles Petrographic Province of Central Australia is about 800 m thick (Goode, 1976). This petrographic province is particularly interesting as it comprises about 20 major and many more minor, layered intrusions. They crop out over an area of about 2500 km^2 (Ballhaus, 1993; Nesbitt *et al.*, 1970). Large bodies of anorthosite have also been described from the Bell Rock Range, Blackstone Range and Hinckley Range complexes of the Giles Petrographic Province.

There is probably more than one simple process responsible for the origin of the anorthosite layers in the various layered igneous complexes. In recent years considerable research effort has been expended on trying to understand the origin and evolution of the economically important platinum, chromium and titanium deposits found in these layered complexes. Several single and multiple parental magma hypotheses have been proposed. The single parental magma theories explain the origin and evolution of some small layered bodies, like the Skaergaard Intrusion of East Greenland. All that is required to produce such layered intrusions is a mechanism for concentrating plagioclase in an actively convecting magma chamber. Large layered intrusive complexes often contain evidence of multiple intrusive events.

The Bushveld Complex seems to consist of four main intrusive bodies that were originally fed from at least seven different conduits (Harmer and Sharpe, 1985, p. 813). Gruenewaldt and others (1985, p. 806) have suggested that this complex evolved from two or more discrete parental magmas. It is likely that these magmas had different silica, magnesium, chromium and incompatible element abundances. The mixing of these magmas was probably important in producing some of the distinctive rocks of this complex, such as the Merensky Reef with its high content of platinum group elements.

Special materials or processes are likely to be required to account for complexes with particularly thick layers of anorthosite, or complexes that contain so much anorthosite that its abundance cannot be accounted for by the differentiation of normal basaltic or picrobasaltic magmas. Such bodies of anorthosite were probably generated by the emplacement of plagioclase-rich mushes that evolved in a deeper magma chamber similar to the one proposed to account for the origin of the massif-type anorthosites (Ashwal, 1993, p. 261).

Anorthosites from the ocean basins

Anorthosites are occasionally dredged from the ocean floor. Most of these samples came from the Atlantic or Indian oceans. These anorthosites probably make up only a small part of the oceanic crust. The idea that layered complexes evolve in oceanic crust is supported by data from ophiolite complexes. Many of these complexes contain thin anorthositic or leuco-gabbroic layers. An interesting suite of 48 plutonic rocks has been collected from the Mid-Cayman Rise in the Caribbean Sea (Elthon, 1987). Five samples are anorthosites, while others include plagioclase-rich leuco-gabbros and leuco-troctolites. Most of the anorthosites from the oceanic crust are interpreted as plagioclase cumulates that formed by gravitational sinking of plagioclase, assisted by adcumulus growth. The lack of cryptic variation in the oceanic crustal layered rocks, suggests that they developed in magma chambers that were frequently replenished by an influx of new primitive magma.

Cognate and accidental inclusions

Fragments, or segregations, of plagioclase-rich material occur in a remarkably wide range of igneous rocks, found in many different tectonic settings (Ashwal, 1993, pp. 281–300). Anorthositic inclusions have been reported from oceanic islands such as Iceland and Hawaii (Clague and Chen, 1986; Kristmannsdottir, 1971), and many island arcs and continental margin arcs. Most of these materials are: (1) cognate inclusions; (2) products of the commingling between conventional magmas and a plagioclase-rich mush; or (3) accidental inclusions of older plagioclase-rich rocks. It is often difficult to establish how a particular anorthositic inclusion formed.

A remarkable suite of plagioclase megacrysts and anorthositic inclusions occurs in the well exposed 'big feldspar dykes' of southern Greenland (Bridgwater and Harry, 1968). Up to 80 per cent of the volume of some dykes is made up of plagioclase megacrysts and individual megacrysts may be up to 1 m in length. The crowding of these megacrysts and the plagioclase-rich inclusions into the upper parts of some larger gabbroic dykes suggests that the included materials floated upwards in their host magma (Emeleus and Upton, 1976, p. 161). Two types of inclusions are common. They are a coarse-grained, massive anorthosite and a strongly foliated anorthosite. The compositions of the plagioclases in the various inclusions are usually broadly similar to those found in their host rocks. This suggests that there is a genetic link between these materials. Because plagioclase megacrysts and anorthositic inclusions are so abundant in one area in southern Greenland it has been proposed that the 'big feldspar dykes' represent a regional commingling of a congealing, or congealed, anorthositic mush with a series of related but more primitive magmas.

The microgabbroic rocks of the Beaver Bay Complex on the northwestern shore of Lake Superior in Minnesota contain another particularly spectacular display of anorthositic inclusions. These inclusions range in size from less than 1 cm in diameter to huge blocks that are hundreds of metres across (Morrison et al., 1983). The Beaver Bay Complex has a sill-like form and the anorthositic inclusions are particularly concentrated near its roof. This suggests that they may have floated in their host magma. Textural and isotopic evidence shows that these inclusions are accidental and represent fragments from an older anorthositic body.

Extraterrestrial anorthosites

In 1969 most petrologists were astounded by the discovery that anorthosites were an important component in the lunar highlands. Until then most of the speculation on the composition of the lunar highlands had presumed that they contained either granitic rocks or primitive chondritic materials. The discovery of large volumes of ancient anorthositic crust forced petrologists to re-evaluate their ideas

on the first stable crust that evolved on Earth and the other terrestrial planets. Lunar anorthosites were collected by both the Apollo and Luna missions. Lately several meteorites have been discovered that contain anorthositic breccias of lunar or possibly mercurian origin.

The most common pristine lunar highland rock is the 'ferroan anorthosite'. These curiously named rocks usually contain less than 0.5 per cent iron as FeO (Ashwal, 1993, p. 318). According to Heiken et al. (1991, pp. 221) roughly 50 per cent of the outer highland crust contains this type of anorthosite. Although they are usually severely brecciated, the few pristine samples are coarse-grained rocks, with large subhedral to euhedral calcium-rich plagioclase crystals (An_{94-97}) surrounded by smaller anhedral crystals of pyroxene or olivine. The pyroxene is usually a low-calcium pigeonite. It may display exsolution lamellae of augite set in an orthopyroxene host. Most of the large samples of this anorthosite contain small amounts of olivine. Some samples contain accessory amounts of ilmenite, aluminium-rich chromite, kamacite (Fe-Ni metal), troilite and a silica phase (cristobalite or tridymite).

Geochemically these primitive lunar anorthosites are particularly interesting as they have a strong positive europium anomaly. This is utterly unlike normal mare basalts that display a strong negative europium anomaly (see Section 7.12). Another geochemical feature of the lunar anorthosites is their remarkably low sodium and potassium contents. It seems likely that these depletions were induced by pre-accretionary events. A few pristine samples of lunar anorthosites have cumulate textures. This is what one would expect because it is usually proposed that the lunar anorthosites are plagioclase cumulate rocks. The final textures and compositions of many of these rocks seem to have been determined by adcumulus growth.

Petrologists have generally supported the idea that the lunar anorthosites formed as flotation cumulates in a primeval magma/lava ocean. The principal evidence in support of this idea is (1) the global distribution of anorthosite on the lunar surface and (2) the lack of complementary mafic phases as this implies that the initial body of magma/lava was thicker than the depth of excavation of the large impact features that now crater the lunar surface. There has been much speculation about the composition of this primeval magma/lava ocean. It was clearly able to crystallise plagioclase and various mafic phases, thus it was probably picrobasaltic or basaltic in composition.

Superficially the heavily cratered surface of Mercury resembles that of the lunar highlands. It has been proposed that the old mercurian surface also contains anorthositic materials. This idea is supported by the limited amount of spectral reflectance data currently available. Mercury is also like the Moon in that its surface is pocked by many large impact craters that are surrounded by bright, light coloured diverging lines of ejecta. On the Moon these ejecta rays usually contain fragments of anorthosite. The mercurian ejecta rays probably contain similar materials excavated from its crust.

The surface of Venus is mainly covered by young features and there is a paucity of primeval heavily cratered areas like those found on the Moon and Mercury, thus it is unlikely that it will contain any large areas of ancient anorthositic crust. About half the surface of Mars is heavily cratered but much of it is covered by wind-blown materials and regolith so that there is no direct evidence of primeval Martian anorthosites. Mars is noted for its huge volcanoes. Future research may well reveal that they evolved above huge magma chambers where cumulate processes produced cumulates including anorthositic rocks. Similar processes may have operated beneath the volcanoes of Venus.

6.14   Non-silicate igneous rocks

The solid inner core of the Earth is its largest body of non-silicate igneous rocks. As was noted in Section 3.3 this material probably has a similar modal and chemical composition as some iron meteorites. Several non-silicate igneous rocks crop out on the surface of the Earth. They include the carbonatites, chromitites, ilmenitites, magnetitites, nelsonites, phoscorites and volcanic flows of materials like native sulphur. The carbonatites and phoscorites belong to a special group of rocks that will be reviewed in Chapter 12. Most of the economically important chromitites, ilmenitites and magnetitites occur as units within layered igneous complexes, such as the Great Dyke of Zimbabwe or the Bushveld and Molopo Farms Complex of southern Africa. Many layered complexes contain small concentrations of sulphides and arsenides. Detailed descriptions of these non-silicate igneous materials can be found in most books on ore petrology.

For many years the origin of the huge nickel-copper sulphide deposits of Sudbury, Ontario in Canada has been a topic of heated debate. The sulphides occur at the base of a lopolith that contains norites, leuco-norites and granophyres. Most petrologists regard the sulphides as accumulations of immiscible metallic sulphide droplets at the base of a magma chamber. Other petrologists, notably Dietz (1964), have suggested that the nickel-rich rocks developed as the result of the impact of a large nickel-iron meteorite. In other areas, such as the Kalgoorlie area of Western Australia, one can examine massive nickel sulphides that segregated from picrobasaltic (komatiitic) magmas.

The El Laco magnetitite lava flows of the highlands of northern Chile are especially intriguing. They are mainly flows, but in some areas this rock is intrusive into other volcanic materials. Individual flows are typically highly vesicular, show contorted flow banding, and display ropy surfaces. In the field they have the general appearance of basaltic lava flows (Haggerty, 1970; Parks, 1961; Rogers, 1968). Their essential minerals are magnetite and hematite, with accessory amounts of feldspar, calcic pyroxene, apatite, calcite and an iron-phosphate mineral. These rocks usually contain up to 98 per cent iron oxide. Both magnetite and hematite may occur as euhedral crystals that are up to several centimetres in diameter. Secondary hematite and maghemite develop extensively as an oxidation product of magnetite. Goethite forms as a late stage vesicle and veinlet infilling.

Another important group of non-silicate rocks contains iron-titanium oxide and apatite. These nelsonites were first described from Nelson County, Virginia, USA (Philpotts, 1967). They usually occur in small intrusive bodies associated with massif type anorthosites, or alkaline rocks such as those found in many carbonatite complexes (see Chapter 12). According to Philpotts (1967, p. 303), nelsonites have a remarkably constant composition of two-thirds by volume iron-titanium oxides and one-third apatite. They are usually associated with dykes rich in ferromagnesian minerals. Philpotts (1967, p. 311) has shown experimentally that nelsonite can evolve as an immiscible liquid separate from a dioritic magma that is rich in iron, titanium and phosphorus. He also claims that such magmas are likely to form during the evolution of the ferrodioritic (jotunite) rocks that are often found with massif-type anorthosites (Philpotts, 1990, p. 306).

The emission of molten sulphur from fumaroles or volcanic vents is a comparatively rare event on Earth. For example, in the period 1975–1985 only three small emissions were recorded, and two came from the same vent (McClelland *et al.*, 1989, pp. 265 and 516). The vent that produced the two flows and a deposit of sulphur pyroclasts occurs in the summit crater of Shinmoe-dake. It is a crater in the southeastern sector of the Kirishima Volcanic Complex, Kyushu, Japan. The second emission of molten sulphur was reported from Irazu, the highest volcano in Costa Rica in Central America. Many spectacular eruptions of sulphur have been reported from Io, the innermost of the four Galilean satellites of Jupiter. This planet is extremely volcanically active and large areas of its surface are covered by young sulphur flows (see Plates 4.1 and 6.3).

6.15 Magmatic differentiation as a planetary process

The petrology and volcanology of every planet is excitingly different. These differences are important as they force petrologists to expand and fine-tune their ideas on how the processes of magmatic differentiation work under changing physical conditions. For example, the discovery of: (1) the circular, steep-sided volcanic domes on Venus; (2) the volcanically active mid-oceanic ridge system and its complementary subduction zones on Earth; (3) the primeval anorthosites on the Moon; (4) the huge domes and giant volcanoes on Mars; and (5) the sulphur-rich volcanic activity on Io, have all irrevocably changed the way petrologists visualise the problem of the diversity in magmatic rocks. As each planet seems to be at a different stage in its evolutionary development the study of these remarkably diverse volcano-magmatic features enables petrologists to refine their ideas on the past and even the future development of Earth. This important idea is explored in many parts of this book and it is reviewed in Section 13.5.

References

Ashwal, L.D., 1993. *Anorthosites*. Springer-Verlag, Berlin, 422 pp.

Ballhaus, C.G., 1993. Petrology of the layered mafic/ultramafic Giles Complex, Western Musgrave Block, Western Australia: igneous stratigraphy, mineralogy and petrogenesis of the Jameson Range, Murray Range, Blackstone Range, Hinckley Range, Bell Rock Range, south Mount West and Latitude Hill intrusions. *Australian Geological Survey Organisation*, Record 1992/73, 75 pp.

Becker, G.H., 1897. Fractional crystallization of rocks. *American Journal of Science*, 4, pp. 257–261.

Bowen, N.L., 1922. The reaction principle in petrogenesis. *Journal of Geology*, 30, pp. 177–198.

Bowen, N.L., 1928. *The Evolution of Igneous Rocks*, Princeton University Press, Princeton, New Jersey, 334 pp.

Bridgwater, D. and Harry, W.T., 1968. Anorthosite xenoliths and plagioclase megacrysts in Precambrian intrusions in South Greenland. *Bulletin Grönlands Geologiske Undersögelse*, 77, 243 pp.

Bridgwater, D., Keto, L., McGregor, V.R. and Myers, J.S., 1976. Archean gneiss complex of Greenland. In Escher, A. and Watt, W.S. (editors) *Geology of Greenland*, pp. 18–75, Grönlands Geologiske Undersögelse, Copenhagen, 603 pp.

Bunsen, R., 1851. *Ueber die Prozesse der vulcanischen Gesteinsbildungen Islands*, Annalen der Physik, Leipzig, 2nd series, editor J.C. Poggendorff, 83, pp. 197–272.

Chayes, F., 1975. *Statistical Petrology*, Carnegie Institution of Washington Yearbook, 74, pp. 542–550.

Clague, D.A. and Chen, C.H., 1986. Ocean crust xenoliths from Hualalai volcano, Hawaii. *Geological Society of America*, Abstracts Program, 18, p. 565.

de Waard, D., 1968. Annotated bibliography of anorthosite petrogenesis. In Isachsen, Y.W. (editor), *Origin of Anorthosites and Related Rocks* pp. 1–11. New York State Museum and Science Service, Memoir 18, 466 pp.

Dietz, R.S., 1964. Sudbury structure as an astrobleme. *Journal of Geology*, 72 (4), pp. 412–434.

Dunham, A.C. and Strasser-King, V.E.H., 1981. Petrology of the Great Whin Sill in the Throckley Borehole, Northumberland. *Report of the Geological Institute*, United Kingdom, 81/4, pp. 1–32.

Elthon, D., 1987. Petrology of gabbroic rocks from the Mid-Cayman Rise spreading centre. *Journal of Geophysical Research*, 92, pp. 658–682.

Emeleus, C.H. and Upton, B.G.J., 1976. The Gardar period in southern Greenland. In Escher, A. and Watt, W.S. (editors) *Geology of Greenland*, pp. 152–181. Grönlands Geologiske Undersögelse, Copenhagen, 603 pp.

Geike, A., 1897. *The Ancient Volcanoes of Great Britain*, Volume 2, Macmillan and Company, Limited, London, 492 pp.

Goldschmidt, V.M., 1922. Stammestypen der Eruptivgesteine, *(Norske) Videnskaps Skrifter* I, Matisk Naturvidenskapelig, Klasse, 10, 6 pp.

Goode, A.D.T., 1976. Small-scale primary cumulus igneous layering in the Kalka layered intrusion, Central Australia. *Journal of Petrology*, 17, pp. 379–397.

Gruenewaldt, G. von, Sharpe, M.R. and Hatton, C.J., 1985. The Bushveld Complex: Introduction and review. *Economic Geology*, 80, pp. 803–812.

Haggerty, S.E., 1970. The Laco magnetite lava flow, Chile. *Carnegie Institution of Washington*, Yearbook 68, pp. 329–330.

Harker, A., 1909. *The Natural History of Igneous Rocks.* Methuen, London, 377 pp.

Harmer, R.E. and Sharpe, M.R., 1985. Field relations and strontium isotope systematics of the marginal rocks of the Eastern Bushveld Complex. *Economic Geology*, 80, pp. 813–837.

Hatton, C.J. and Gruenewaldt, G. von, 1990. Early Precambrian layered intrusions. In Hall, R.P. and Hughes, D.J. (editors) *Early Precambrian Basic Magmatism*, pp. 56–82, Blackie, Glasgow, 486 pp.

Heiken, G., Vaniman, D. and French, B.M., 1991. *Lunar Sourcebook: A Users Guide to the Moon.* Cambridge University Press, Cambridge, 736 pp.

Hess, H.H., 1939. Extreme fractional crystallization of a basaltic magma; the Stillwater igneous complex. *Transactions of the American Geophysical Union*, 3, pp. 430–432.

Hildreth, E.W., 1979. The Bishop Tuff: evidence for the origin of compositional zonation in silicate magma chambers. *Geological Society of America*, Special Paper, 180, pp. 43–75.

Hildreth, E.W., 1981. Gradients in silicic magma chambers: implications for lithospheric magmatism. *Journal of Geophysical Research*, 86 (B11), pp. 10153–10192.

Hunt, T.S., 1862. Descriptive catalogue of a collection of the crystalline rocks of Canada: descriptive catalogue of a collection of the economic minerals of Canada sent to the London International Exhibition. Montreal. *Geological Survey of Canada*, 1862, pp. 61–83.

Hunter, D.R., 1975. *The Regional Geological Setting of the Bushveld Complex.* Economic Geology Research Unit, University of the Witwatersrand, Johannesburg, South Africa, 18 pp.

Irvine, T.N., 1980. Magmatic infiltration metasomatism, double-diffusive fractional crystallization, and adcumulus growth in the Muskox intrusion and other layered intrusions. In Hargraves, R.B. (editor) *Physics of Magmatic Processes*, Princeton University Press, Princeton, 586 pp.

Isachsen, Y.W. (editor), 1969. Origin of Anorthosites and Related Rocks. *New York State Museum and Science Service*, Memoir 18, 466 pp.

Kjarsgaard, B.A. and Hamilton, D.L., 1988. Liquid immiscibility and the origin of alkali-poor carbonatites. *Mineralogical Magazine*, 52, pp. 43–55.

Kristmannsdottir, H., 1971. Anorthosite inclusions in Tertiary dolerite from the island groups Hrappsey and Purkey, West Iceland. *Journal of Geology*, 79, pp. 741–748.

Leelanandam, C., 1987. Proterozoic anorthosite massifs: an overview. *Indian Journal of Geology*, 59, pp. 179–194.

Le Maitre, R.W., Bateman, P., Dudek, A., Keller, J., Lameyre, J., Le Bas, M.J., Sabine, P.A., Schmidt, R., Sørensen, H., Streckeisen, A., Woolley, A.R. and Zanettin, B., 1989. *A Classification of Igneous Rocks and Glossary of Terms: Recommendations of the International Union of Geological Sciences Subcommission on the systematics of igneous rocks.* Blackwell Scientific Publications, Oxford, 193 pp.

McBirney, A.R., 1980. Mixing and unmixing of magmas. *Journal of Volcanology and Geothermal Research*, 7, pp. 357–371.

McBirney, A.R., 1993. Differentiated rocks of the Galapagos hotspot. In Prichard, H.M., Alabaster, T., Harris, N.B.W. and Neary, C.R. (editors), *Magmatic Processes and Plate Tectonics*, pp. 61–69. Geological Society of London, Special Publication Number 76, 526 pp.

McBirney, A.R., 1995. Mechanisms of differentiation in the Skaergaard Intrusion. *Journal of the Geological Society*, London, 152, pp. 421–435.

McBirney, A.R. and Nakamura, Y., 1974. Immiscibility in late-stage magmas of the Skaergaard intrusion. *Carnegie Institution of Washington*, Yearbook, 73, pp. 348–352.

McClelland, L., Simkin, T., Summers, M., Nielsen, E. and Stein, T.C., 1989. *Global Volcanism 1975–1985*, Prentice Hall, Englewood Cliffs, New Jersey, 655 pp.

Morrison, D.A., Ashwal, L.D., Phinney, W.C., Shih, C.Y. and Wooden, J.L., 1983. Pre-Keweenawan anorthosite inclusions in the Keweenawan Beaver Bay and Duluth Complexes, northeastern Minnesota. *Bulletin of the Geological Society of America*, 94, pp. 206–221.

Murray, C.G., 1972. Zoned ultramafic complexes of Alaskan type: feeder pipes of andesite volcanoes. *The*

Geological Society of America, Memoir 132, pp. 313–335.

Nesbitt, R.W., Goode, A.D.T., Moore, A.C. and Hopwood, T.P., 1970. The Giles Complex, Central Australia: a stratified sequence of mafic and ultramafic intrusions. In Visser, D.J.L. and Gruenewaldt, G. von (editors) *Symposium on the Bushveld Igneous Complex and other layered intrusions*, pp. 547–564. The Geological Society of South Africa, Special Publication 1, 763 pp.

Nielsen, R.L., 1990. A numerical approach to in situ fractionation: application to differentiation in open magma systems. *EOS Transactions*, American Geophysical Union Fall Meeting, 71 (43), p. 1665.

Olmsted, J.F., 1969. Petrology of the Mineral Lake Intrusion, Northwestern Wisconsin. In Isachsen, Y.W. (editor), *Origin of Anorthosites and Related Rocks*, pp. 149–161, New York State Museum and Science Service, Memoir 18, 466 pp.

Osborne, E.F., 1979. The reaction principle. In Yoder, H.S. (editor) *The Evolution of Igneous Rocks: Fiftieth Anniversary Perspectives*, pp. 133–169, Princeton University Press, Princeton, New Jersey, 588 pp.

Parks, C.F., 1961. A magnetite 'flow' in northern Chile. *Economic Geology*, 56, pp. 431–436.

Parsons, I. (editor), 1987. *Origins of Igneous Layering*. D. Reidel Publishing Company, Dordrecht, 666 pp.

Philpotts, A.R., 1967. Origin of certain iron-titanium oxide and apatite rocks. *Economic Geology*, 62, pp. 303–315.

Philpotts, A.R., 1990. *Principles of Igneous and Metamorphic Petrology*, Prentice Hall, Englewood Cliffs, New Jersey, 498 pp.

Roedder, E., 1951. Low temperature liquid immiscibility in the system $K_2O–FeO–Al_2O_3–SiO_2$. *American Mineralogist*, 36, pp. 282–286.

Roedder, E., 1979. Silicate liquid immiscibility in magmas. In Yoder, H.S. (editor) *The Evolution of Igneous Rocks: Fiftieth Anniversary Perspectives*, pp. 15–57,

Princeton University Press, Princeton, New Jersey, 588 pp.

Rogers, D.P., 1968. The extrusive iron oxide deposits, 'El Laco', Chile. *Geological Society of America*, Abstracts Programs, pp. 252–253.

Turner, J.S. and Gustafson, L.B., 1978. The flow of hot saline solutions from vents in the sea-floor – some implications for exhalative massive sulphides and other ore deposits. *Economic Geology*, 73, pp. 1082–1100.

Scrope, G.P., 1825. *Consideration on volcanoes, the probable cause of their phenomena, the laws which determine their march, the disposition of their products, and their connection with the present state and past history of the globe, leading to the establishment of a new theory of the Earth*. W. Phillips, London, 270 pp.

Sorensen, H.S. and Wilson, J.R. 1995. A strontium and neodymium isotopic investigation of the Fongen-Hyllingen layered intrusion, Norway. *Journal of Petrology*, 36 (1), pp. 161–187.

Sparks, R.S.J., Huppert, H.E., Koyaguchi, T. and Hallworth, M.A., 1993. Origin of modal and rhythmic igneous layering by sedimentation in a convecting magma chamber. *Nature*, 361, pp. 246–249.

Turner, J.S. and Chen, C.F., 1974. Two-dimensional effects of double-diffusive convection. *Journal of Fluid Mechanics*, 63, pp. 577–593.

Wager, L.R. and Brown, G.M., 1968. *Layered Igneous Rocks*, Oliver and Boyd, Edinburgh, 588 pp.

Wager, L.R., Brown, G.M. and Wadsworth, W.J., 1960. Types of igneous cumulates. *Journal of Petrology*, 1, pp. 73–85.

Walker, D. and De Long, S.E., 1982. Soret separation of mid-ocean ridge basalt magma. *Contribution to Mineralogy and Petrology*, 79, pp. 231–240.

Wilson, M. 1993. Magmatic differentiation. *Journal of the Geological Society*, London, 150, pp. 611–624.

Windley, B.F., 1969. Anorthosites of southern West Greenland. *American Association of Petroleum Geologists*, Memoir 12, pp. 899–915.

CHAPTER 7 Picrobasalts and basalts

From the Earth today to the small asteroid Vesta 4.5 billion years in the past, on Mars, Venus and Mercury, the generation of basalts has spanned the history of the Solar System...

Basaltic Volcanism Study Project, 1981, p. xxvii.

7.1 Abundance and distribution of the picrobasalts and basalts

The picrobasalts and basalts are a chemically coherent group of volcanic rocks that erupted throughout geologic time. According to the Basaltic Volcanism Study Project (1981, p. xxvii), this type of magmatism is a fundamental process in the evolution of the terrestrial planets. When one examines picrobasalts and basalts from different tectonic settings one discovers that their modal and major element compositions are broadly similar, but closer inspection usually reveals subtle differences. These differences generally become magnified during magmatic differentiation. All the common liquid lines of descent appear to diverge from the picrobasaltic-basaltic field. Magmas that begin in this field but are silica oversaturated usually evolve towards the rhyolite field, whereas magmas that are silica undersaturated traverse either the midalkaline or the alkaline fields and terminate in the phonolite field.

Basalt is an old term and it is generally used to describe fine-grained, or glassy, basic volcanic rocks. They usually contain plagioclase and pyroxene, also lesser amounts of iron-titanium oxides, and other minerals like olivine, quartz or nepheline. Quartz and nepheline are incompatible phases and are not found in the same rock. Picrobasalt is a newer term introduced by Le Maitre in 1984 (p. 245). It is used to describe fine-grained, or glassy, ultrabasic volcanic rocks that normally contain plagioclase, pyroxene, olivine and lesser amounts of iron-titanium oxides. According to the IUGS subcommission on the systematics of igneous rocks (Le Maitre *et al.*, 1989, p. 28) basalts contain between 45 and 52 per cent silica and less than 5 per cent total alkalis.

Picrobasalts occupy a field beside the basalts in the TAS classification (see Fig. 2.7). They contain between 41 and 45 per cent silica and less than 3 per cent total alkalis. Picrobasaltic or basaltic rocks that contain more than 18 per cent MgO and 1–2 per cent total alkalies are called picrites. The komatiites and meimechites are also high-magnesium rocks but they contain less than 1 per cent total alkalis.

When considering the origin and evolution of the picrobasalts and basalts one soon discovers that there are many ways of classifying them. They can be grouped into their rock association, tectonic setting, or the thermal environment in which they originated. A particular basalt may belong to more than one of these categories. For example, lavas and pyroclastic rocks that belong to the midalkaline association may erupt in several different tectonic settings and rocks belonging to two, or even three, distinct rock associations may crop out in a small area such as Hawaii or the Strait of Taiwan.

Volcanic rocks from the Hawaiian Islands have played an important part in the development of ideas on the evolution of basaltic rocks. In 1935 Powers recognised two rock associations on Hawaii, namely an early primitive suite and a group of later alkaline rocks. Subsequently, Tilley (1950, p. 42) used a total alkali silica diagram to show that these two groups of rocks were chemically distinct. He called the primitive suite the tholeiitic series, and the other rocks the alkalic series. MacDonald (1968, p. 481) recognised three petrographic suites, and pointed out that a single straight line on a total alkali silica diagram can separate the rocks of the tholeiitic suite from rocks of the other two suites, a midalkaline suite and a later alkaline (or nephelinitic) suite (see Fig. 7.1).

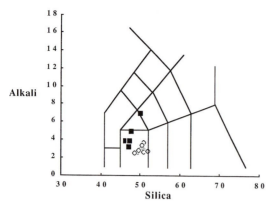

Fig. 7.1 A total alkali silica (TAS) diagram that shows selected basaltic rocks from Hawaii. The dark squares are the alkalic rocks whereas the open diamonds are the tholeiitic rocks. Chemical data obtained from the Basaltic Volcanism Study Project, 1981, pp. 166–167.

Hawaiian tholeiitic basalts typically contain olivine phenocrysts (Fo_{89} to Fo_{70}), and their essential minerals are plagioclase (An_{68} to An_{50}) and pyroxene (augite + enstatite ± pigeonite). The basalts of the midalkaline suite contain olivine (Fo_{87} to Fo_{50}), both as a phenocryst and in the groundmass, plagioclase (An_{68} to An_{39}, with the more sodic minerals crystallising in the groundmass), and a clinopyroxene (see Table 1.1). The latter is usually a diopside that contains more Ca, Fe and Ti than the clinopyroxenes of the tholeiitic basalts. Some basalts of the midalkaline suite also contain kaersutite (amphibole), and phlogopite.

7.2   Major element chemistry of the picrobasalts and basalts

Manson (1967) was the first to study thoroughly the abundance of major elements in the picrobasalts and basalts. He separated this group of rocks from the other igneous rocks by using an empirical chemical screen that contained 33 parameters (see Table 7.1). Limits were, for example, set on various major oxides (i.e. SiO_2 less than 56 per cent, Al_2O_3 between 10.5 and 22.0 per cent), and on the abundance of various normative minerals (i.e. normative quartz less than 12.5 per cent). The idea of using a chemical screen to define a rock, or to define a type of magma, is important, because this approach assumes that at any particular time there is likely to be a consensus among petrologists about the major oxide, or normative, limits they would set to define a particular igneous rock or magma. To obtain such consensus, one normally has to set broad limits for individual parameters, but despite this a rock or magma can easily be defined if the number of parameters is large.

Manson discovered a complete chemical gradation between the different types of basaltic rocks. The only exception was a small group of leucite-bearing rocks (Manson, 1967, p. 227). If these aberrant rocks are plotted on a TAS classification diagram they plot outside the basalt and picrobasalt fields. They form a cluster in the leucite tephrite field.

7.3   Simple basalt system

The CIPW normative minerals purport to be the crystalline phases that are likely to crystallise, under low pressure conditions, from a magma of basaltic chemical composition. Norms are particularly important in igneous petrology because they provide a link between natural rocks and analogous experimental systems. They enable one to compare glassy (hyaline) rocks with holocrystalline rocks, and they often reveal subtle intra-elemental differences as shown in Table 7.1. This table shows that the tholeiitic basalts are likely to contain normative quartz. Alkali basalts normally contain normative nepheline. Calc-alkali basalts have high total normative feldspar contents, whereas the normal mid-oceanic basalts are likely to contain neither normative quartz nor normative nepheline.

In 1962 Yoder and Tilley introduced the idea of the basalt tetrahedron. The end-members in this system are forsterite (Fo), diopside (Di), nepheline (Ne) and quartz (Qz). They provide a straightforward framework for evaluating ideas on the origin and evolution of the various basaltic rocks (see Fig. 7.2). A closer look at this system reveals that albite (Ab) lies on the line nepheline-quartz (i.e. Ne + 2Qz = Ab), and enstatite (En) lies between forsterite and quartz (i.e. Fo + Qz = 2En). The tetrahedron is cut by two planes. One is the critical plane of silica undersaturation and it is defined by forsterite-diopside-albite. The other is the plane of silica saturation and it is defined by enstatite-diopside-albite. Basaltic rocks that plot on the nepheline side of the plane forsterite-diopside-albite are undersaturated in silica, while those that plot on the quartz side of the plane enstatite-diopside-albite are oversaturated in silica.

Table 7.1 Major element composition and CIPW norms of selected picrobasalts and basalts.

	Mean picrobasalt Le Maitre, 1984, p. 250	Mean basalt Le Maitre, 1984, p. 250	Mean basalt Manson, 1967, p. 236	Tholeiitic basalt Le Maitre, 1976, p. 626	Calc-alkali basalt Nockolds & Le Bas, 1977, p. 312	MOR basalt Mid-Atlantic Ridge Melson & Thompson, 1971, p. 429	Alkali basalt (screened mean) Chayes, 1975, p. 548
SiO_2	43.55	49.45	49.2	49.58	51.31	49.21	45.68
TiO_2	2.03	1.8	1.9	1.98	0.88	1.39	2.58
Al_2O_3	10.72	15.35	15.8	14.79	18.6	15.81	14.68
Fe_2O_3	4.31	3.28	3	3.38	2.91	2.21	3.58
FeO	8.58	7.95	8	8.03	5.8	7.19	8.14
MnO	0.21	0.15	0.17	0.18	0.15	0.16	0.17
MgO	18.08	7.71	6.6	7.3	5.95	8.53	8.92
CaO	10.24	11.01	10	10.36	10.3	11.14	10.44
Na_2O	1.44	2.57	2.7	2.37	2.93	2.71	3.11
K_2O	0.51	0.52	1	0.43	0.74	0.26	1.3
P_2O_5	0.33	0.21	0.33	0.24	0.12	0.15	0.52
H_2O^+	0	0	0.9	0.91	0.3	0	0.87
H_2O^-	0	0	0	0.5	0	0	0
Total	100	100	99.6	100.08	99.99	98.76	99.99
CIPW Norms							
Quartz	0	0	0	2.29	0.83	0	0
Orthoclase	3.02	3.06	5.99	2.58	4.39	1.56	7.75
Albite	11.32	21.74	23.15	20.33	24.86	23.22	17.86
Anorthite	21.29	28.81	28.4	28.85	35.52	30.6	22.45
Nepheline	0.47	0	0	0	0	0	4.7
Diopside	21.71	19.75	15.97	17.53	12.08	19.65	21.14
Hypersthene	0	15.63	15.59	19.05	16.13	11.34	0
Olivine	31.33	2.33	2.03	0	0	7.35	14.68
Magnetite	6.24	4.76	4.42	4.98	4.23	3.25	5.25
Ilmenite	3.85	3.42	3.67	3.82	1.68	2.68	4.95
Apatite	0.77	0.5	0.78	0.57	0.28	0.35	1.22

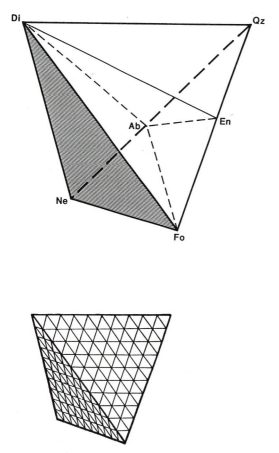

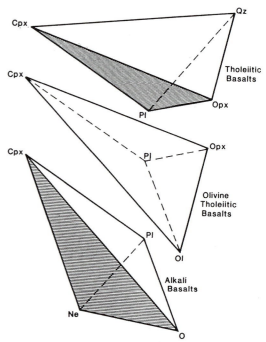

Fig. 7.3 An exploded view of the generalised simple basalt system showing the main types of basaltic rocks (adapted from Yoder and Tilley, 1962, p. 352).

Fig. 7.2 Schematic representation of the basalt tetrahedron (adapted from Yoder and Tilley, 1962, p. 350). All basaltic rocks that plot on the quartz side of the plane diopside-enstatite-albite are oversaturated in silica, whereas those samples that plot on the nepheline side of the plane diopside-forsterite-albite are silica undersaturated.

The saturated basaltic rocks lie between these two important planes.

Yoder and Tilley (1962, p. 352) also introduced what they called the simple basalt system. In this system they substituted the normal basalt minerals, olivine, clinopyroxene, orthopyroxene and plagioclase, for the pure end-members, forsterite, diopside, enstatite and albite. Their generalised system was then used as a classification of the basaltic rocks (see Fig. 7.3). Basalts that contained normative quartz and hypersthene are called oversaturated tholeiites. Those basalts that contained normative hypersthene, but no normative quartz or olivine, are called saturated tholeiites. The name olivine tholeiite is given to the basalts that contained both nor-

mative hypersthene and olivine. Such rocks are usually midalkaline basalts. The name olivine basalt is given to the rocks that contained normative olivine but do not contain normative hypersthene or nepheline. These basalts plot on the critical plane of silica undersaturation and are always midalkaline basalts. The basalts that contained normative olivine and nepheline are called alkali basalts. Depending on their normative nepheline content these rocks may be members of either the midalkaline suite or the alkaline suite.

7.4 Komatiites and related rocks

Nine different types of basalt were investigated by the Basaltic Volcanism Study Project (1981, pp. 2–4). The survey begins with a review of Archean volcanism. Particular attention is focused on the magnesium-rich picrobasalts and basalts that characterise the greenstone belts that developed at this time. Many of these rocks are special because they belong to the komatiitic magmatic suite (Viljoen and Viljoen, 1969;

Arndt and Nisbet, 1982). Their petrological importance was first recognised by the Viljoen brothers who discovered excellent examples of these rocks in the Komati river valley of the Barberton area in northeastern South Africa. Similar rocks have now been found on all the continents, except Antarctica (Hall and Hughes, 1990). It is frequently difficult to piece together the tectonic environment in which these rocks erupted. This is because many greenstone belts display 'horrendous structural complexity' (de Wit and Ashwal, 1986, p. 17).

Komatiites are typically picrobasalts or basalts that contain more than 18 per cent MgO (recalculated on an anhydrous basis), less than 1 per cent $(Na_2O + K_2O)$, and less than 1 per cent TiO_2 (see Fig. 7.4 on p. 168). It is often convenient to divide the komatiites into aluminium-depleted and aluminium-undepleted groups. The aluminium-depleted komatiites are mainly older rocks (about 3.4 giga-years), such as the rocks from the Barberton Province (see Table 7.4). Komatiites of the aluminium-undepleted group typically occur in the younger greenstone belts (about 2.7 giga-years), such as those found in the Munro Township Province in Canada. A survey of komatiites has shown that their abundance has been dwindling since the Archean. The last substantial recorded eruption was about 80 million years ago. It was on the island of Gorgona that lies about 50 km off the southwest coast of Colombia (Echeverria, 1982, pp. 199–209).

Many komatiites contain pillow structures and individual flows may display chilled tops. A particularly distinctive feature of many komatiites is their spinifex texture. Rocks with this texture contain a matted jumble of large, skeletal, platy, bladed or acicular crystals of olivine or pyroxene. This texture usually develops in the upper parts of flows or along the margins of dykes, sills or former lava lakes. In rocks with an olivine spinifex texture the olivine has a plate-like or lattice habit. It forms complex grains made up of a multitude of individual plates of olivine that are aligned parallel to one another. Pyroxene spinifex texture usually contains skeletal needle-like crystals of clinopyroxene (augite and pigeonite) arranged in complex sheaths perpendicular to the margin of the flow. Many spinifex-textured flows display prominent petrographic and chemical layering. The upper layer is usually spinifex-textured and it overlies an olivine cumulate layer.

In most komatiites the primary phases are at least partly replaced by minerals such as albite, chlorite, epidote, serpentine, talc and tremolite. Although such alteration is common there is robust petrographic and geochemical evidence that the bulk major element composition of many of these rocks has not been significantly changed by post-eruption metamorphism. Taken collectively the various petrographic, structural and textural features show that many komatiites were once wholly liquid and congealed from a primitive magma. It is also likely that on being erupted komatiitic lavas were disinclined to nucleate and had a strong tendency towards heterogeneous nucleation. Experimental studies on the melting relations of ultramafic rocks have shown that komatiitic magmas are likely to differ from basaltic magmas in having a much larger temperature interval (about 500°C) between their liquidus and solidus (Arndt, 1976).

The eruption temperature of komatiites that contain about 30 per cent MgO is probably approximately 1600°C (Nisbet, 1982). This suggests that komatiitic magmas are only likely to develop in those parts of the upper mantle where the temperature is abnormally high. Some petrologists regard this observation as evidence of high mantle temperatures in the early Precambrian (Bickle, 1990, p.127). Others have suggested that komatiitic magmas are more likely to segregate out of an upwelling mantle plume where the temperature at its centre is about 350°C hotter than the surrounding materials. This idea is supported by Campbell *et al.* (1989) who published a paper entitled 'melting in an Archean mantle plume, heads it's basalts tails it's komatiites'. They suggest that komatiitic magma is generated in the centre and tail of a mantle plume as it traverses the mantle. Once this magma has formed it pursues an axial route through the head of the plume and it is rapidly propelled up to the surface. Basaltic and picrobasaltic magmas that contain less magnesium would tend to evolve in the cooler peripheral zones of the plume-head.

Experimental data has revealed that komatiitic magmas are likely to begin to segregate from the other materials in a mantle plume at depths of at least 400 km, and then as they rise to higher levels they become superheated (Wei *et al.*, 1990). Superheating usually breaks down the network structures that are normally found in silicate liquids, and when these liquids eventually cool crystal nucleation becomes sluggish (Lofgren, 1983). This sluggish nucleation is likely to be an important factor in the development of the distinctive spinifex texture. Thick olivine-rich flows that contain economic quantities of nickel sulphide often occur at the base of Archean magnesium-rich volcanic sequences. One area where economically important komatiite-hosted iron-nickel ore deposits crop out is the Kambalda Dome, south of Kalgoorlie in the Eastern Goldfields region of Western Australia.

The meimechites are a rare group of rocks that are chemically akin to the komatiites. They are ultrabasic to basic volcanic rocks that contain more than 18 per cent MgO, but differ from normal komatiites by containing more than 1 per cent TiO_2 (Le Maitre et al., 1989, p. 27). A typical meimechite is a porphyritic rock composed of abundant, often altered, olivine phenocrysts set in a groundmass of olivine, clinopyroxene, iron-titanium oxides and glass. The meimechites were originally described from the Meimecha Valley in northern Siberia (Moor and Sheinman, 1946, p. 141). These rocks will be examined in more detail in the chapter on alkaline rocks because they usually contain normative leucite and nepheline (see Table 11.4).

7.5 Mid-oceanic ridge lavas

Basaltic rocks (Hekinian, 1982; Saunders and Norry, 1989; Floyd, 1991) are continually being extruded from or intruded into the 65 000 km long, segmented, mid-oceanic ridge system. Most volcanic and subvolcanic activity is concentrated in a narrow strip along the crest of the ridge that is called the neovolcanic zone. The magmatic processes that operate within the neovolcanic zone produce less than $3\,km^2$ of new crust per annum. This is equivalent to 5 per cent of the mid-oceanic ridge system widening by just 1 m per year. The morphology of the neovolcanic zone changes from areas with high to those with lower spreading rates. Rates of spreading are usually quoted as half-rates, or the rate at which a plate moves away from a spreading ridge. In a few segments of the mid-oceanic ridge system the spreading half-rates are up to 80 mm per year or 80 km per million years.

Away from the influence of fracture zones, mantle plumes, marginal basins and areas of very slow spreading the oceanic crust is exceptionally uniform in thickness. It has a mean thickness of 7 km. The crust is abnormally thin near large-offset fracture zones and areas of very slow spreading. It is anomalously thick in areas associated with active mantle plumes. Near large-offset fracture zones magma generation and movement is likely to be locally inhibited

at sites where cold lithosphere abuts an active spreading axis. Secondly magma is likely to become focused into discrete centres as it ascends towards the surface and it only spreads out along the axis of the ridge when it is near the surface. This implies that magma emplaced into an axial magma chamber is likely to decrease in volume towards the distal ends of the various segments of the neovolcanic zone.

In areas like the East Pacific Rise where the spreading half-rate is over 60 mm per year, volcanic activity is dominated by the eruption of lava from a linear trough that develops along the crest of the axial high. The trough is essentially an axial summit caldera that arises when magma drains out of the axial zone instigating summit collapse (MacDonald et al., 1993). This remarkable type of caldera may be tens of kilometres long, yet only 50 to 500 m wide. Axial highs are elevated because of the buoyancy of the upwelling magma and the thermal expansion of the rocks along the edges of the spreading plates.

MacDonald et al. (1993, p. 30) have likened the axial high in fast-spreading areas to 'a long, skinny, magma filled balloon whose diameter is a sensitive measure of magma supply'. Apparently the elevation of the axial high is not directly related to the accumulation of volcanic products. When an axial high splits, cools and moves off the axis of the ridge it ceases to be a topographic feature. The newly formed segment of oceanic crust is likely to have the same mean thickness, about 7 km, as normal oceanic crust.

In areas where the spreading rate is of intermediate velocity large volcanic edifices are often constructed. This is because the magma supply is periodic enabling new crust to cool, thicken, and become strong enough to support the weight of a large volcanic edifice. With further spreading these volcanoes may be severed with the parts drifting away from one another on different lithospheric plates. In zones with slow spreading rates the neovolcanic zone is usually discontinuous. If spreading slows down even more, and the half-rate drops below about 8 mm per year, little magma rises into the spreading centre and it generates a thinner than normal layer of oceanic crust (Bown and White, 1994, p. 444).

Areas of slow spreading are often characterised by the appearance of rift valleys. Detailed bathymetric charts show that the neovolcanic zone within the inner floor of these rifts is usually dotted with many small volcanoes that are only about 60 m high. Smith and Cann (1992) have proposed that the products from a large number of these volcanoes are likely to coalesce and construct the upper crustal layer found on slow-spreading ridges. Oceanic ridge volcanism is unlike other types of volcanism because it is usually episodic yet it continues for most of the lifetime of an ocean basin. Oceanic basins are likely to survive for more than 150 000 000 years.

In areas of very slow spreading, such as the Mid-Cayman rise in the Caribbean Sea, the volume of magma rising up beneath the neovolcanic zone is so small that it loses much of its heat by conduction. This forces the zone of partial melting to move to depths of between 20 and 40 km. Low volume magmas that equilibrate at these depths usually differ from normal mid-oceanic ridge basalts as they are richer in sodium and have lower CaO/Al_2O_3 ratios (Bown and White, 1994, p. 446).

Currently the whole neovolcanic zone produces about 20 km^3 of new magma each year. Low-velocity seismic anomalies are found beneath all active mid-oceanic ridges (Su *et al.*, 1992, p. 149). They normally extend down to a depth of about 300 km, but in some areas they can be traced into the lower mantle. According to McKenzie and Bickle (1988, p. 660) the primary magma produced beneath the mid-oceanic ridge system is normally a tholeiitic basalt that contains about 10 per cent MgO (see Table 7.1). Magma of this composition probably forms at a temperature of 1280°C, from a 13.5 per cent partial melt, at a depth of about 15 km (0.48 GPa). Bown and White (1994, p. 444) have proposed a slightly higher temperature of 1300°C in the source region. They also suggest temperatures of between 1360°C and 1400°C for the source areas of magmas that form near a major mantle plume, such as the one now beneath Iceland.

Along most of the length of the mid-oceanic ridge system melting results from the passive pulling apart of lithospheric plates and the consequent upwelling and decompression of mantle materials. Because the interior of the mantle is hotter than the solidus of mantle materials at atmospheric pressure, the upward movement of these materials usually results in their partial melting when they rise to a depth of less than about 40 km. Such upwelling is generally sufficiently fast for heat conduction to be small and the volume of material partially melted is controlled by the heat content of the upwelling mantle materials. The melting process beneath the mid-oceanic ridges can generate enough magma to produce an average thickness of oceanic crust of about 7 km, increasing to about 25 km where the ridge system coincides with a mantle plume.

MORBs and MORBs

When first studied the basaltic rocks of the mid-oceanic ridges and ocean floor were considered homogeneous in composition. They were called low-potassium, tholeiitic basalts (Engel and Engel, 1963). Subsequent studies have revealed that several discrete rock types erupt in the neovolcanic zone. These rocks are now usually referred to as N-MORB, T-MORB, E-MORB, ferrobasalt and nepheline-normative alkali basalt. To account for the origin and evolution of these rocks, one has to examine the processes that control magma genesis in this tectonic setting. Most normal mid-oceanic ridge basalt magmas (N-MORB) equilibrate with the solid phases in the upper mantle at depths of about 15 km beneath active spreading ridges (see Table 7.1). Another major group of magmas is generated at higher temperatures and/or at greater depths in areas where plumes coincide with spreading ridges. These are the enriched mid-oceanic ridge basaltic magmas or E-MORB magmas. Some picritic or picrobasaltic magmas are generated at even higher temperatures in this tectonic setting.

Magmas also may be generated beneath spreading ridges located near, but not over, mantle plumes. These are the transitional or T-MORB magmas. Such magmas have compositions intermediate between the normal and enriched types. Regional heterogeneities within the upper mantle also may influence the compositions of magmas generated at spreading ridges. Local variations in the amount of partial melting and the depth of magma segregation also influence the composition of the magma, particularly its incompatible component abundance. Normally the greatest amount of partial melting occurs at the shallowest depths. Even in this environment it usually does not exceed 25 per cent. The nepheline-normative mid-oceanic ridge basalts generally contain somewhat higher concentrations of sodium but normal contents of silica. Such rocks are probably derived from magmas generated by a low degree of partial melting at pressures greater than those required to produce normal mid-oceanic ridge basalts. Some mid-oceanic ridge lavas are interpreted as being products of the mixing of magmas generated at different pressures in an upwelling body of mantle materials.

It is also likely that the compositions of many mid-oceanic ridge lavas have been modified by low pressure fractional crystallisation and magma mixing in high-level magma chambers. In some areas ferroba-saltic magmas are produced by fractional crystallisa-tion in a closed system that essentially excludes oxygen. With further differentiation this magma may separate into two separate immiscible magmas. One is rich in iron but basic in overall composition and the other much more silicic in composition.

The geochemistry and petrology of very few areas of the mid-oceanic ridges is known in the same detail as many land areas. Detailed submarine research such as that carried out in the FAMOUS (French-American Mid-Ocean Undersea Study) area south-west of the Azores or along the East Pacific Rise between $12°00'N$ and $12°30'N$ (Reynolds *et al.*, 1992) is expensive, but especially important. It shows how new segments of oceanic-floor are constructed and how older magmatic rocks are moved sideways to form bands of rocks that are preserved in their proper sequence in time. Detailed submarine studies also provide essential information on the abundance and distribution of off-axis volcanic activity.

The basaltic rocks collected from mid-oceanic ridges seldom crystallise under equilibrium condi-tions. This is because they are extruded into cold sea-water where they cool rapidly. Water has a greater thermal conductivity and specific heat than air. The margins of flows and pillows are usually quenched to glass and even the interiors of the thicker flows have textures that suggest high rates of cooling. Many mid-oceanic ridge basalts seem por-phyritic, but the megacrysts they contain are seldom in equilibrium with the rest of the rock (Natland, 1991, p. 65).

Rapid cooling is probably the main reason under-water lava forms into distinctive pillow-like shapes. When the lava encounters sea-water a chilled rind forms on its outer surface that is flexible enough to change shape as the lava flows. If more lava feeds into the flow pressure inside the rind may cause bul-bous protrusions to form, become detached from the main flow and form separate pillows. If this process is repeated it may produce piles of pillows hundreds of metres thick. Individual pillows usually range in size from about 10 cm to over 1 m in diameter. After solidification they frequently consist of a glassy rim enclosing a more crystalline interior with radial cool-ing joints. Drilling into the upper volcanic layer of the ocean-floor has revealed that it usually contains massive sheet flows, interlayered with pillowed flows. Some pillows fragment and produce local accumula-tions that are generally described as volcanic breccias.

7.6 Flood lavas

Large areas of the continents are covered by broad, flat, lava flows of enormous volume. These flood lavas have been linked with the break-up of conti-nents, climatic change and even the mass extinction of biological species. Normally these immense piles of lava are about a kilometre thick, but locally they may be much thicker. Originally some of these flood lavas covered areas of $1–2\,000\,000\,km^2$ and individual flows incorporated several hundred km^3 of rock.

The ages of these huge bodies of lava extend back for at least 2.7 giga-years to when the extensive Kli-priviersberg Flood Lavas of the Ventersdorp Super-group were erupted in northern South Africa (Bowen *et al.*, 1986, pp. 303–305). Other well documented continental flood lava provinces include the North Australian (or Antrim Plateau) Province (575 million years), the Siberian Platform Province (230 million years), the Karoo Province of southern Africa (195 million years), the Parana Basin Province of eastern South America (130 million years), Deccan Province of India (60 million years), the Thulean (or North Atlantic) Superprovince of the British Isles, Faeroe Islands, Greenland, Baffin Island and the continental margins of northwestern Europe and eastern Green-land (55 million years), the Ethiopian-Yemen Pro-vince (34 million years to the present), the Columbia River Province of Miocene age (17–8 million years) and the less voluminous and younger lavas of the adjoining Snake River Province (Table 7.2).

The Klipriviersberg Flood Lavas are the basal unit in the Ventersdorp Supergroup. They lie directly over the gold-bearing sediments of the Witwatersrand Supergroup. Most lavas are now confined to the Wit-watersrand Basin, where they crop out in a huge area that extends from Evander in the east to Johannes-burg in the north and Welkom in the southwest. Because of their propinquity to the auriferous Wit-watersrand sediments their petrology and geochem-istry have been thoroughly investigated (Bowen *et al.*, 1986; Linton *et al.*, 1993; Myers *et al.*, 1989). When these lavas erupted they extended over an area in excess of $30\,000\,km^2$, and their volume was probably between $23\,000\,km^3$ and $82\,000\,km^3$ (Linton *et al.*, 1993, p. 12). A clear-cut geochemical stratigraphy has been established that divides these lavas into eight units. The basal unit is picritic to meimechitic in composition, whereas most of the overlying flood lavas are basaltic andesites (or basaltic icelandites sensu stricto) (Myers *et al.*, 1989, p. 5).

Table 7.2 Major element compositions of flood lavas from the Columbia River Province. After Basaltic Volcanism Study Project, 1981, p. 82.

Rock type Number	Basalt CP-1	Basalt CP-2	Basalt CP-3	Basalt CP-6	Basalt CP-8	Basalt CP-9	Basalt CP-10	Basalt Mean, n 7	Icelandite CP-4	Icelandite CP-5	Icelandite CP-7	Icelandite Mean, n 3
SiO_2	48.35	48.14	48.71	50.17	50.84	49.98	47.91	49.16	53.92	55.4	52.59	53.97
TiO_2	1.57	2.71	0.96	3.15	1.55	3.49	3.26	2.38	1.83	1.96	2.8	2.2
Al_2O_3	15.49	16.19	16.91	13.23	14.71	12.64	13.23	14.63	14.23	13.67	13.09	13.66
Fe_2O_3	3.26	3.06	2.44	3.52	1.6	2.3	1.84	2.57	2.27	4.98	7.25	4.83
FeO	8.05	9.56	7.77	10.71	8.75	12.67	12.49	10	9.1	7.16	5.88	7.38
MnO	0.17	0.18	0.2	0.22	0.19	0.22	0.24	0.2	0.2	0.23	0.23	0.22
MgO	7.03	5.39	7.86	4.41	6.99	4.27	5.8	5.96	4.25	3.34	2.67	3.42
CaO	9.92	8.66	11.06	8.2	10.48	8.35	9.99	9.52	8.42	6.95	6.01	7.13
Na_2O	2.76	3.3	2.48	2.85	2.3	2.46	2.38	2.65	2.92	2.41	3.18	2.84
K_2O	0.51	0.9	0.28	1.26	0.66	1.33	0.9	0.83	1.35	1.81	2.52	1.89
P_2O_5	0.24	0.36	0.18	0.67	0.25	0.56	0.78	0.43	0.31	0.32	0.85	0.49
H_2O total	1.52	0.98	0.4	0.85	0.53	0.67	0.5	0.78	0.83	1.33	2.56	1.57
CO_2	0.05	0.11	0.04	0.05	0.06	0.04	0.08	0.06	0.04	0.05	0.08	0.06
Total	98.92	99.54	99.29	99.29	98.91	98.98	99.4	99.19	99.67	99.61	99.71	99.66
CIPW Norms												
Quartz	0	0	0	4.76	1.95	4.64	0	0.18	6.69	15.36	12.29	11.43
Orthoclase	3.1	5.43	1.67	7.56	3.97	7.99	5.38	5.02	8.07	10.88	15.33	11.39
Albite	23.99	28.51	21.23	24.5	19.8	21.18	20.38	22.95	25.01	20.76	27.71	24.5
Anorthite	29.15	27.27	34.59	19.9	28.34	19.86	23.03	26.18	22	21.52	14.42	19.32
Diopside	16.07	11.69	16.02	14.09	18.71	15.56	18.24	15.86	15.14	9.38	8.29	11.04
Hypersthene	16.39	10.6	13.88	16.31	21.27	19.28	22.01	20.32	15.49	10.2	3.58	9.72
Olivine	2.79	6.88	6.75	0	0	0	0.13	0	0	0	0	0
Magnetite	4.87	4.54	3.59	5.2	2.37	3.4	2.71	3.82	3.34	7.34	10.85	7.16
Ilmenite	3.07	4.22	1.85	6.09	3	6.76	6.28	4.64	3.53	3.8	5.49	4.27
Apatite	0.57	0.86	0.42	1.59	0.59	1.33	1.84	1.03	0.73	0.76	2.04	1.17

It is frequently suggested that the origin and evolution of many large flood lava provinces is directly related to the arrival of the head of an extensive mantle plume at the base of the lithosphere (Morgan, 1981; Campbell and Griffiths, 1990). This hypothesis is supported by the recognition that the eruption of flood lavas is often the first event in a series of magmatic events that eventually result in the construction of linear chains of volcanoes. It is also normal for the eruption of flood lavas to take place extremely rapidly.

The complex Thulean Superprovince is associated with the opening of the North Atlantic, the Deccan Province is associated with the detachment of the Seychelles Platform from the western part of India and the Parana Basin Province is linked to the separation of South America from southwestern Africa. Erlank *et al.* (1984) have shown conclusively that the magmatic rocks of the Etendeka Plateau of northwestern Namibia are the missing eastern edge of the Parana Basin Province of Brazil. If the magmatic rocks of the Etendeka Plateau are excluded the lavas of the Parana Basin Province (Serra Geral Formation) cover an area of approximately $1\,200\,000\,\mathrm{km}^2$. If comagmatic dykes, sills and outliers are considered the size of this province increases to about $2\,000\,000\,\mathrm{km}^2$. These lavas have an average thickness of $650\,\mathrm{m}$, and a volume of approximately $650\,000\,\mathrm{km}^3$.

Most of the lavas of the Parana Basin Province occur in southwestern Brazil with smaller outcrops in Uruguay, Paraguay and Argentina. According to Bellieni *et al.* (1983, p. 370) the volcanic rocks of the Parana Basin Province consist of tholeiitic basalts, basaltic icelandites (sensu stricto), icelandites (sensu stricto), dacites and rhyolites. The dominant rock is basaltic icelandite. Both the Karoo and Parana Basin provinces contain high-titanium and low-titanium lavas. In both areas the high-titanium lavas erupted in cratonic areas underlain by ancient lithosphere, whereas the low-titanium lavas typically erupted in areas containing younger rocks.

Plume-heads and flood lavas

The introduction of a huge, possibly $2000\,\mathrm{km}$ in diameter, plume-head beneath a lithospheric plate is likely to cause thermal erosion at its base and doming, stretching and fracturing of the overlying rocks. These processes are likely to promote decompression melting in the plume-head resulting in the rapid onset of flood volcanism. In continental areas the thermal erosion of the lithosphere is likely to be coupled with assimilation resulting in noticeable changes in the chemical composition of the ascending magma. The lithosphere beneath the Karoo Province is known from the study of xenoliths to be highly complex and chemically inhomogeneous. This has resulted in the generation of diverse magmas (Cox, 1988, p. 239).

According to Bristow (1988, p. 33) the focus of magmatic activity and volcanism in the Karoo Province was located along lines of pre-existing structural weakness between old cratonic nuclei. The cratonic areas were probably underlain by thick roots of lithospheric materials and rising magma was shepherded between the cratonic blocks with their extensive roots (Ballard and Pollack, 1987, p. 253). The vents from which flood lavas issue are often difficult to locate, but are likely to be fissures that are hidden by later flows. Great thicknesses of Karoo lavas are exposed in the spectacular Drakensberg Mountains of the Kingdom of Lesotho and adjoining areas of South Africa (see Plate 7.1). Field research in this area has revealed that dykes are likely to continue feeding magma into a flow until cooling and congelation impedes its process. The magma then breaks through the congealing flow and generates a new flow above. A single inconspicuous dyke probably can feed several, discrete, superimposed flows.

The Ethiopian-Yemen Province is still volcanically active. It is on the Afro-Arabian dome (Gass, 1970, p. 286) that splits into three at the Afar triple junction. This is the meeting place of the African Rift System, the Red Sea neovolcanic zone and Gulf of Aden rift and neovolcanic zone. In the Ethiopian-Yemen Province the subaerial volcanic pile is normally between 500 and $1500\,\mathrm{m}$ thick, but locally it may attain a thickness of $3000\,\mathrm{m}$. Prior to the opening of the Red Sea the basalts of the Yemen Plateau were joined with those of Ethiopia. The province contains about $300\,000\,\mathrm{km}^3$ of Tertiary flood volcanic rocks. They are mainly basalts, but the sequence also contains some 10 per cent of trachytic and rhyolitic pyroclastic flow deposits. Most of the rocks of the Ethiopian-Yemen Province were extruded during three episodes of vigorous activity, between 32 and 21 million years, between 13 and 9 million years, and between 4.5 and 1.5 million years (Mohr and Zanettin, 1988). The volcanically active Afar triple junction and spreading zone is particularly interesting because it is part of the Ethiopian-Yemen continental flood basalt province, it is directly linked to two zones of active sea-floor spreading, it is directly connected to the active African continental rift system to the south and south-west, and it appears to lie above a mantle plume that has been active, but approximately stationary relative to the crust, for over 35 million years.

A feature of many continental flood basalt provinces is the close association of lavas with large numbers of basaltic dykes and sills. This observation has led to the proposal that some extensive Proterozoic dyke swarms may now be all that remains of what had once been extensive fields of continental flood lavas. Most continental flood basalts are low-MgO, quartz normative tholeiites and they often occur interbedded with more differentiated basaltic icelandites and icelandites (or tholeiitic basaltic andesites and tholeiitic andesites). According to Yoder (1988, p. 151) an important reason why continental flood basalts differ from mid-oceanic ridge basalts is that magma finds it more difficult to penetrate a thicker and usually cooler continental lithosphere than to break through a divergent plate boundary that is fracturing under tension. The impounding of magma beneath a continent may promote fractional crystallisation in a closed magmatic system resulting in the generation of iron-rich residual magmas. In the mid-oceanic ridge environment the passage of magma is generally unimpeded and this enables the magma to retain a more primitive, magnesium-rich composition.

Magee and Head (1995, pp. 1549–1550) have compared the processes for generating flood lavas on Venus and Earth. They suggest that the volume of flood lava associated with the huge rift zone that borders Lada Terra on Venus was enhanced by the rifting process that encouraged decompression melting. It has also been proposed that the flood lavas of Venus are likely to concentrate in areas of thinned, rifted, lithosphere. This is because in such areas the volume of magma that segregates from an upwelling diapir is larger because the process of decompression melting is prolonged by the diapir rising closer to the surface.

Oceanic flood lavas

After surveying the abundance, distribution and origin of continental flood lavas it is instructive to speculate on whether similar large floods of lava can be found in the oceanic environment. The Thulean Superprovince shows that such lavas can erupt in both settings. In the Icelandic segment of this superprovince a dwindling type of flood volcanism is still active. As we have already noted in Section 4.4 in 1783 the Lakagigar eruption in Iceland opened up a 10 km long fissure, and for 50 days basaltic lava was extruded at a rate of about 2300 m^3/s (Thorarinsson, 1969). This huge eruption released not only a flood of lava, but also so much volcanic ash, gas and aerosols that it caused a bluish haze to settle over Iceland and much of Europe (Wood, 1992).

Oceanic flood lavas are likely to form huge oceanic plateaus, such as the Ontong Java Plateau, the Manihiki Plateau and the Kerguelen-Heard Plateau (Storey et al., 1992). These plateaus are broad almost flat-topped, elevated parts of the ocean. They usually grade into linear aseismic ridges or seamount chains. The large oceanic plateaus should be separated from the aseismic ridges because the former were constructed in a comparatively short time. It is likely that the magmas used to form the large oceanic plateaus were produced by partial melting in the heads of huge mantle plumes (see Section 3.5). The eruption of these enormous reservoirs of magma produced the largest volcanic outbreak of the past 200 million years.

The Ontong Java Plateau is estimated to contain about 25 times more lava than the continental flood lava that erupted in the Deccan Province. This probably means that the Ontong Java mantle plume rose beneath a thinner layer of lithosphere thus experiencing a longer period of decompression melting. This increased the size of the plume-head, resulting in the generation of a megaplume. The emplacement of a megaplume beneath an area of oceanic lithosphere is likely to initiate widespread crustal heating and prodigious volcanic activity. According to Caldeira and Rampino (1991) such an event would have global repercussions. It might locally increase the rate of oceanic floor spreading, boost global climatic change and trigger a worldwide rise in sea level.

It is interesting to speculate about what is likely to happen when a large oceanic plateau collides with an island arc or a continental margin. The oceanic plateau probably will append itself to the island arc or continental margin. The Ontong Java Plateau is currently colliding with the Solomon Island Arc. Its leading edge has been obducted onto the arc, but its huge mass is still pushing the arc and refashioning the shape of the local subduction system. The Ontong

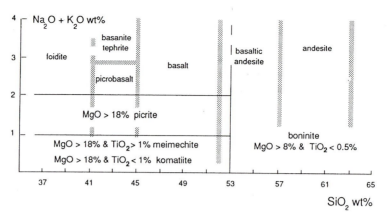

Fig. 7.4 Classification and nomenclature of the high-Mg volcanic rocks, boninite, komatiite, meimechite and picrite. After Le Maitre *et al.*, 1989, p. 27.

Java Plateau is about $12\,000\,000\,km^2$ in area. If one was to assume that $25\,000\,000\,km^2$ of oceanic plateau is produced every 200 million years then one might assume that over a period of 3.5 giga-years about $440\,000\,000\,km^2$ is produced. This is larger than the area currently covered by all the oceans and marginal seas (about $360\,000\,000\,km^2$). This calculation shows that one has always to be mindful of large oceanic plateaus when one speculates about the evolution of the continents.

In this Section we have established that often during the evolution of the Earth there have been enormous eruptions of lavas. Some petrologists like to use the term large igneous province, or LIP, to describe the areas covered by these flood lavas (Coffin and Eldholm, 1993). This phrase is valuable because it can be used to describe both large volcanic and large plutonic provinces. Extensive comagmatic dyke swarms, large bodies of layered igneous rocks and large oceanic plateaus of magmatic origin all evolved from huge volumes of magma. Future research will show whether the rocks from these diverse large igneous provinces can all be linked together as products of mantle megaplumes.

Flood lavas are characteristic of all the terrestrial planets. For example most of the surface of Venus is covered by flood lavas and these lavas and other volcanic landforms suggest that mantle plumes are a particularly important mechanism for transporting internal heat to the surface of this planet. The mare basins of the Moon are all filled with old flood lavas. Mars with its huge lava fields and its immense volcanoes abounds in large igneous provinces.

7.7   Picrites

Picrites are a broad group of volcanic rocks that contain more than 18 per cent MgO, less than 53 per cent SiO_2 and 1–2 per cent total alkalies (see Fig. 7.4). They include basalts, picrobasalts and some foidites. Volcanic rocks that contained more than 18 per cent MgO, less than 53 per cent SiO_2 and over 2 per cent total alkalies may be called alkalic picrites (Cox *et al.*, 1965, p. 148). According to Tröger (1935) a typical picrite contains 51 per cent olivine, 37 per cent clinopyroxene, 8 per cent bytownitic plagioclase, and lesser amounts of amphibole, phlogopite, iron-titanium oxides, apatite, picotite and analcime. Picrites normally contain more olivine than basalts, but not enough olivine and other mafic minerals for them to be classified as ultramafic rocks. Many picrites are picrobasalts as they contain between 41 and 45 per cent silica. Tables 7.3 and 7.4 show the major element and CIPW normative compositions of three typical picrites. None of these rocks are ultramafic rocks, as they all contain over 20 per cent normative feldspar.

The picrites occupy a special place in igneous petrology because experimental studies have revealed that picritic melts are likely to be produced by moderate amounts of partial melting of a lherzolitic source rock at the ambient pressures typical of the low velocity zone in the upper mantle (Jaques and Green, 1980, p. 287). Marsh and Eales (1984, p. 27) have proposed that the huge volumes of Lesotho-type flood lavas in the Karoo Province evolved

Table 7.3 Mean major element composition and CIPW norms of some Hawaiian lavas. After the Basaltic Volcanism Study Project, 1981, pp. 166–167.

	Picrite Mean ($n = 2$)	Alkali basalt Total ($n = 5$)	Tholeiitic basalt Mean ($n = 15$)
SiO_2	47.17	47.46	50.7
TiO_2	1.65	2.71	2.48
Al_2O_3	9.05	15.27	13.2
Fe_2O_3	2.66	4.11	2.54
FeO	9.04	7.92	8.83
MnO	0.17	0.19	0.17
MgO	19.96	6.89	8.02
CaO	7.71	9.61	10.57
Na_2O	1.54	3.35	2.32
K_2O	0.33	1.12	0.48
P_2O_5	0.19	0.51	0.26
H_2Ototal	0.27	0.71	0.27
CO_2	0.06	0.06	0.08
Total	99.77	99.9	99.93
CIPW Norms			
Quartz	0	0	2.54
Orthoclase	1.96	6.69	2.85
Albite	13.1	27.25	19.72
Anorthite	16.9	23.58	24.3
Nepheline	0	0.76	0
Diopside	16.15	17.27	21.59
Hypersthene	19.47	0	19.94
Olivine	24.92	13.95	0
Magnetite	3.89	4.08	3.71
Ilmenite	3.16	5.22	4.74
Apatite	0.45	1.2	0.61

from primary picritic magmas. Minor amounts of picrite have been described from many continental flood lava provinces. Bristow (1984, p. 117) has, for example, described the various high-Mg rocks of the Karoo Province. Many of these rocks contain more than 2 per cent alkalies and probably should be named alkalic picrites.

Enormous volumes of picritic lava erupted during the initial stages in the development of the rift between West Greenland and Baffin Island (Clarke and Pedersen, 1976, p. 381). These lavas are a part of the complex Thulean Superprovince (see Table 7.4). According to Gill *et al.* (1995, p. 35) the flood lavas and in particular the picrites of this superprovince can be linked to two discrete mantle plumes, an early Baffin Bay plume and a later Icelandic plume. This proposal explains why large volumes of high temperature picritic magma could erupt in West Greenland and Baffin Island that would have been located at the outer margins of a single plume head

(White and McKenzie, 1989). In the two-plume model picritic magmas segregate and remained undifferentiated in the higher temperature axial zones of the separate plumes.

Picrites often crop out as minor intrusions. Many such intrusions have been discovered in the Hebridean Subprovince of the British Tertiary Province (Sabine and Sutherland, 1982, pp. 532–533). Some of these picrites such as the picritic layer at the base of the Garbh Eilean sill in the Shiant Isles are olivine cumulates. Other intrusions have chilled salvages and high temperature metamorphic aureoles. They probably evolved from a primary, or close to primary, picritic magma (Drever and Johnston, 1967).

7.8 Continental rifts and paleorifts

Magmatism and basaltic volcanism is an integral part of continental extension and rifting. There is usually no clear cut chemical distinction between the continental flood volcanic rocks and the continental rift volcanic rocks, except that the latter are likely to include more rocks that belong to the midalkaline and alkaline suites. Currently the Afar triple-junction area of Ethiopia contains both flood lavas and continental rifting. Continental rifting is usually linked to the incipient, or abortive, break-up of continental or super-continental plates. Some areas where continental rifting is currently active include the East African Rift System, the Rhine Graben in western Europe, the Baikal Rift in Siberia and the Rio Grande Rift of New Mexico. Large continental rifts such as the East African Rift System contain huge volumes of volcanic rocks, but other rifts such as the Rio Grande Rift have produced much smaller volumes of volcanic rocks.

The East African Rift System contains enormous volumes of midalkaline and alkaline volcanic rocks (see Sections 10.5 and 11.4), the Oslo Paleorift contains a wide range of midalkaline volcanic and plutonic rocks (see Section 10.9), whereas most of the basaltic rocks of the Rio Grande Rift are hypersthene and olivine normative tholeiitic basalts (Basaltic Volcanism Study Project, 1981, p. 115). Such rift and paleorift systems usually contain large volumes of basalt, and a wide range of midalkaline and alkaline rocks that will be examined in Chapters 10 and 11.

Table 7.4 Selected major element compositions and CIPW norms of komatiites and picrites.

	Komatiite W. Australia 3.5 Ga, Cattell & Taylor, 1990, p. 12	Komatiite South Africa 3.5 Ga, Cattell & Taylor, 1990, p. 12	Komatiite W. Australia 2.7 Ga, Cattell & Taylor, 1990, p. 12	Picrite Zimbabwe Cox *et al.* 1965, p. 148	Picrite Baffin Island Clarke, 1970 No 5, Table 1
SiO_2	46.62	46.63	48.47	44.97	45.86
TiO_2	0.8	0.31	0.33	1.48	0.77
Al_2O_3	5.46	3.3	7.81	5.74	10.98
Fe_2O_3	14.78	14.73	10.43	3.45	11.59
FeO	*	*	*	10.18	*
MnO	0.25	0.18	0.18	0.18	0.18
MgO	23.45	31.34	24.18	24.15	20.03
CaO	8.49	3.3	8.17	6.26	9.35
Na_2O	0.06	0.09	0.39	0.91	1.06
K_2O	0.02	0.05	0.03	0.75	0.08
P_2O_5	0.07	0.07	0.01	0.23	0.09
H_2O^+	0	0	0	0.87	0
H_2O^-	0	0	0	0.29	0
CO_2	0	0	0	0.5	0
Total	100	100	100	99.96	99.99
CIPW Norms					
Orthoclase	0.12	0.29	0.18	4.51	0.47
Albite	0.5	0.75	3.26	7.83	8.88
Anorthite	14.39	8.35	19.26	9.52	24.71
Diopside	21.44	5.87	16.52	16.31	16.49
Hypersthene	31.29	40.29	32.62	18.08	11.45
Olivine	26.36	39.48	24.31	35.23	33.01
Magnetite	4.24	4.23	3	5.1	3.33
Ilmenite	1.5	0.58	0.62	2.87	1.45
Apatite	0.16	0.16	0.23	0.55	0.21

*Total iron as Fe_2O_3.

7.9   Intraplate basalts

The oceanic intraplate basalts usually crop out on oceanic islands and seamounts. Chains of intraplate volcanoes are found also in many continental areas. During the past 35 million years a double chain of central-type volcanoes has formed in eastern Australia. The evolution of these volcanoes is particularly intriguing as they evolved concurrently with another two chains of volcanoes that lie off the eastern shore of Australia in the Tasman Sea (Johnson, 1989).

Madeira, Canary, Cape Verde, Saint Helena, Tristan da Cunha, Reunion, Mauritius, Comoros, Tahiti, Hawaii and Guadaloupe are all well-known intraplate volcanic islands, or groups of islands. In the Western Pacific there are many islands and seamounts that represent extinct intraplate volcanoes. As we have already noted the volcanically productive Hawaiian islands contain rocks that belong to the subalkaline tholeiitic, midalkaline and alkaline suites

(see Table 7.3). On most small-volume, oceanic intraplate islands the rocks mainly belong to the mid-alkaline or alkaline suites.

Present-day intraplate volcanism is usually related to local, somewhat small-scale mantle plume activity. Often the trace of these mantle plumes can be tracked for considerable distances. The Louisville seamounts that cut across the South Pacific to the east of New Zealand are admirable examples of this process. Some petrologists consider this mantle plume to be a relic of the megaplume that produced the Ontong Java Plateau.

Geochemical investigations have revealed that the basalts from the smaller oceanic islands and seamounts are more akin to basalts from the small-volume continental magmatic provinces than normal mid-oceanic ridge basalts. According to McKenzie and O'Nions (1995, p. 154) the basalts from the smaller oceanic islands and seamounts usually contain isotopic signatures that show that at least some of their constituent materials were resident in the continental lithosphere long before the present oceanic basins formed. To explain this anomaly it has been suggested that the critical continental components become entrained in the upper mantle convective system. Once entrained they remain in the outer boundary layers of the convective system for several cycles. This makes these critical components available for incorporation into magmas that are generated beneath the smaller oceanic islands and seamounts.

7.10   Subduction system basalts

Basalts are commonly extruded on island arcs and continental margin arcs (Smellie, 1994). The volume of magma generated by subduction systems is huge. After the mid-oceanic ridge system the various subduction systems rank second in the amount of magma they produce. The compositions of the materials erupted are very variable as they range from the tholeiitic suite, via the calc-alkaline suite, to mid-alkaline suite. Most of the basalts that erupt in this tectonic setting belong to the high-alumina or calc-alkaline suite (see Table 7.1). Generally basalts make up only 20 per cent of the total volume of volcanic rocks extruded above subduction systems. The dominant volcanic rocks are usually basaltic andesites and andesites. These rocks are examined in more detail in Chapters 8 and 9 that survey the magmas generated at convergent plate margins.

7.11   Basaltic meteorites

Some achondritic meteorites (Dodd, 1981; McSween, 1987, p. 101–155), such as the eucrites and shergottites, have basaltic modal compositions (see Table 2.5). Other meteorites such as the howardites and mesosiderites contain a mixture of clasts that include fragments of basalt. Eucrites are the most abundant type of basaltic meteorite, yet by some curious quirk the name given to these meteorites that look like terrestrial basalts is derived from the Greek word 'eukritos' that means easily distinguished. Eucrites are usually composed of only two primary minerals, Ca-rich plagioclase (An_{80-90}) and the Ca-poor pyroxene (pigeonite). Their main claim to fame is that they are the oldest known basalts, with a mean age of 4.54 giga-years. It is usually postulated that they originally erupted on the surface of a large asteroid.

Remote sensing has confirmed that basaltic rocks crop out on the surface of the belt asteroids, such as the large asteroid Vesta. In contrast to the eucrites, the rare shergottites are far younger (circa 1300 million years), and they are usually considered to have originated on the surface of Mars (see Section 2.11).

7.12   Lunar picrobasalts and basalts

If picrobasalts and basalts from the maria are compared with the normal subalkaline basalts of Earth they are found to be depleted in silicon, aluminium, sodium and potassium. Individual mare lavas contain variable amounts of titanium with TiO_2 varying between 0.5 per cent and 13 per cent (see Table 2.6). The essential minerals in mare picrobasalts and basalts are clinopyroxene, plagioclase and iron-titanium oxides. Olivine is often an important minor phase. The mineralogy of lunar rocks is less complex than their counterparts on Earth because they lack hydrous phases.

The mare are contained within the older and more densely-cratered areas of the lunar highlands. Most highland rocks are older than 4.0 giga-years, and have been pulverised by large numbers of impact events. Over 60 per cent of the samples of the lunar highlands are breccias. Most of the remaining samples are of rocks produced by impact melting. A few lunar highland rocks have been called lunar highland basalts. They are readily distinguished from mare basalts because they have much higher Al_2O_3/CaO ratios (Heiken *et al.*, 1991).

7.13   Petrogenesis of the picrobasaltic and basaltic rocks

Experimental studies have revealed that eutectic-like partial melting of typical upper mantle source rocks such as lherzolite (see Table 3.1) yields the common types of primitive picrobasaltic and basaltic magmas. Differences in composition can be replicated by changing the pressure or the abundance and composition of the volatile and incompatible components in the source material at the time of melt segregation. Generally high pressures and sources enriched in CO_2 produce alkali picrobasaltic and basaltic magmas. Lower pressures, often accompanied by H_2O-rich sources, produce the subalkalic varieties of picrobasaltic and basaltic magmas (Yoder, 1976, p. 205).

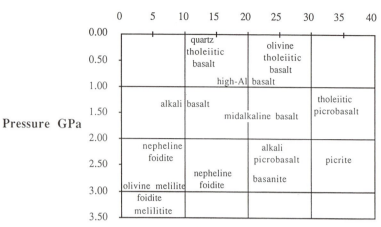

Fig. 7.5 A simplified petrogenic grid that attempts to show the pressures at which various picrobasaltic, basaltic and alkaline magmas segregate from a standard upper mantle rock. Partly based on Green, 1970, p. 26.

Green (1970, p. 26) introduced a simplified petrogenic grid that shows the physical condition in which various picrobasaltic and basaltic magmas are likely to form with the partial melting of a standard upper mantle rock (see Fig. 7.5).

The partial melting of natural multiphase upper mantle rocks probably proceeds via several different invariant points and the magmas produced at each of these points would have uniform, but discrete, major element compositions. Changes in pressure not only change the phases present in the mantle rocks, they also change the composition of the initial magma generated by partial melting. This idea can be illustrated by examining the system forsterite-nepheline-silica that occupies the base of the basalt tetrahedron. At low pressures the melt produced in this system is oversaturated in silica, whereas at high pressures the magma is saturated in nepheline therefore undersaturated in silica (see Fig. 7.6).

7.14 Main Lava Series from the Isle of Skye

The rocks of the Main Lava Series from Skye in the Inner Hebrides of Scotland are a well-known and historically important group of rocks (Emeleus and Gyopari, 1992; Harker, 1904; Thompson et al., 1982). Their origin and evolution will be described in extra detail because so much is now known about these rocks. In the past the typical basic

rocks of this series have been called transitional basalts because they lie astride the critical plane of silica undersaturation (Yoder and Tilley, 1962). Some rocks are nepheline-normative while others are hypersthene-normative.

The primary magnesium-rich (13 to 15 per cent MgO) magmas of the Main Lava Series were probably generated by decompression partial melting in a body of abnormally hot, effectively anhydrous, mantle material (Scarrow and Cox, 1995; Thompson et al., 1982). When this partial melting took place the source rocks were part of the Icelandic mantle plume (see Section 3.5). The melting process began in the garnet stability field at a depth of about 110 km, and continued to a depth of about 60 km. It is likely that most primary magma segregated at depths of between 90 and 70 km. At 60 km the presence of the more rigid continental lithosphere deterred the further rise of the mantle plume and this impeded further magma generation. The temperature in the source area was probably in the range 1510°C to 1390°C (Scarrow and Cox, 1995, pp. 17–18). The nepheline-normative primary magma segregated at greater depths than the hypersthene-normative magma. Most magmas segregated at the top of the zone of partial melting. While rising towards the surface these magmas are estimated to have cooled at a rate of about three degrees Celsius per kilometre. Their eruption temperature was about 1200°C. While in transit to the surface the composition of most magmas changes because of the crystallisation and removal of some olivine and a little magnesiochromite or chromite.

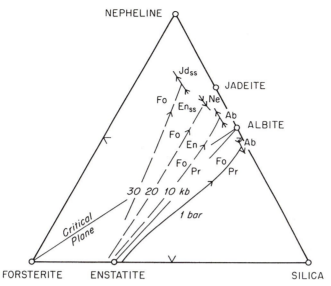

Fig. 7.6 The system forsterite-nepheline-silica that occupies the base of the basalt tetrahedron. Note at low pressures the melt produced in this system is oversaturated in silica, whereas at high pressures the magma is saturated in nepheline and undersaturated in silica. After Yoder, 1976, p. 132, and others.

References

Arndt, N.T., 1976. Melting relations of ultramafic lava (komatiites) at 1 atmosphere and high pressure. *Carnegie Institution of Washington Yearbook*, 75, pp. 555–562.

Arndt, N.T. and Nisbet, E.G. (editors), 1982. *Komatiites*. George Allen and Unwin, London, 526 pp.

Ballard, S. and Pollack, H.N., 1987. Diversion of the heat by Archean cratons: a model for southern Africa. *Earth and Planetary Science Letters*, 85, pp. 253–264.

Basaltic Volcanism Study Project, 1981. B*asaltic Volcanism on the Terrestrial Planets*, Pergamon Press, Incorporated, New York, 1286 pp.

Bellieni, G., Brotzu, P., Comin-Chiaramonti, M., Ernesto, A.J., Melfi, I.G., Piccirillo, E.M. and Stolfa, D., 1983. Petrological and paleomagnetic data on the plateau basalts to rhyolite sequences of the Southern Parana Basin (Brazil). *Anual Academia Brazil, Ciencia*, 55 (4), pp. 355–383.

Bickle, M.J., 1990. Mantle evolution. In Hall, R.P. and Hughes, D.J., (editors) *Early Precambrian Basic Magmatism*, pp. 111–135, Blackie, Glasgow, 486 pp.

Bowen, T.B., Marsh, J.S., Bowen, M.P. and Eales, H.V., 1986. Volcanic Rocks of the Witwatersrand Triad, South Africa. *Precambrian Research*, 31, pp. 297–324.

Bown J.W. and White, R.S., 1994. Variation with spreading rate of oceanic crustal thickness and geochemistry. *Earth and Planetary Science Letters*, 121, pp. 435–449.

Bristow, J.W., 1984. Picritic rocks of the North Lebombo and South-East Zimbabwe. In Erlank, A.J. (editor) *Petrogenesis of the volcanic rocks of the Karoo Province*, pp. 105–123, Special Publication of the Geological Society of South Africa, Number 13, 395 pp.

Bristow, J.W., 1988. Flood basalts in new perspective. *Nuclear Active*, 39, pp. 26–36.

Caldeira, K. and Rampino, M.R., 1991. The Mid-Cretaceous Super Plume, carbon dioxide and global warming. *Geophysical Research Letters*, 18 (6), pp. 987–990.

Campbell, I.H. and Griffiths, R.W., 1990. Implications of mantle plume structure for the evolution of flood basalts. *Earth and Planetary Science Letters*, 99, pp. 79–93.

Campbell, I.H., Griffiths, R. and Hill, R.I., 1989. Melting in an Archean mantle plume: heads its basalts tails its komatiites. *Nature*, 339, pp. 697–699.

Cattell, A.C. and Taylor, R.N., 1990. Archaean basic magmas. In Hall, R.P. and Hughes, D.J. (editors) *Early Precambrian Basic Magmatism*, pp. 11–39, Blackie, Glasgow, 486 pp.

Chayes, F., 1975. Statistical Petrology. *Carnegie Institution of Washington Yearbook*, 74, pp. 542–550.

Clarke, D.B., 1970. Tertiary basalts from Baffin Island: possible primary magmas from the mantle. *Contributions to Mineralogy and Petrology*, 25, pp. 203–224.

Clarke, D.B. and Pedersen, A.K., 1976. Tertiary volcanic province of West Greenland. In Escher, A. and Watt, W.S. (editors) *Geology of Greenland*, pp. 364–385, Geological Survey of Greenland, 603 pp.

Coffin, M.F. and Eldholm, O., 1993. Large Igneous Provinces. *Scientific American*, 269(4), pp. 26–33.

Cox, K.G. 1988. The Karoo Province. In McDougall, J.D. (editor) *Continental Flood Basalts*, pp. 239–271, Kluwer Academic Publishers, Dordrecht, 341 pp.

Cox, K.G., Johnson, R.L., Monkman, L.J., Stillman, C.J., Vail, J.R. and Wood, D.N., 1965. The Geology of the Nuanetsi Igneous Province. *Philosophical Transactions of the Royal Society of London*, A 257, pp. 71–218.

de Wit, M.J. and Ashwal, L.D., 1986. Summary of Technical Sessions. In de Wit, M.J. and Ashwal, L.D. (editors) *Workshop on Tectonic Evolution of Greenstone Belts*, pp. 13–30, Lunar and Planetary Institute, Houston, Texas, 227 pp.

Dodd, R.T., 1981. *Meteorites: A Petrologic-Chemical Synthesis*. Cambridge University Press, Cambridge, 368 pp.

Drever, H.I. and Johnston, R., 1967. Picritic minor intrusions. In Wyllie, P.J. (editor) *Ultramafic and Related Rocks*, pp. 71–82, John Wiley and Sons, New York, 464 pp.

Echeverria, L.M., 1982. Komatiites from Gorgona Island, Colombia. In Arndt, N.T. and Nisbet, E.G. (editors), *Komatiites*. Chapter 15, pp. 199–209, George Allen and Unwin, London, 526 pp.

Emeleus, C.H. and Gyopari, M.C., 1992. The Isle of Skye. In *British Tertiary Volcanic Province*, pp. 13–68, U.K. Geological Conservation Review Series, Chapman and Hall, London, 259 pp.

Engel, C.G. and Engel, A.E.J. 1963. Basalts dredged from the north-eastern Pacific Ocean. *Science*, 140, pp. 1321–1324.

Erlank, A.J. (editor), 1984. *Petrogenesis of the Volcanic Rocks of the Karoo Province*. Special Publication of the Geological Society of South Africa, 13, 395 pp.

Erlank, A.J., Marsh, A.R., Duncan, R., Miller, R.M., Hawkesworth, P.J. and Rex, D.C., 1984. Geochemistry and petrogenesis of the Etendeka volcanic rocks from SWA/Namibia. In Erlank, A.J. (editor), *Petrogenesis of the Volcanic Rocks of the Karoo Province*, pp. 195–

245, Special Publication of the Geological Society of South Africa. 13, 395 pp.

Floyd, P.A. (editor), 1991. *Oceanic Basalts*. Blackie, Glasgow, 456 pp.

Gass, I.G., 1970. Tectonic and magmatic evolution of the Afro-Arabian dome. In Clifford, T.N. and Gass, I.G. (editors) *African Magmatism and Tectonics*. pp. 285–300, Oliver and Boyd, Edinburgh, 461 pp.

Gill, R.C.O., Holm, P.M. and Nielsen, T.F.D., 1995. Was a short-lived Baffin Bay plume active prior to initiation of the present Icelandic plume? Clues from the high-Mg picrites of West Greenland. *Lithos*, 34, pp. 27–39.

Green, D.H., 1970. The origin of basaltic and nephelinitic magmas. *Transactions of the Leicester Literary and Philosophical Society*, 64, pp. 26–54.

Hall, R.P. and Hughes, D.J. (editors), 1990. *Early Precambrian Basic Magmatism*. Blackie, Glasgow, 486 pp.

Harker, A., 1904. The Tertiary Igneous Rocks of Skye. *Memoirs of the Geological Survey of the United Kingdom*, 48 pp.

Heiken, G., Vaniman, D. and French, B.M., 1991. *Lunar Sourcebook: A User's Guide to the Moon*. Cambridge University Press, New York, 736 pp.

Hekinian, R., 1982. *Petrology of the Ocean Floor*. Elsevier Scientific Publishing Company, Amsterdam, 393 pp.

Jaques, A.L. and Green, D.H., 1980. Anhydrous melting of peridotite at 0–15 Kb pressure and the genesis of tholeiitic basalts. *Contribution to Mineralogy and Petrology*, 73, pp. 287–310.

Johnson, R.W. (editor), 1989. *Intraplate Volcanism in Eastern Australia and New Zealand*. Cambridge University Press, Cambridge, 408 pp.

Le Maitre, R.W., 1976. The chemical variability of some common igneous rocks. *Journal of Petrology*, 17 (4), pp. 589–637.

Le Maitre, R.W., 1984. A proposal by the IUGS Subcommission on the Systematics of Igneous Rocks for a chemical classification of volcanic rocks based on the total alkali silica (TAS) diagram. *Australian Journal of Earth Sciences*, 31, pp. 243–255.

Le Maitre, R.W., Bateman, P., Dubek, A., Keller, J., Lameyre, J., Le Bas, M.J., Sabine, P.A., Schmid, R., Sorensen, H., Streckeisen, A., Woolley, A.R. and Zanettin, B., 1989. *A Classification of Igneous Rocks and Glossary of Terms: Recommendations of the International Union of Geological Sciences Subcommission on the Systematics of Igneous Rocks*. Blackwell Scientific Publications, Oxford, 193 pp.

Linton, P.L., McCarthy, T.S. and Brown, C.A., 1993. *The sequential eruption and tectonic history of the Klipriviersberg Group as illustrated by the distribution of geochemical units.* University of the Witwatersrand, Economic Research Unit, Information Circular Number 271, 18 pp.

Lofgren, G.E., 1983. Effect of heterogeneous nucleation on basaltic textures: a dynamic crystallisation study. *Journal of Petrology*, 24, pp. 229–255.

MacDonald, G.A., 1968. Composition and Origin of Hawaiian lavas. *Geological Society of America*, Memoir, 116, pp. 477–522.

MacDonald, K.C., Scheirer, D.S., Carbotte, S. and Fox, P.J., 1993. It's only topography: Part 2. *GSA Today*, 3 (2), pp. 30–35.

Magee, K.P. and Head, J.W., 1995. The role of rifting in the generation of melt: implications for the origin and evolution of Lada Terra-Lavinia Planitia region of Venus. *Journal of Geophysical Research*, 100 (E1), pp. 1527–1552.

Manson, V. 1967. Geochemistry of basaltic rocks: major elements. In Hess, H.H. and Poldervaart, A. (editor), *Basalts: The Poldervaart Treatise on Rocks of Basaltic Composition*, pp. 215–269, Volume 1, Interscience Publishers, New York, 482 pp.

Marsh, J.S. and Eales, H.V., 1984. The chemistry and petrogenesis of igneous rocks of the Karoo Central Area, Southern Africa. In Erlank, A.J. (editor), *Petrogenesis of the volcanic rocks of the Karoo Province*, pp. 27–67, Special Publication of the Geological Society of South Africa, Number 13, 395 pp.

McKenzie, D. and Bickle, M.J., 1988. The volume and composition of melt generated by extension of the lithosphere. *Journal of Petrology*, 29 (3), 625–679.

McKenzie, D. and O'Nions, R.K., 1995. The source regions of ocean island basalts. *Journal of Petrology*, 36 (1), pp. 133–159.

McSween, H.Y., 1987. *Meteorites and their Parent Planets.* Cambridge University Press, Cambridge, 237 pp.

Melson, W.G. and Thompson, G., 1971. Petrology of a transform fault zone and adjacent ridge segments. *Philosophical Transactions of the Royal Society of London*, A 268, pp. 423–441.

Mohr, P. and Zanettin, B., 1988. The Ethiopian Basalt Province. In McDougall, J.D. (editor), *Continental Flood Basalts*, pp. 63–110, Kluwer Academic Publishers, Dordrecht, 341 pp.

Moor G.G. and Sheinman, Y.M., 1946. Porody iz severnoi okrainy Sibirskoi platformy (transliterated from Russian). *Doklady Akademii Nauk SSSR*, Leningrad, Volume 51 (2), pp. 141–144.

Morgan, W.J., 1981. Hotspot tracks and the opening of the Atlantic and Indian Oceans. In Emiliani, C. (editor) *The Sea*, pp. 443–487, Volume 7, The Oceanic Lithosphere, John Wiley and Sons, New York, 1738 pp.

Myers, R.E., McCarthy, T.S., Bunyard, M., Cawthorn, R.G., Falatsa, T.M., Hewitt, T., Linton, P., Meyers, J.M., Palmer, K.J. and Spencer, R., 1989. *Geochemical Stratigraphy of the Klipriviersberg Group Volcanics.* University of the Witwatersrand, Economic Research Unit, Information Circular Number 215, 26 pp.

Natland, J. 1991. Mineralogy and crystallisation of oceanic basalts. In Floyd, P.A. (editor), *Oceanic Basalts*, pp. 63–93, Blackie, Glasgow, 456 pp.

Nisbet, E.G., 1982. The tectonic setting and petrogenesis of komatiites. In Arndt, N.T. and Nisbet, E.G. (editors), *Komatiites*, pp. 501–520, George Allen and Unwin, London, 526 pp.

Nockolds, S.R. and Le Bas, M.J., 1977. Average calc-alkali basalt. *Geological Magazine*, 144 (4), pp. 311–312.

Powers, H.A., 1935. Differentiation of Hawaiian lavas. *American Journal of Science*, 5th series, 30, pp. 57–71.

Reynolds, J.R., Langmuir, C.H., Bender, J.F., Kastens, K.A. and Ryan, W.B.F., 1992. Spatial and temporal variability in the geochemistry of basalts from the East Pacific Rise. *Nature*, 359, pp. 493–499.

Sabine, P.A. and Sutherland, D.S., 1982. Petrography of British rocks. In Sutherland, D.S. (editor) *Igneous Rocks of the British Isles*, pp. 479–544, John Wiley and Sons, Chichester, 645 pp.

Saunders, A.D. and Norry, M.J. (editors), 1989. *Magmatism in the Ocean Basin.* Geological Society of London, Special Publication 42, Blackwell Scientific Publications, Oxford, 398 pp.

Scarrow, J.H. and Cox, K.G., 1995. Basalts generated by decompressive adiabatic melting of a mantle plume: a case study from the Isle of Skye, NW Scotland. *Journal of Petrology*, 36 (1), pp. 3–22.

Smellie, J.L. (editor), 1994. Volcanism associated with extension at consuming plate margins. *Geological Society of London*, Special Publication Number 81, 293 pp.

Smith, D.K and Cann, J.R., 1992. The role of seamount volcanism in crustal construction at the Mid-Atlantic Ridge (24°–30°N). *Journal of Geophysical Research*, 97, pp. 1645–1658.

Storey, M., Kent, R.W., Saunders, A.D., Salters, V.J., Hergt, J., Whitechurch, H., Sevigny, J.H., Thirlwall, M.F., Leat, P., Ghose, N.C. and Gifford, M., 1992. Lower Cretaceous Volcanic Rocks on Continental Margins and their relationship to the Kerguelen Plateau. *Proceedings of the Ocean Drilling Program*, Scientific Results, 120, pp. 33–53.

Su, W-J., Woodward, R.L. and Dziewonski, A.M., 1992. Deep origin of mid-ocean-ridge seismic velocity anomalies. *Nature*, 360, pp. 149–152.

Thompson, R.N., Gibson, I.L., Marriner, G.F., Mattey, D.P. and Morrison, M.A., 1982. Trace element evidence of multi-stage mantle fusion and polybaric fractional crystallisation in the Palaeocene lavas of Skye, N.W. Scotland. *Journal of Petrology*, 21, pp. 265–293.

Thorarinsson, S., 1969. The Lakagigar eruption of 1783. *Bulletin Volcanologique*, 33, pp. 910–929.

Tilley, C.E. 1950. Some aspects of magmatic evolution. *Quarterly Journal of the Geological Society* (London), 106, pp. 37–61.

Tröger, W.E., 1935. *Spezielle Petrographie der Eruptivgesteine: Ein Nomenklatur – Kompedium*. Schweizerbart'sche Verlangsbuchhandlung, Stuttgart, Germany, 360 pp.

Viljoen, M.J. and Viljoen, R.P., 1969. The geology and geochemistry of the lower Ultramafic Unit of the Onverwacht Group and a proposed new class of igneous rocks. *Geological Society of South Africa*, Special Publication, 2, pp. 55–86.

Wei, J.F., Tronnes, R.G. and Scarf, C.M., 1990. Phase relations of alumina-undepleted and alumina-depleted komatiites at pressures of 4–12 GPa. *Journal of Geophysical Research*, 95, pp. 15 817–15 828.

White, R. and McKenzie, D., 1989. Magmatism at rift zones: the generation of volcanic continental margins and flood basalts. *Journal of Geophysical Research*, 94 B, pp. 7685–7729.

Wood, C.A., 1992. Climatic effects of the 1783 Laki eruption. in Harington, C.R. (editor), *The Year Without Summer?* Canadian Museum of Nature, Ottawa, 576 pp.

Yoder, H.S., 1976. *Generation of Basaltic Magma*. (US) National Academy of Sciences, Washington, D.C., 265 pp.

Yoder, H.S., 1988. The great basaltic 'floods'. *South African Journal of Geology*, 91 (Alex. L. du Toit Memorial Lectures, Number 20), pp. 139–156.

Yoder, H.S. and Tilley, C.E., 1962. Origin of basalt magmas: An experimental study of natural and synthetic rock systems. *Journal of Petrology*, 3 (3), pp. 342–532.

CHAPTER 8

Convergent plate margins and their magmas

The tectonic and magmatic processes which have shaped the post-Archean continents are dominated by convergent plate tectonics.

Hamilton, 1995, p. 13.

8.1 When plates converge

Along convergent plate margins the oceanic lithosphere is subducted at velocities that range between 10 and a 110 km per million years. This is where diverse magmas evolve and spectacular chains of volcanoes thrive (Gill, 1981; Smellie, 1994; Thorpe, 1982). It is also where new continental crust evolves and the process of subduction inserts new chemical heterogeneities and complexities into the upper mantle.

Oceanic plates form on mid-oceanic ridges and as they cool they contract and become progressively denser and thicker. This results in their gradual subsidence as they pull away from their parental spreading ridge. Later during subduction the bulk density of the descending oceanic crust increases greatly as increased pressures and temperatures cause the basaltic and gabbroic rocks to transform into denser eclogites. This transformation is the principal driving force in subduction systems.

Convergent plate boundaries are complex because the individual parts of a single, continuous island arc or continental margin arc are likely to evolve in many different ways. The rocks from some convergent plate boundaries contain evidence of repeated episodes of collision, aggregation, rifting and volcanism. They may contain unrelated rock sequences or terranes that have been thrust together. Oceanic trenches are the normal surface expression of convergent plate boundaries. In active subduction systems the downgoing slab is cold and heat flow normally remains low over the trench and forearc. Volcanism activity is usually concentrated in linear belts or volcanic arcs. They lie parallel to the trench system. Heat flow increases swiftly as the volcanic arc is approached and it may remain moderately high well into the

back-arc-basin. Most volcanoes and volcanic products are concentrated in the volcanic arc and there is a marked decrease in their abundance towards the back-arc-basin. To obtain some idea of the scale of these features one should recognise that in Japan the main volcanic arcs run parallel to the oceanic trench and that the distance between these major linear features varies between 150 to 300 km.

Subduction is continually introducing cold oceanic lithospheric materials into the upper mantle thus cooling the upper mantle. Both subducting and overriding plates are usually seismically coupled to depths that vary between 30 to 50 km (Hamilton, 1995, p. 3). At greater depths the subducting plate usually lacks strength and deforms internally. It may be traced to depths of at least 300 km. A plate that is being subducted bends into a broad curve. This bending begins on the seaward side of the oceanic trench. The plate then dips gently beneath the accretionary wedge that builds up in front of the overriding plate. Most of the materials that accumulate in the accretionary wedge are scraped from the top of the subducting plate. They typically consist of a melange of sedimentary materials.

Generally the sediments that are conveyed on the subducting plate remain undeformed until they are only about 30 km from the trough of the oceanic trench. Profiles obtained from reflection-seismic studies show that when these abyssal sediments reach the oceanic trench they are normally complexly imbricated and thickened by thrusting. The sediments that are subducted become even more highly sheared with their response to their changing tectonic environment being more like that of a highly viscous fluid than a solid rock. The subduction of seamounts and other large topographic features is likely to distort the shape of the overriding accretionary wedge. Seismic activity also increases as these protuberances collide with the overriding plate.

The principal bend in the subducting plate is frequently called the subduction hinge. It is normally located beneath the overriding plate where it takes the form of a gentle flexure that gradually changes the dip of the subducting slab. In most subduction systems the mean dip of subducting slab ranges from 30° to 70°. As a convergent plate boundary evolves the subduction hinge is likely to recede resulting in the advance of the overriding plate. An examination of the convergent plate boundaries that border the Pacific Ocean readily confirms this idea. The Pacific Ocean is currently shrinking as the plates that flank it move towards it. There is a concomitant expansion in the area of the other oceans. This means that along most convergent margins the overriding plates are being pulled towards a retreating subduction hinge. Extension is the dominant tectonic milieu in the overriding plate and this favours the injection of magma.

Mantle-wedges and seismic tomography

In a convergent plate setting most primary magmas are generated in the mantle-wedge above the subducting plate. Seismic tomography provides images of the thermal structure and distribution of magma within subduction systems (Zhao and Hasegawa, 1993). Beneath the northern part of Japan's largest island seismic tomography has revealed the shape of a high-seismic-velocity slab of cold material that is currently being subducted. The slab is about 85 km thick. Immediately above it in the mantle-wedge there is a layer that has a seismic velocity that is normal for the upper mantle. Above this normal layer and still inside the mantle-wedge there is an irregular zone with an anomalously low seismic velocity. Such zones usually conform to the shape of a layer that is subparallel to the subducting slab. It is usually about 50 km thick and it is often traced to a depth of at least 200 km. This anomalous layer is interpreted as an extensive upward flow of hot, probably fertile mantle material. Partial melting is likely to occur within this hot material as it rises to lower pressures. At a depth of about 60 km there is usually a change in the layer's shape. Its dip steepens and it connects with the low seismic velocity materials normally located beneath the volcanic arc.

A synthesis of the data obtained from seismic tomography studies of modern subduction systems

and various investigations of the materials present in defunct subduction systems has revealed that primary magmas are mainly generated in those parts of a subduction system where there is reaction between volatiles emanating from a subducting plate and hot fertile mantle materials moving up through the mantle-wedge. This is probably what is currently happening beneath central Japan. In this area the cooler layer of mantle material that lies atop the subducting plate is being pulled away from the rest of the mantle-wedge and hot, normally fertile mantle material is rising to fill the potential void. Water is the main volatile that facilitates partial melting in subduction systems. It lowers the melting temperatures of the phases in the rocks it permeates. This magma generating process usually takes place at lower temperatures than those that operate in intraplate settings or beneath mid-oceanic ridges. When evaluating the melting process in subduction systems it is apposite to record that this partial melting process is likely to stop unless the fusible materials in the mantle-wedge are continually renewed.

In a typical subduction system most of the volcanoes develop about 100 km above the subducting plate. This distance is usually the same irrespective of the chemical composition or physical properties of the upper plate. The precise position of the volcanoes is normally determined by the shape of the magma generating and distribution system within the mantle wedge and it is not normally related to the position where the subducting plate releases most of its volatiles.

Cordillera and subduction systems

The surface landforms above many active subduction systems are often loosely described as fold-mountains. This term seems to imply that rocks within these mountain belts have been folded by compressional forces. An examination of the geology of an area such as Peru that contains these landforms reveals a very different tectonic setting. In a broad tract of land, extending from the coast in the west, across the Western Cordillera and the Altiplano to the crest of the Eastern Cordillera, most of the young deformation is extensional (Hamilton, 1995, p. 11). This suggests that the overriding plates at convergent plate margins are usually areas of extension and not compression. Studies of the first-motions of earthquakes in volcanic arcs usually reveal that the dominant stress is extensional, perpendicular to the trend of the arc. Further irrefutable evidence of extension is provided by the many island arcs that border the Pacific Ocean. They have retreating subduction hinges and actively spreading back-arc-basins.

It is sometimes suggested that part of the magma generated in subduction systems is produced by the partial melting of the materials being subducted. Peacock *et al.* (1994, p. 241) evaluated this process and concluded that it was only likely to be substantial if local shearing generates abnormally high rates of heating or if young, hot oceanic lithosphere was being subducted. The volcanic rocks of the Austral Volcanic Zone of southern Chile have been suggested as a possible example of rocks generated by the second of these processes. In this area subduction is taking place at a rate of about 2 cm per year and the material being subducted is relatively hot MOR basalt. The volume and composition of the sedimentary rocks being subducted at different sites is extremely variable. In the Mariana Trench, south of Japan, all the incoming sediment is subducted and even the base of the overriding plate is being actively eroded. At other sites, such as the oceanic trench linked to the subduction system beneath the Lesser Antilles in the eastern Caribbean most of the incoming sediments accumulate on the leading edge of the overriding plate.

A clearer perception of the processes that operate during the subduction of volatile-rich, hydrothermally altered materials can be deduced from the study of the rocks displayed in sections through defunct subduction complexes. Excellent examples of this type of material are exposed on Santa Catalina Island near Long Beach, California. These rocks show that large volumes of water-rich fluid are released during the prograde metamorphism induced by subduction. They also reveal that the mineral chlorite can stabilise and transport volatiles to a depth of at least 100 km (Belout, 1991, p. 416).

According to Peacock (1990, pp. 329–337) the expulsion of large volumes of pore water and gas at shallow depths influence the thermal and rheological characteristics of the accretionary prism. At greater depths, below about 45 km, water and carbon dioxide are released by prograde metamorphic reactions. Amphibole-bearing oceanic crust that is being subducted is only expected to begin to melt if it remains or becomes abnormally hot. In normal, cooler and older slabs partial melting is mainly confined to the overlying mantle-wedge.

Serpentinite diapirs are emplaced in some forearc settings. In the Mariana arc system such serpentinite diapirs are considered to represent hydrated and mobilised mantle-wedge materials. Their origin was probably triggered by the release of volatiles from the underlying subducting slab.

Some late Mesozoic and younger subduction complexes contain large volumes of blueschists. These distinctive metamorphic rocks crop out in areas as far apart as Alaska, southern Europe, California, Japan and New Caledonia in the southwest Pacific. The characteristic minerals in these rocks equilibrated under conditions of high pressure and low temperature. Such singular conditions are likely to exist when a slab of cold ocean-floor material is subducted at a fast rate. In some subduction complexes there is evidence that temperatures were less than 200°C down to depths of up to 30 km.

8.2　Diverse magmas in subduction systems

A diverse group of extrusive and intrusive rocks can evolve within a subduction system. Generally there is a broad correlation between their composition and the thickness and composition of the lithosphere in which they evolved. Basalts of tholeiitic lineage are usually found to erupt from young intraoceanic island arcs. In more mature oceanic island arcs the typical volcanic rocks are of the calc-alkali lineage. They range in composition from calc-alkali basalts (Nockolds and Le Bas, 1977, p. 313) via basaltic andesites and andesites to dacites (see Tables 7.1 and 8.1). The proportion of intermediate rocks of calc-alkali lineage is much greater in mature island arcs and continental margin arcs where the crust is thicker. When subduction occurs beneath a segment of continental crust or beneath a thick deposit of terrigenous sedimentary rocks igneous magmas of silicic composition are often produced. It is generally in this geological setting that one finds large volumes of dacites and rhyolites at the surface and granodiorites and granites emplaced into the crust.

Tholeiitic and boninitic lineages

Most of the volcanic rocks produced during the early stages in the evolution of intraoceanic island

Table 8.1 Mean major element compositions and CIPW norms of the subalkalic suite. After Le Maitre, 1984, pp. 243–255.

Name	Basalt	Basaltic andesite	Andesite	Dacite	Rhyolite
SiO_2	49.45	54.21	59.95	66.17	74.07
TiO_2	1.8	1.29	0.82	0.6	0.27
Al_2O_3	15.35	16.37	17.03	15.93	13.23
Fe_2O_3	3.28	3.38	3.18	2.47	1.5
FeO	7.95	6.58	3.78	2.47	1.17
MnO	0.15	0.18	0.13	0.1	0.06
MgO	7.71	4.49	3.21	1.68	0.33
CaO	11.01	8.8	6.56	4.19	1.05
Na_2O	2.57	2.95	3.46	3.82	3.95
K_2O	0.52	1.04	1.69	2.39	43.31
P_2O_5	0.21	0.21	0.19	0.18	0.06
Total	100	100	100	100	100
CIPW norms					
Quartz	0	6.99	14.76	23.1	31.76
Corundum	0	0	0	0	0.31
Orthoclase	3.06	6.15	9.97	14.13	25.46
Albite	21.74	24.92	29.26	32.36	33.43
Anorthite	28.82	28.37	25.97	19.24	4.81
Diopside	19.75	11.31	4.29	0.3	0
Hypersthene	15.63	14.41	9.13	5.73	1.4
Olivine	2.33	0	0	0	0
Magnetite	4.76	4.9	4.62	3.58	2.18
Ilmenite	3.42	2.46	1.56	1.14	0.51
Apatite	0.5	0.5	0.45	0.42	0.14

arcs are deeply buried within the growing volcanic pile. When these older volcanic sequences are examined in areas such as the island arcs of the western Pacific they are usually found to contain rocks of both the tholeiitic and boninitic lineages. Rocks of the former lineage are frequently called island-arc tholeiites. They are normally more abundant than the high-Mg and low-Ti basaltic andesites and andesites of the boninitic lineage (see Table 8.2). Petersen (1890, p. 25) introduced the term boninite to describe volcanic rocks from the Bonin (or Ogaswara Gunto) islands to the north of the Marianas in the northwestern Pacific Ocean (see Crawford, 1989). The boninites he described were glassy lavas that contained about 54.4 per cent SiO_2 and 12.8 per cent MgO. They exhibited many geochemical characteristics of primary magmas, with their high magnesium, nickel and chromium contents, yet they are also enriched in silica relative to the basaltic rocks. According to Le Maitre *et al.* (1989, p. 27) a typical boninite contains more than 53 per cent SiO_2, more than 8 per cent MgO and less than 0.5 per cent TiO_2.

In 1976 Peccerillo and Taylor divided the volcanic rocks that evolved in subduction systems into four groups using a K_2O versus SiO_2 diagram (see Fig. 8.1). Their groups are: (1) low-potassium series; (2) calc-alkali series; (3) high-potassium series; and (4) shoshonite series. Island-arc tholeiites and boninites are all subsumed in their low-potassium series. The andesites that commonly evolve in normal subduction settings plot as intermediate members of the calc-alkali series. Most of the rocks that belong to Peccerillo and Taylor's high-potassium and shoshonite groups are extruded in the back-arc area behind the volcanic arc. They are often reported from areas where there is evidence of crustal extension. These rocks belong to the midalkaline or alkaline lineages and will be considered in Chapters 10 and 11. Table 8.3 contains selected major element data on primitive rocks of the low-potassium, calc-alkali, high-potassium and shoshonite series from the Japanese volcanic arcs (Tatsumi and Eggins, 1995, p. 65).

Many areas of active subduction contain magmatic rocks that display a systematic increase in their

Table 8.2 Major element compositions and CIPW norms of selected boninitic rocks. After Crawford *et al.*, 1989, p. 9.

Number	B-1	B-2	B-3	B-4	B-5	Mean
SiO_2	53.71	54.75	56.2	58.5	59.46	56.52
TiO_2	0.21	0.26	0.13	0.43	0.16	0.24
Al_2O_3	7.77	13.28	10.57	13.3	11.16	11.22
Fe_2O_3	2.33	2.4	2.02	2.5	1.37	2.12
FeO	7.03	5.33	6.23	3.95	5.9	5.69
MnO	0.17	0.13	0.16	0.12	0.13	0.14
MgO	18.49	10.72	11.19	9.47	11.26	12.23
CaO	4.64	8.11	7.44	6.13	4.97	6.26
Na_2O	0.83	1.5	1.54	2.61	2.19	1.73
K_2O	0.19	0.91	0.4	1.28	0.64	0.68
P_2O_5	0.03	0.05	0.02	0.13	0.04	0.05
H_2O^+	2.65	2.95	3.02	1.37	2.92	2.58
H_2O^-	1.64	n.d	0.93	0.22	n.d.	0.93
CO_2	<0.1	0.08	<0.1	n.d.	0.04	0.05
Total	99.69	100.47	99.85	100.01	100.24	100.44
CIPW norms						
Quartz	7.38	8.93	12.66	11.48	14.17	10.95
Orthoclase	1.18	5.52	2.47	7.69	3.89	4.15
Albite	7.37	13.03	13.6	22.45	19.06	15.12
Anorthite	17.75	27.54	21.64	21.14	19.27	21.53
Diopside	4.93	10.72	13.41	7.06	4.67	8.14
Hypersthene	57.35	30.05	32.85	25.35	36.48	36.34
Magnetite	3.55	3.58	3.06	3.69	2.05	3.18
Ilmenite	0.42	0.51	0.26	0.83	0.31	0.47
Apatite	0.07	0.12	0.05	0.31	0.1	0.12
	100	100	100	100	100	100

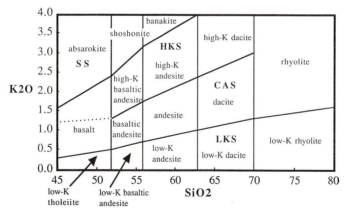

Fig. 8.1 A K_2O versus SiO_2 diagram that is sometimes used to classify volcanic rocks from divergent plate margins. SS = shoshonite series, HKS = high-potassium series, CAS = calc-alkali series and LKS = low-potassium series. After Peccerillo and Taylor, 1975, p. 66.

incompatible element content (particularly potassium, rubidium and lanthanum) as one proceeds across the upper plate from the volcanic arc into the back-arc zone (see Fig. 8.2). This trend is clearly displayed by the volcanic rocks of the Sunda Arc in Indonesia (Whitford and Nicholls, 1976, pp. 66–67). Earlier Buddington (1927) had discovered a similar geochemical trend in the plutonic rocks in the compound batholiths of the Coast Range of southeastern Alaska (see Section 9.8).

Table 8.3 Major element compositions and CIPW norms of selected primitive rocks of the low-K, calc-alkali, high-K and shoshonite series from the Japanese arcs. After Tatsumi and Eggins, 1995, p. 65.

	Low-K	Calc-alkali	High-K	Shoshonitic
SiO_2	49.1	50.57	49.21	49.81
TiO_2	0.65	0.51	1.03	1.71
Al_2O_3	15.09	14.48	15.18	17.54
FeO total	8.97	8.36	9.14	7.12
MnO	0.19	0.17	0.25	0.16
MgO	13.34	12.64	10.72	9.68
CaO	10.6	10.57	11.04	7.92
Na_2O	1.77	1.95	1.97	1.73
K_2O	0.2	0.65	1.41	3.56
P_2O_5	0.09	0.1	0.05	0.77
Total	100	100	100	100
CIPW norms				
Orthoclase	1.18	3.85	8.34	21.06
Albite	15.01	16.53	16.7	14.66
Anorthite	32.71	28.91	28.48	29.62
Diopside	15.53	18.4	20.97	3.74
Hypersthene	18.46	17.9	4.28	10.75
Olivine	13.31	11.02	16.76	13.25
Magnetite	2.35	2.19	2.39	1.86
Ilmenite	1.24	0.97	1.96	3.26
Apatite	0.21	0.23	0.12	1.8

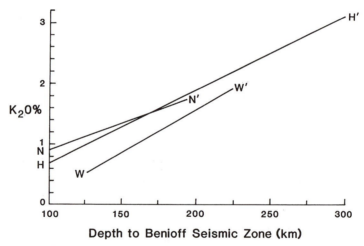

Depth to Benioff Seismic Zone (km)

Fig. 8.2 This diagram shows the relationship between K_2O and depth to Wadati-Benioff seismic zone for volcanic rocks (where SiO_2 equals 55 per cent) from various subduction systems. Trend lines NN′and HH′ are averaged results for many island arcs compiled by Nielson and Stoiber (1973) and Dickinson (1968), respectively. WW′ is the trend line for the Sunda Arc in Indonesia (Whitford and Nicholls, 1976, p. 67).

8.3   Petrogenesis in a subduction setting

In a normal subduction system the subducting plate is a slab of cold oceanic lithosphere. It has an upper fractured layer of metabasaltic rocks interspersed with variable amounts of pelagic and terrigenous sedimentary materials. A middle layer that contains various gabbroic rocks and a lower layer of depleted mantle peridotite. As the slab moves down these various rocks are metamorphosed. The normal metamorphic sequence is zeolite facies, greenschist facies, amphibolite facies and eclogite facies. During subduction some heat is generated locally by shear-strain or friction heating.

The subducted rocks gradually dehydrate and release an aqueous fluid. This process continues down to a depth of about 250 km. On being expelled most of this fluid permeates the overlying mantle-wedge. If the fluid encounters hot, upward flowing mantle material the fluid lowers its melting point, density and viscosity. This encourages partial melting and buoyancy assists in the segregation and accumulation of a primary magma. As this magma rises, pressure decreases, thus encouraging more partial melting. The amount of magma generated by decompression melting is usually related to the thickness of the lithosphere. Decompression melting is greatest beneath young intraoceanic arcs.

According to Pearce and Parkinson (1993, p. 398) between 15 and 30 per cent of the mantle material is likely to melt during magma generation in a subduction setting. A primary magma in equilibrium with the phases present in the mantle-wedge and sustained at a pressure of about 2.8 GPa is likely to be picrobasaltic composition. Local variations in composition may be produced by small changes in temperature, the composition of the mantle material rising into the mantle-wedge, the ambient pressure and the composition of the migratory fluid, or fluids, that react with the mantle phases.

The fluids that migrate into the mantle-wedge are likely to contain variable amounts of elements such as carbon, fluorine, silica, sulphur, chlorine, potassium, rubidium, strontium, thorium and uranium. During the life of a particular mantle-wedge the upwelling mantle materials are likely to change both in composition (or relative fertility) and temperature and in the evolution of a normal subduction system the compositions of the primary magmas are likely to change in a systematic way producing the initial magmas that belong to the tholeiitic, boninitic and calc-alkalic lineages. The boninitic magmas may have been derived from depleted source rocks that have previously yielded a magma by partial melting. According to Tatsumi and Eggins (1995, p. 139) they most probably develop at somewhat shallow depths in the uppermost mantle under water-undersaturated conditions.

Less frequently, magmas that equilibrate at considerably higher pressures will be released from the mantle-wedge. These magmas are likely to be the primogenitors of magmas of the midalkaline and alkaline lineages. It has also been suggested that quartz eclogite from a subducting slab may occasionally become mixed into the hotter rising mantle material, leading to the generation of magmas that contain more silica and have a somewhat higher than normal K/Na ratio (Ringwood, 1974).

When picrobasaltic magmas reach the boundary zone between the mantle and crust their upward progress into the crust is likely to be impeded by their high densities. On being detained at the mantle–crust interface, cooling and fractional crystallisation is likely to occur resulting in the generation of a more buoyant magma. The various changes in the chemical composition of these magmas as they migrate upwards will normally depend on the thickness and composition of the crust traversed. In intraoceanic island arcs the crust will be thin and essentially basaltic in composition, whereas in areas where the subduction system operates under a mature continental margin the crust is likely to be much thicker and contain various sialic rocks. It is anticipated that when a normal picrobasaltic-basaltic magma tries to traverse a thick segment of continental crust it will become at least temporarily impounded at one or more levels. While impounded it is likely to cool, undergo fractional crystallisation, commingle with other magmas, and react with at least some of the wall rocks it encounters. These processes acting together, or alone, are likely to yield a variety of differentiated magmas that under normal circumstances have compositions within the basaltic andesite to rhyolite range (see Table 8.1).

Pearce and Parkinson (1993, p. 398) have tried to work out the chemical composition of the source materials in subduction systems operating in settings that range from intraoceanic arcs to continental margin arcs. They found that in island arcs with active back-arc-basins such as the Scotia, Tonga, Kermadec and Mariana arcs the source rocks are usually slightly depleted, particularly in the highly incompatible elements. It is likely that this depletion resulted from an earlier extraction of magma to produce the rocks of the back-arc-basin. In intraoceanic arcs that lack back-arc-basins the source rocks are likely to be fertile and similar to the materials that yield MOR basalts. The source rocks that partially melt beneath continental margin arcs are likely to be fertile or even enriched in incompatible elements. Many rocks that display enrichment in incompatible elements occur at arc-transform fault intersections or in areas undergoing rifting. It is also often proposed that this enrichment may be due to reaction between magma generated in the mantle-wedge and enriched subcontinental lithosphere.

8.4 Andesites from non-subduction settings

Basaltic andesites and andesites are the characteristic rocks of the volcanic chains that develop along convergent plate margins. Rocks that are chemically similar and plot in the same fields on the TAS diagram (Le Maitre *et al.*, 1989, p. 28) also erupt in non-subduction, or anorogenic, tectonic environments. These rocks have been called flood icelandites, icelandites and oceanic andesites (see Fig. 2.14). In those large areas of the continents that are covered by flood basalts there are usually also extensive outcrops of basaltic icelandites (see Table 7.4). These rocks are all comagmatic and belong to the tholeiitic lineage. In many flood basalt provinces the volume of basaltic icelandite seems to have been augmented by the assimilation of sialic crustal materials.

The term icelandite was coined by Carmichael (1964, p. 442), to distinguish between andesites that evolve in subduction settings and andesites (sensu lato) that erupt from central-type volcanoes in Iceland. He revealed that the intermediate rocks, from volcanoes like Thingmuli in Iceland, were 'poorer in alumina and richer in iron' than typical subduction system andesites (see Fig. 2.14). The oceanic andesites were first described from the aseismic Ninety East Ridge in the Indian Ocean. These rocks are compositionally similar to the basaltic icelandites and icelandites of Iceland. According to Thompson *et al.* (1973, p. 1020) the rocks from the Ninety East Ridge probably evolved from the fractional crystallisation of a tholeiitic basaltic magma that was impounded in a series of shallow magma chambers beneath the ridge.

Some andesitic rocks are unusual because they are generated by impact melting. Their genesis is triggered by a hypervelocity impact. A well-known example of an andesitic impact-melt-sheet is found surrounding the Manicouagan astrobleme site in Quebec. These rocks show that andesites can originate from the impact melting of surface materials. Impact melting was probably important in the Hadean and early Archean on all the terrestrial planets.

8.5 Geochemistry of the andesites

The average andesite contains about 17.2 per cent Al_2O_3 and 7.1 per cent (Fe_2O_3 + FeO), whereas the average icelandite contains considerably less Al_2O_3 (14.4 per cent) and considerably more (Fe_2O_3 + FeO) with a mean of 9.2 per cent. The Al_2O_3/(Fe_2O_3 + FeO) ratio is characteristically less than two for the icelandites and more than two for the andesites. Andesites do not usually contain any visible quartz, yet they normally contain over 10 per cent normative quartz (see Table 8.1). When compared with the MOR basalts the andesites are generally depleted in scandium, vanadium, chromium, cobalt and nickel. These are elements that are preferentially concentrated in the ferromagnesian minerals. This suggests that andesitic magma is probably a differentiation product of a more primitive magma from which the early formed ferromagnesian minerals have been removed. In most normal andesites the chalcophile elements (copper, zinc and molybdenum) have low, but variable concentrations (see Fig. 2.20). High concentrations of these elements are found in the porphyry copper-lead-zinc-molybdenum-silver-gold ore deposits. Gill (1981, p. 160) concluded that the strontium, lead and neodymium isotopic ratios found in most subduction system andesites show that their magmas were generated from a mixture of both mantle and sialic crustal sources. Andesites with high $^{87}Sr/^{86}Sr$ ratios (i.e. higher than 0.704) are usually extruded above areas with thick, somewhat old continental crust.

8.6 Eruption style of andesitic volcanoes

Many andesite volcanoes build remarkably symmetrical cones (see Plate 8.1). If the public is asked to describe a volcano they usually describe a similar conical landform. Such structures are usually produced by a series of eruptions of pyroclasts and lavas from a central vent. In an eruptive cycle the pyroclastic rocks are usually extruded first because this explosively released material is derived from the upper fraction of a gas-charged magma. The lavas are extruded later and develop from a more degassed magmatic material. Lavas of basaltic andesite composition generally form extensive flows whereas the more viscous silicic andesites tend to produce shorter stubby flows or even lava domes.

Lava domes are particularly characteristic of dacitic and rhyolitic lavas. Andesite eruptions usually register about two on the volcanic explosivity index. When a volcano that had been erupting basaltic andesites and andesites starts to erupt more silicic materials, such as dacite, there is generally a substantial increase in its explosivity.

8.7 Mineral deposits at convergent plate boundaries

Along the convergent plate boundaries that border the Pacific Ocean there are many large porphyry ore deposits. Most of these deposits were emplaced during the last 150 million years. They normally occur within, or in the country rocks that surround, small intrusive bodies. The latter are often interpreted as the altered subvolcanic remains of earlier andesitic, or dacitic, volcanoes. Generally the ores form during the waning stages of volcanic activity when a silicic magma is crystallising in the upper layers of a compositionally zoned magma-chamber. Aqueous, often saline, fluids are released into the crystallised carapace of the magma-chamber and into the surrounding country rocks. Depending on the initial chemical composition of the fluid, metal sulphides or oxides crystallise amidst the silicate phases. Some silicate minerals thrive in this fluid-rich milieu and grow large thus producing rocks with a porphyritic or even a pegmatitic texture. These textural features explain why these rocks and their related ore deposits are frequently called porphyries. Whether such rocks should be classified as igneous rocks or metasomatic rocks remains the subject of lively discussion.

The rocks surrounding porphyry ore deposits are usually fractured or brecciated. This breaking up of the country rocks seems to result from a late-stage build-up of fluid pressures in the magma chamber. Once fracturing has taken place the pressure in the magma chamber falls, the fluid boils and more material is transported into the fracture system. Ore minerals containing copper and tin are likely to occur in veins near the remains of the old magma chamber, whereas gold, silver and manganese generally occur in veins that cut the pyroclastic rocks of the caldera facies. Active subduction systems frequently change their positions. This is readily revealed by an examination of the distribution of porphyry ore deposits. For example, during the past 90 million years the zone of copper-molybdenum mineralisation in the cordillera of South America has migrated about 400 km east from Chile into Argentina.

According to Sillitoe (1976) the composition of porphyry ore deposits often varies systematically across island arcs and active continental margins. As these deposits are usually associated with subduction systems it is likely that at least some of their sulphur and ore metals are derived from sulphide minerals that have been subducted. These minerals would have originally been deposited on the ocean-floor by the action of black smokers or other hydrothermal systems. Some deposits, such as the tin-bearing porphyries of Bolivia, contain remarkably high concentrations of a single metal. Such geochemical anomalies may arise if the magma that produced the subvolcanic porphyry system had previously assimilated crustal rocks enriched in that metal.

Many other types of ore deposits, particularly epithermal vein deposits, that contain metals such as copper, zinc, niobium, molybdenum, silver, tin, antimony, tantalum, tungsten, gold, mercury, lead, thorium and uranium, develop in island arcs and continental margin arcs (Mitchell and Carlile, 1984). Many of these deposits are situated on basement antiforms or domes and their location is mainly related to structural controls and the availability of suitable source rocks from which the metals can be scavenged. Other deposits are regarded as more directly linked to the cooling of late-stage, water and potassium-rich magmas. As such magmas cool they release enormous volumes of hot fluid, this promotes widespread alteration, the scavenging and transport of metals, and the evolution of various types of hydrothermal veins including pegmatites. Müller and Groves (1995, p. 119) have shown that there are spatial and probably genetic associations between copper-gold mineralisation and high-potassium suites of igneous rocks that crop out in what they call late oceanic arc, continental arc and post-collisional arc settings. They claim that many large volcanic-hosted gold deposits are intimately related to potassic igneous rocks (see Section 11.6).

8.8 Rhyolite–granite terminus

Rhyolites and granites are exceptional rocks because silicic magma occupies the principal low temperature trough in petrogeny's residua system (Bowen, 1937). This means that all the common magmatic lineages, such as the tholeiitic, calc-alkali and midalkalic, have

Fig. 8.3 The 1886 eruption of Mount Tarawera on North Island, New Zealand blasted a chasm through a cluster of rhyolitic volcanic domes revealing their shape and internal structure (see Fig. 8.7).

liquid lines of descent that terminate in the rhyolite field (see Figs 8.3–8.6). Several, but not all, of the many magmatic lineages that terminate in the rhyolite field evolve within normal subduction systems. Most calc-alkali dacitic and rhyolitic magmas develop in subduction systems beneath continental margin arcs.

Rhyolitic magma may evolve even during the death throes of a subduction system when an ocean-basin closes and continental blocks collide. The High Himalayan leucogranites and their retinue of dykes form a discontinuous belt that extends for nearly 3000 km along the High Himalayas from northern Pakistan to Bhutan. They formed after the collision between the Indian and Eurasian plates about 50 million years ago. The collision caused rapid crustal thickening, with the stacking of crustal slabs. Widespread structural readjustments led to the partial melting of part of the micaceous metasedimentary component in the rocks of the Higher Himalayan Crystalline Series (see Section 9.13). A representative example of these silicic rocks is the Manaslu leucogranite of Central Nepal (Guillot *et al.*, 1993).

Rhyolites, or as they are sometimes called oceanic rhyolites, also erupt from sites close to spreading ridges in those few areas where the oceanic crust is thick. Iceland and the Galapagos Islands are excellent examples of areas where this type of rhyolite is currently being produced (McBirney, 1993). It is likely that in these areas the volume of rhyolitic magma was augmented by the remelting of pockets of silicic rock that had earlier evolved by magmatic differentiation in the thickened oceanic crust. In both Iceland and the Galapagos Islands it is probable that most of the heat needed for remelting was supplied by active mantle plumes.

For many decades it has been acknowledged that rocks of the granite-granodiorite association are remarkably abundant in the upper continental crust. In more recent times it has been recognised that in many continental areas there are huge volumes of silicic volcanic rocks, particularly rhyolites. Many earlier estimates of the abundance of rhyolite miscalculated the size of the various silicic pyroclastic deposits that were ejected during cataclysmic plinian or ultraplinian eruptions.

Fig. 8.4 Deadman Creek Dome is the southernmost of the domes in the Inyo Crater Chain in the Long Valley–Mono Basin Province of eastern California. This rhyolitic dome is about 550 years old and it represents the final, relatively passive, vent-filling stage in the Deadman Creek eruptions.

The largest continuous outcrop of rhyolite is in the Sierra Madre Occidental Province of western Mexico and New Mexico. In this province rhyolitic materials cover an area of more than $250\,000\,km^2$ to a mean depth of about 1 km. The well exposed and mainly undisrupted silicic rocks of the Mogollon-Datil volcanic field of southwestern New Mexico forms an outlier of the larger Mexican province (Elston, 1989, pp. 43–46). The Mogollon-Datil volcanic field covers an area of about $40\,000\,km^2$. It lies to the south of the more extensive volcanic outcrops of the Basin and Range Province. During the last 40 million years the tectonics of western Mexico and the southwestern USA have changed from subduction of the Farallon Plate to an extensional regime within a continental plate. Between 40 and 20 million years ago large volumes of intermediate to silicic rocks were extruded. When plotted on a TAS diagram these rocks occupy a field that lies astride the midalkaline–subalkaline boundary (Davis *et al.*, 1993, p. 476). The midalkaline signature of many of these rocks is not too unexpected as some of their parental magmas were generated up to 900 km from the oceanic trench at a time (about 37 million years) when the trench was in collision with a mid-oceanic ridge.

In 73 000 BC the ultraplinian eruption of Toba Volcano in northern Sumatra produced a sheet of rhyolitic tephra that blanketed an area in excess of $5\,000\,000\,km^2$. Other vast outpourings of rhyolitic and dacitic rocks are found in the Cenozoic Taupo Zone of North Island, New Zealand. The high viscosity of rhyolitic magmas and lavas profoundly influences their eruptive style. Dacitic magma was responsible for the huge ultraplinian eruption of Mount Pinatubo in the Philippines on June 15, 1991. This event produced an eruption column that was over $35\,000\,m$ high with an umbrella-shaped cloud that covered a huge area. Pyroclastic flows cascaded out of the eruption column and covered an area of about a $100\,km^2$. When the ash cleared a new large caldera was revealed. Large rhyolitic pyroclastic flow, surge and air-fall deposits are usually ejected during cataclysmic plinian or ultraplinian eruptions and the eruptive orifice that survives this

Fig. 8.5 Bedded ash-fall and ash-flow deposits of the El Cajete Member of the Valles Rhyolite, including the El Cajete Pumice, from the Valles Resurgent Caldera in New Mexico. The Valles Rhyolite erupted after the caldera moat volcanism. It is thus younger than the voluminous (600 cm^3) Bandelier Tuff. The latter erupted during two major cycles, 1.4 and 1.1 million years ago.

type of eruption is usually either a large caldera, or an even larger volcano-tectonic depression.

8.9   Petrography and geochemistry of the dacites and rhyolites

Most volcanic silicic rocks crop out as pyroclastic rocks, obsidian, pumice or pitchstone. Lavas of this composition are usually porphyritic with flow-banded or eutaxitic structures. Quartz is the most common phenocryst. Hornblende and biotite are the characteristic ferromagnesian minerals in normal calc-alkali rhyolites, but in the peralkaline rhyolites, the sodium-rich minerals arfvedsonite, riebeckite, aegirine and aenigmatite are commonly observed.

When the mean major element compositions of dacite and rhyolite are compared the former is normally found to be enriched in total iron, magnesium, calcium and aluminium, while the latter is enriched in silica and potassium. These differences are well illus-

trated by their normative mineral compositions (see Table 8.1). The rhyolites are strongly enriched in quartz and orthoclase, whereas the dacites are usually enriched in plagioclase, hypersthene and the iron-titanium oxides.

8.10   Landforms shaped by silicic eruptions

The high viscosity of normal dacitic and rhyolitic magmas and lavas profoundly influences their style of eruption (see Section 4.3). Such eruptions are usually explosive and result in the extrusion of extensive pyroclastic deposits. Many dacitic and rhyolitic lavas crop out in volcanic domes or coulee. For example, the Mount Tarawera volcanic centre on North Island, New Zealand contains both coulee and volcanic domes (see Fig. 8.7). An examination of this volcanic centre is particularly informative as the eruption of 1886 blasted a chasm through a

Fig. 8.6 Rhyolitic pyroclastic deposits of the Bishop Tuff as seen from Power-house Road in the Owens River Gorge. This gorge is incised into the volcanic tableland to the south of Long Valley. The Bishop Tuff erupted from Long Valley Caldera circa 0.71 million years ago. It now crops out over an area of about 900 Km^2 in eastern California.

cluster of volcanic domes revealing their shape in three-dimensions and their internal structure (see Fig. 8.3 and Fig. 8.7). The Mono Domes of the Long Valley-Mono Basin Province of eastern California are another famous cluster of rhyolitic domes (see Fig. 8.4; Bailey *et al.*, 1989). Volcanic domes, cryptodomes and coulee are often extruded quietly, immediately prior to large explosive events that expel extensive pyroclastic flow, surge and air-fall deposits.

8.11   Mount Saint Helens Volcano – an active andesite–dacite volcano

A variety of basaltic, basaltic andesitic, andesitic, dacitic and rhyodacitic rocks have erupted from Mount Saint Helens Volcano during the past 40 000 years (Mullineaux and Crandell, 1981; Foxworthy and Hill, 1982). Before 1980 it had a symmetrical

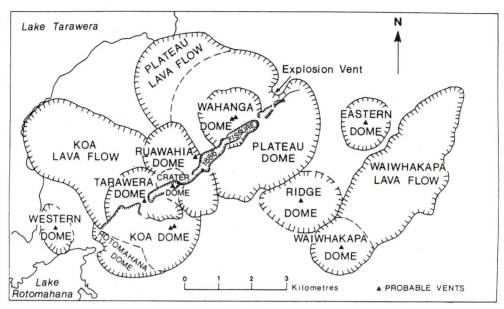

Fig. 8.7 Sketch map of rhyolite domes and flows of the Tarawera Complex cut by a line of craters that developed during the 1886 fissure eruption, central North Island, New Zealand (see Fig. 8.3). Adapted from Cole, 1970.

cone that was almost entirely constructed within the past 2200 years. The upper 400 m of the cone was removed on May 18, 1980, by a 2.7 km³ landslide. This released pressure on a growing cryptodome, resulting in a massive debris avalanche and lateral blast. A huge plinian eruption column developed. It spread dacitic pyroclastic flows and air-fall deposits over a wide area. The explosive eruption devastated approximately 600 km². Debris avalanches buried the upper 24 km of the North Fork Toutle valley to a depth of some 50 m.

This volcano lies about 250 km east of the convergent boundary between the Juan de Fuca Plate and the North American Plate. Beneath it the crust is between 40 and 46 km thick and the subducted slab dips at approximately 45°. According to Smith and Leeman (1993, p. 283) the basalts that have issued from Mount Saint Helens Volcano were of two distinct types with distinct incompatible element abundance patterns. To explain these chemical characteristics they claim that the volcano contains a complex plumbing system that can prevent mixing between different batches of magma. Smith and Leeman (1993, p. 290) also suggest that the dacitic materials that erupt from this volcano are derived from magma supplied from a mid-crustal magma chamber at depths of 7–11 km beneath the volcano. Whether the primary magma that evolved to produce the dacitic and andesitic magmas, assimi-

lated some crust rocks, reacted with subcrustal lithospheric materials or incorporated young and hot subducted materials is still being debated.

References

Bailey, R.A., Miller, C.D. and Sieh, K., 1989. Long Valley caldera and Mono-Inyo craters volcanic chain, eastern California. *New Mexico Bureau of Mines Resources Memoir*, 47, 227–254.

Belout, G.E., 1991. Field-based evidence for devolatization in subduction zones: implications for arc magmatism. *Science*, 251, pp. 413–416.

Bowen, N.L., 1937. Recent high-temperature research on silicates and its significance in igneous geology. *American Journal of Science*, 33, pp. 1–21.

Buddington, A.F., 1927. Coast range intrusives of southeastern Alaska. *Journal of Geology*, 35 (3), pp. 224–246.

Bullard, F.M., 1977. *Volcanoes of the Earth*. University of Queensland Press, St. Lucia, Queensland, Australia, 579 pp.

Carmichael, I.S.E., 1964. The petrology of Thingmuli, a Tertiary volcano in Eastern Iceland, *Journal of Petrology*, 5 (3), pp. 435–460.

Cole, J.W., 1970. Structure and eruptive history of the Tarawera Volcanic Complex. *New Zealand Journal of Geology and Geophysics*, 13 (4), pp. 879–902.

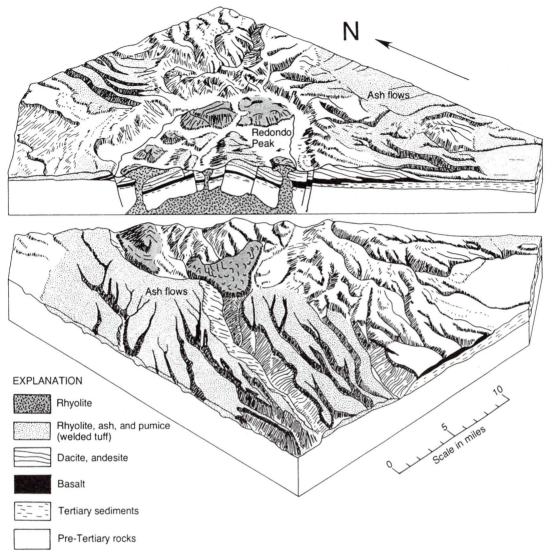

N

Ash flows

Redondo
Peak

Ash flows

EXPLANATION

Rhyolite

Rhyolite, ash, and pumice
(welded tuff)

Dacite, andesite

Basalt

Tertiary sediments

Pre-Tertiary rocks

10

5

Scale in miles

0

Fig. 8.8 Block diagram showing the Valles Resurgent Caldera in the Jemez Mountains, New Mexico (see Fig. 8.5). Note the up domed caldera floor and the surrounding ring of rhyolitic domes. Adapted from Bullard, 1977, p. 107.

Crawford, A.J. (editor), 1989. *Boninites.* Unwin Hyman, London, 465 pp.

Crawford, A.J., Falloon, T.J. and Green, D.H., 1989. Classification, petrogenesis and tectonic setting of boninites. In Crawford, A.J. (editor), *Boninites*, pp. 1–49, Unwin Hyman, London, 465 pp.

Davis, J.M., Elston, W.E. and Hawkesworth, C.J., 1993. Basic and intermediate volcanism of the Mogollon-Datil volcanic field: implications for mid-Tertiary tectonic transitions in southwestern New Mexico, USA. In Prichard, H.M., Alabaster, T., Harris,

N.B.W. and Neary, C.R. (editors), *Magmatic Processes and Plate Tectonics*, pp. 469–488, Geological Society of London, Special Publication Number 76, 526 pp.

Dickinson, W.R., 1968. Circum-Pacific andesite types. *Journal of Geophysical Research*, 73, pp. 2261–2269.

Elston, W.E., 1989. Overview of the Mogollon-Datil volcanic field. *New Mexico Bureau of Mines and Mineral Resources*, Memoir Number 46, pp. 43–46.

Foxworthy, B.L. and Hill, M., 1982. *Volcanic Eruptions of 1980 at Mount Saint Helens: The First 100 Days*. U.S. Geological Survey Professional Paper 1249, 125 pp.

Gill, J.B., 1981. *Orogenic Andesites and Plate Tectonics*, Springer-Verlag, Berlin, 390 pp.

Guillot, S., Pecher, A., Rochette, P. and Le Fort, P., 1993. The emplacement of the Manaslu granite of Central Nepal: field and magnetic susceptibility constraints. In Treloar, P.J. and Searle, M.P. (editors), *Himalayan Tectonics*, pp. 413–428, Geological Society of London, Special Publication Number 74, 640 pp.

Hamilton, W.B., 1995. Subduction systems and magmatism. In Smellie, J.L. (editor), *Volcanism Associated with Extension at Consuming Plate Margins*, pp. 3–28, Geological Society of London, Special Publication Number 81, 293 pp.

Le Maitre, R.W., 1984. A proposal by the IUGS Subcommission on the Systematics of Igneous Rocks for a chemical classification of volcanic rocks based on the total alkali silica (TAS) diagram. *Australian Journal of Earth Sciences*, 31 (2), pp. 243–255.

Le Maitre, R.W., Bateman, P., Dudek, A., Keller, J., Lameyre, J., Le Bas, M.J., Sabine, P.A., Schmidt, R., Sørensen, H., Streckeisen, A., Woolley, A.R. and Zanettin, B., 1989. *A Classification of Igneous Rocks and Glossary of Terms: Recommendations of the International Union of Geological Sciences Subcommission on the systematics of igneous rocks.* Blackwell Scientific Publications, Oxford, 193 pp.

McBirney, A.R., 1993. Differentiated rocks of the Galapagos hotspot. In Prichard, H.M., Alabaster, T., Harris, N.B.W. and Neary, C.R. (editors), *Magmatic Processes and Plate Tectonics*, pp. 61–69, Geological Society of London, Special Publication Number 76, 526 pp.

Mitchell, A.H.G. and Carlile, J.C., 1984. Arc-related gold-copper mineralization, basement domes and crustal extension. *Journal of Southeast Asian Earth Sciences*, 10 (1/2), pp. 39–50.

Müller, D. and Groves, D.I., 1995. *Potassic Igneous Rocks and Associated Gold-Copper Mineralization.* Springer-Verlag, Berlin, 210 pp.

Mullineaux, D.R. and Crandell, D.R., 1981. *The eruptive history of Mount Saint Helens.* United States Geological Survey Professional Paper 1250, pp. 3–15.

Nielson, D.R. and Stoiber, R.E., 1973. Relationship of potassium content in andesite lavas and depth to seismic zone. *Journal of Geophysical Research*, 78, pp. 6887–6892.

Nockolds, S.R. and Le Bas, M.J., 1977. Average calc-alkali basalt. *Geological Magazine*, 114 (4), pp. 311–312.

Peacock, S.M. 1990. Fluid processes in subduction zones. *Science*, 248, pp. 329–337.

Peacock, S.M., Rushmer, T. and Thompson, A.B., 1994. Partial melting of subducting oceanic crust. *Earth and Planetary Science Letters*, 121, pp. 227–244.

Pearce, J.A. and Parkinson, I.J., 1993. Trace element models for mantle melting: application to volcanic arc petrogenesis. In Prichard, H.M., Alabaster, T., Harris, N.B.W. and Neary, C.R. (editors), *Magmatic Processes and Plate Tectonics*, pp. 373–403, Geological Society of London, Special Publication Number 76, 526 pp.

Petersen, J., 1890. Beitrage zur Petrographie von Sulphur Island, Peel Island, Hachijo und Mijakeshima. *Jahrbuch der Hamburgischen Wissenschaftlichen Anstalten*, Hamburg, 8, pp. 1–58.

Peccerillo, A. and Taylor, S.R., 1975. Geochemistry of Upper Cretaceous Volcanic Rocks from the Pontic Chain, Northern Turkey. *Bulletin Volcanologique*, 39 (4), pp. 1–13.

Peccerillo, A. and Taylor, S.R., 1976. Geochemistry of the Eocene calc-alkaline volcanic rocks from the Kastamonu area, northern Turkey. *Contributions to Mineralogy and Petrology*, 58, pp. 63–81.

Ringwood, A.E., 1974. The petrological evolution of island arc systems. *Journal of the Geological Society of London*, 130, pp. 183–204.

Sillitoe, R.H., 1976. Andean mineralisation: a model for the metallogy of convergent plate margins. In Strong, D.F. (editor), *Metallogeny and Plate Tectonics*, pp. 59–100, Geological Society of Canada, Special Paper 14.

Smellie, J.L. (editor), 1994., *Volcanism Associated with Extension at Consuming Plate Margins.* Geological Society of London Special Publication Number 81, 293 pp.

Smith, D.R., and Leeman, W.P., 1993. The origin of Mount Saint Helens andesites. *Journal of Volcanology and Geothermal Research*, 55, pp. 271–303.

Tatsumi, Y. and Eggins, S., 1995. *Subduction Zone Magmatism.* Blackwell Science, Cambridge, USA, 211 pp.

Thompson, G., Bryan, W.B. and Frey, F.A., 1973. Petrology and geochemistry of basalts and related rocks from the DSDP leg 22 sites 214 and 216, Ninety-East Ridge, Indian Ocean. *EOS: Transactions of the American Geophysical Union*, 54 (11), pp. 1019–1021.

Thorpe, R.S. (editor), 1982. *Andesites: Orogenic Andesites and Related Rocks.* John Wiley and Sons, Chichester, U.K., 724 pp.

Whitford, D.J. and Nicholls, I.A., 1976. Potassium variation in lavas across the Sunda Arc in Java and Bali. In Johnson, R.W. (editor), *Volcanism in Australasia*, pp. 63–75, Elsevier Scientific Publishing Company, Amsterdam, The Netherlands, 405 pp.

Zhao, D. and Hasegawa, A., 1993. P wave tomographic imaging of the crust and upper mantle beneath the Japan islands. *Journal of Geophysical Research*, 98, pp. 4333–4353.

CHAPTER 9 — Granites – the crusty enigma

Complexity is to be expected, for in the system above a subducting slab many sites of melting and contamination are possible.

Fyfe, 1988, p. 341.

9.1 Granites and the granitic rocks

Historically the development of ideas on the evolution of the continental crust and on the origin of the granitic rocks have been intricately intertwined. As the Earth has evolved so have the granitic rocks and they are now the pre-eminent type of rock in the upper continental crust. According to Le Maitre *et al.* (1989, pp. 14–16) granite is a plutonic rock that contains between 20 to 60 per cent quartz and alkali feldspar makes up between 35 and 90 per cent of its total feldspar content. The granitic rocks include alkali feldspar granite, granite (sensu stricto), granodiorite and tonalite (see Fig. 2.2). On the plutonic QAPF diagram these rocks occupy an impressively large field that includes all the rocks that contain between 20 to 60 per cent quartz. Chemically the alkali feldspar granites and granites are equivalent to the rhyolites, the granodiorites are similar to the dacites, while most tonalites are akin to the andesites (see Section 2.7 and Table 9.1).

If one tries to use chemical criteria to develop a detailed classification of the silicic rocks one soon discovers the great merit of Shand's (1947) idea of separating igneous rocks into subaluminous, metaluminous, peraluminous and peralkaline groups (see Section 2.9). Granitic rocks usually belong to either the metaluminous, peraluminous or peralkaline groups. The tonalites and granodiorites are typically metaluminous. Most granites (sensu stricto) are peraluminous, whereas many alkali feldspar granites are peralkaline. The excess of alumina in the peraluminous granitic rocks usually results in the crystallisation of minerals such as biotite, muscovite, corundum, garnet, topaz and tourmaline. In the peralkaline rocks the excess in alkalies usually results in the crystallisation of the soda-bearing pyroxenes and amphiboles (see Table 9.1).

Before proceeding it is timely to ponder why we are surveying the granitic rocks as a discrete and special group of rocks. More particularly, it may seem strange to separate them from the rhyolites and dacites. Discussions on rhyolite petrogenesis are usually about how diverse primary magmas follow separate liquid lines of descent yet terminate in the rhyolite field (see Section 8.8). While this idea is important in evaluating the origin of the granitic rocks, the granitic rocks and their exocontact zones can usually be induced to reveal more varied and valuable petrogenetic information than the rhyolites and dacites. This is because the granitic rocks contain a variety of phases and enclaves (inclusions), display complex and occasionally unusual textures and often contain unmistakable evidence that they formed from the melting of crustal rocks. One might summarise the contrasting attributes of rhyolites and granites by recording that rhyolites typically provide information about cataclysmic volcanism, whereas the granitic rocks and the metamorphic rocks that surround them are more likely to carry information about crustal evolution. The peraluminous autochthonous granites are, for example, often surrounded by migmatites. They also may contain metasedimentary xenoliths and xenocrysts. Perhaps they should be regarded as a special group of igneous rocks because they are restricted to a singular continental intracrustal setting. Read (1957, p. 373), a pioneer of granite research, once declared that 'it is only by the grace of granitization that we have continents to live on'.

9.2 Watery granites or fiery granites

In the eighteenth century Werner (1749–1817) and the Neptunist school considered all granite to be

Table 9.1 Major element compositions and CIPW norms of the granitic rocks.

Rock type	Granite Le Maitre, 1976, p. 606	Peralkaline granite Nockolds, 1954	Granodiorite Le Maitre, 1976, p. 609	Tonalite Le Maitre, 1976, p. 612
SiO_2	71.3	71.08	66.09	61.52
TiO_2	0.31	0.4	0.54	0.73
Al_2O_3	14.32	11.26	15.73	16.48
Fe_2O_3	1.21	4.28	1.38	1.83
FeO	1.64	2.19	2.73	3.82
MnO	0.05	0.11	0.08	0.08
MgO	0.71	0.25	1.74	2.8
CaO	1.84	0.84	3.83	5.42
Na_2O	3.68	4.92	3.75	3.63
K_2O	4.07	4.21	2.73	2.07
P_2O_5	0.12	0.07	0.18	0.25
H_2O^+	0.64	0.39	0.85	1.04
H_2O^-	0.13	0	0.19	0.2
CO_2	0.05	0	0.08	0.14
Total	100.07	100	99.9	100.01
CIPW norms				
Quartz	29.14	26.11	22.62	16.92
Corundum	0.92	0	0.26	0
Orthoclase	24.56	24.96	16.3	12.46
Albite	31.21	34.6	32.11	31.21
Anorthite	8.06	0	17.54	22.98
Diopside	0	3.19	0	1.52
Hypersthene	3.38	0.82	7.49	9.85
Aegirine	0	6.34	0	0
Magnetite	1.75	3.06	2.02	2.71
Ilmenite	0.58	0.76	1.04	1.42
Apatite	0.28	0.16	0.43	0.59
Calcite	0.12	0	0.19	0.34

coarse-grained crystalline rocks that precipitated out of a vast, primeval body of water (see Section 1.2 for more details and references). During the seventies and eighties of the eighteenth century Hutton postulated that granites crystallised from molten rock materials and was the first to conceptualise the magma-igneous rock system (see Fig. 9.1). It was only after the publication of Lyell's *Principles of Geology* (1830) that it became widely recognised that the granitic rocks formed at different times in Earth history. In 1837 Lyell visited the Oslo region where he had stimulating discussions with Keilhau on the origin and emplacement of the plutonic rocks of that area (Holtedahl, 1963). Keilhau was particularly troubled by the problem of how large bodies of granite, such as the Drammen Granite, could make room for themselves in the solid crust. He claimed that some contact zones in the Oslo region contained evidence of a gradual transition from sedimentary rocks to granite. This led Keilhau to suggest that the older

sedimentary rocks of the Oslo region could be changed into granitic rocks by a slow process he called granitification. Lyell remained convinced that most granitic rocks had crystallised from molten materials that cooled slowly beneath the surface.

In the nineteenth century many geologists claimed that while some granites were magmatic in origin others were not. Several French geologists who were particularly interested in the textural features observed in contact zones postulated that metamorphic processes were likely to produce granitic rocks. Delesse (1861) suggested that some of these metamorphic granites could under suitable thermal and tectonic conditions become mobile and intrude overlying rocks. The multifold ideas on the origin of granite were succinctly summarised by Green (1882, p. 443) who declared that there were 'granites and granites ... some formed in one way and some in another'. In his classification of crystalline rocks he (1882, p. 445) proposed three classes: (1) the volcanic

Fig. 9.1 Leucogranite (light) intruding metasedimentary rocks (dark) in the Mount Rushmore area of the Black Hills, South Dakota.

rocks; (2) the plutonic rocks that included many granites; and (3) the metamorphic rocks that included some granites.

Termier (1904) elaborated the idea of generating granites in situ by metamorphic processes. He suggested that highly energised emanations can ascend from unspecified depths and transform pre-existing rocks into granites. In brief, his argument was that the evolution of granites and the generation of regionally metamorphic rocks were two effects of the same cause. This idea keeps reappearing in a modified form in debates on the thermal and chemical evolution of the continental crust.

9.3 Granitisation, ultrametamorphism and migmatites

In 1907 Sederholm introduced the term migmatite and focused attention on these important rocks that contain both igneous-looking and metamorphic components. He reported many well exposed examples of

these mixed rocks from the contact zones surrounding the granitic rocks of southern Finland. These outcrops and similar rocks from all the continents show that some granites have gradational, or thermally harmonious, contact zones. In most of these contact zones supracrustal rocks appear to be transmuted by imperceptible changes, first into thermally metamorphosed rocks, then into migmatites and finally into granitic rocks. Over the years Sederholm's ideas on the origin of granite changed. In the end he placed much less emphasis on magmatic processes and more on the movement of volatile-rich fluids. He called these fluids ichor. This strange-sounding word has variously been used to describe the blood of the gods, the blood of dragons, and the watery serum that discharges from human wounds. Read (1957, p. 120) states that Sederholm was not attracted by the term 'granitic juices' as it seemed to have a vegetable flavour about it. After careful consideration Sederholm decided to use the stronger Greek word ichor. The main flaw in the idea of using a fluid to transform crustal rocks into granite is that it implies that a small quantity of fluid can introduce sufficient

energy to transform large volumes of rocks into granite. Sederholm's perception of the importance of
hydrothermal systems in modifying the mineralogy
and textures of some granitic rocks and the rocks
that surround them has been confirmed by subsequent research.

In 1909 the Swede, Holmquist, introduced the term
ultrametamorphism to describe an extreme metamorphic, or quasi-igneous, process that ultimately
resulted in the partial fusion of the country rocks.
This idea is important as it links anatexis, high-
grade thermal metamorphism and the development
of thermal anomalies in the continental crust. Wegmann (1935) and a few others attributed the origin of
granite to the migration of ions. They became known
as the dry transformists and they championed the
idea of an advancing migmatite front that moved
ahead of a granitization front.

At this time granitization, either wet or dry, was a
widely accepted process. Proponents of the process
usually stipulated that it resulted in the addition of
sodium and potassium and the removal of iron, magnesium and calcium. The main advocate of the granitization hypothesis was Read. He (1957) eloquently
presented his ideas, and those of the other wet transformists, in an erudite book entitled *The Granite
Controversy*. According to him granitization was a
metasomatic process that produced extensive changes
in the chemical composition of large volumes of rock.
Petrologists usually have difficulty dealing with metasomatic processes because when they examine the
products of this process, the initial rocks have
already been replaced and the fluids that facilitated
the change have disappeared.

The granitization theory reached its zenith in 1955
when most of the participants in a symposium at
Nancy in France supported it (Mehnert, 1968).
Three years later, with the publication of Tuttle
and Bowen's (1958) memoir on the origin of granite
in the light of experimental studies in the system
$NaAlSi_3O_8–KAlSi_3O_8–SiO_2–H_2O$, the views of most
petrologists began to change. They now championed
the idea that magmatic processes are important in the
evolution of most granitic rocks. The experimental
system used by Tuttle and Bowen was trailblazing
because it included water. It is now usually accepted
that silicic magmas are likely to contain between one
and four weight per cent water, and on cooling some
magmas will eventually become water saturated. This
is likely to produce rocks with complex, even pegmatitic, textures. The release of this water is also likely
to establish a long-lasting hydrothermal system that

metasomatically alters the rocks it encounters (Luth,
1976). Some experimental petrologists claim that the
formation of water-saturated liquids of silicic composition is a 'normal consequence of regional metamorphism' (Wyllie, 1983, p. 24). He (p. 25) has also
suggested that silicic magma can be produced by the
partial melting of normal crustal rocks and that the
various granitic rocks can be generated by mixing
silicic magma with different proportions of relict
refractory materials. This idea has been further developed by other petrologists, such as Chappell *et al.*
(1987) who have proposed that substantial amounts
of residual source materials (or restite) can be
entrained in silicic magmas. They postulate that a
comprehensive suite of chemically related, but different, granitic rocks can be produced by mixing
different proportions of restite and silicic magma.

Detailed studies of granitic rocks from many localities show subtle and more robust heterogeneities and
disequilibrium features over a wide range of scales.
These features are usually products of mixing
between the intruding magma and various other
materials. Recent theoretical investigations into
how granitic diapirs move through the crust have
revealed how and why some of this mixing takes
place. Weinberg (1995, p. 162) has shown that granitic diapirs moving through the crust have several
features in common with mantle plumes moving
through the mantle. In both there is likely to be viscous coupling between the hot rising body and the
rocks that surround it. Such an action probably will
lead to some country rocks being dragged upwards
resulting in their decompression and partial melting.
This raises the buoyancy of the entrained materials so
speeding up the ascent of the diapir without consuming its thermal energy. As long as this process continues to operate it will produce an expanding and
accelerating diapir that ingests some the country
rocks it encounters. This runaway process is likely
to slow down and eventually stop when the granitic
diapir reaches a zone of cooler and more viscous
crust. The cooler environment will encourage crystallisation leading to a reduction in buoyancy.

Structural and geophysical patterns

In the 1920s Cloos examined the granitic rocks in a
new way. He attempted to map the structural features of granitic plutons and the rocks that enveloped
them. In his (1923) well-known, but seldom read,

book *Das Batholithenproblem* he traced the interaction between buoyant silicic magmas and their carapace of country rocks. Later Ramberg (1970) greatly increased our knowledge of how plutons are emplaced by devising successful fluid dynamic experiments that simulated these processes. Recent advances in field mapping, image interpretation and the collection and interpretation of geophysical data have all enhanced our perceptions on how granitic rocks are emplaced and how they make room for themselves at different levels within the crust.

Lately there have been several excellent structural investigations of granitic bodies and their tectonic settings. For example, Brun *et al.* (1990) investigated the emplacement history of the small Flamanville granodiorite in Normandy, northwestern France. Emplacement of this pluton was at least partly synchronous with regional deformation. Its internal foliation reveals a gradual increase in strain towards its margins, showing that the pluton probably expanded laterally during emplacement. This idea is supported by the shallow plunging and stretching lineation discovered in the metamorphosed rocks surrounding the pluton.

Geophysical methods may help one figure out the internal structure and three-dimensional form of granitic bodies. Gravimetric, magnetic, radiometric image analysis and reflection seismic techniques have all been used in attempts to discover the shape of granitic bodies at depth. According to Vigneresse (1990, p. 258) gravimetric methods are particularly successful, because they show differences in density between granitic bodies and the rocks that surround them. Unfortunately the resolution of this method deteriorates with depth. This means that the floor of a granitic body determined from gravity data inversion may not correspond exactly with the actual physical contact. Ambiguities usually arise because the basal contact between the granitic rocks and their source rocks is gradational or the granitic rocks are part of a more extensive layered plutonic body. Seismic reflection profiles are more reliable at revealing layering beneath or within granite batholiths. Radiometric image analysis often provides a valuable technique for uncovering the general pattern of variation in composition within a batholith.

Seismic methods have also been particularly valuable in determining the form of silicic magma chambers beneath large active caldera, such as the Yellowstone Caldera (Smith and Braile, 1984, pp. 103–106). Most granitic bodies that have been geophysically investigated are lenticular or sill-shaped.

The Manaslu Leucogranite of Central Nepal in the Himalayas has been mapped as an 8 km thick lenticular slab, or laccolith, that lies on a 12 km thick body of crystalline rocks that have been thrust over the Indian lithospheric plate during continent–continent collision. Guillot and Le Fort (1995, p. 231) have proposed that the High Himalayan leucogranites were generated by anatexis. This process operated at pressures of between 0.6 to 0.8 GPa, and temperatures of between 620 and 730°C. Geochemical studies, particularly strontium isotopic studies, show that there were two main source rocks (metagreywackes and metapelites). The anatectic process was largely controlled by the amount of water present in these rocks. Water-saturated conditions prevailed during the melting of the metagreywackes. This favoured the evolution of the two-mica leucogranites. Water-absent conditions prevailed during the partial melting of the metapelites. This resulted in biotite, plagioclase and monazite fractionation and ultimately in the evolution of tourmaline-bearing leucogranites.

An error of category

Recent reviews of the nature and origin of the granitic rocks (Bonin, 1986; Brown *et al.*, 1988; Shimizu, 1986; Clarke, 1992; Pitcher, 1993) all conclude that granite can be produced in many ways. Much of the controversy that once raged about the origin of granite arose because of a categorical error. All granitic rocks were placed in a single category and it was then presumed that they all formed in the same way. Every granite poses new challenges. Before attempting to respond to these challenges and placing a granite in some currently fashionable classification, one should ponder Pitcher's (1993, p. 28) maxim that 'to each tectonic situation, to each stage of the Wilson cycle, there is likely to be a corresponding type of granitic association'.

9.4 Special ways of classifying granitic rocks

For over two centuries petrologists have tried to understand the evolution of the continental crust, and the part played by silicic magmas in this epic process. In a review paper Barbarin (1990, p. 229) evaluated over 30 different petrogenetic classifications

Table 9.2 Mean major element compositions and CIPW norms of the M-, I-, S- and A-type granitic rocks. After Whalen *et al.*, 1987, p. 409.

Rock type	M-type	I-type	S-type	A-type
SiO_2	67.24	69.17	70.27	73.81
TiO_2	0.49	0.43	0.48	0.26
Al_2O_3	15.18	14.33	14.1	12.4
Fe_2O_3	1.94	1.04	0.56	1.24
FeO	2.35	2.29	2.87	1.58
MnO	0.11	0.07	0.06	0.06
MgO	1.73	1.42	1.42	0.2
CaO	4.27	3.2	2.03	0.75
Na_2O	3.97	3.13	2.41	4.07
K_2O	1.26	3.4	3.96	4.65
P_2O_5	0.09	0.11	0.15	0.04
Total	98.63	98.59	98.31	99.06
CIPW norms				
Quartz	27.44	28.49	33.77	30.5
Corundum	0	0	2.52	0
Orthoclase	7.54	20.37	23.8	27.72
Albite	34.05	26.86	20.75	34.76
Anorthite	20.15	15.22	9.25	1.85
Diopside	0.59	0.12	0	1.41
Hypersthene	6.21	6.32	7.79	1.35
Magnetite	2.86	1.53	0.83	1.82
Ilmenite	0.95	0.83	0.93	0.5
Apatite	0.21	0.26	0.36	0.09

of the granitic rocks. The criteria used in these classifications were remarkably diverse. They included petrographic (various), modal composition, type and abundance of enclaves or inclusions, mafic mineral content, biotite composition, zircon morphology, opaque oxide compositions, major element abundance, trace element abundance, isotopic ratios, nature of cognate ore bodies and tectonic setting. Most of these classifications are neither systematic nor practical. They were usually devised to resolve a local problem or to illustrate a particular tectonic or petrogenetic hypothesis that was popular at the time. Barbarin has attempted to show how these diverse classifications, with their complex and often contradictory nomenclature, can be related to one another. A lucid and valuable critique of the sometimes ambiguous and often puzzling alphabetic classifications of the granitic rocks (see Table 9.2 and Section 9.8) is to be found in Clarke's book on granites (1992, pp. 3–15).

9.5   Modal classifications

As we have already noted in the plutonic rock classification proposed by the IUGS Subcommission on the Systematics of Igneous Rocks, granite (sensu stricto) occupies a large field on the QAP diagram. This liberal definition of granite is easily defended because most of the different types of granite recognised by systematic petrographers fall within this field yet they have broadly similar chemical compositions. This apparent anomaly arises because sodium and potassium are distributed in several different ways among minerals, such as the feldspars and the micas, that usually crystallise in these rocks.

Tuttle and Bowen (1958, p. 79) investigated the abundance of normative albite-orthoclase-quartz in a large collection of granitic rocks. They discovered that the rocks contained similar amounts of all three normative minerals. When plotted on a QAP diagram their data-points form a remarkably coherent group. The boundary between what some petrographers regard as the authentic granites and the monzogranites cuts through the main cluster of data-points. As has already been recorded the granitic rocks include the alkali feldspar granites, the granites (sensu stricto), the granodiorites and the tonalites (see Table 9.1). Granites and granodiorites that contain more than 95 per cent felsic minerals are called leucogranites and leucogranodiorites respectively (Le Maitre, 1989, p. 19). Tonalites that contain more than 90 per cent felsic minerals are called trondhjemites. Trondhjemites are especially abundant in Archean cratonic areas.

To understand the cooling and crystallisation history of a typical granitic rock one has to know a little about the structural complexities of the alkali feldspars. At high temperatures they crystallise as a complete solid solution series, but at lower temperatures the alkali feldspars of intermediate composition occur as intergrowths of the two end-member feldspars. These intergrowths of sodium and potassium feldspar range in size from submicroscopic to megascopic and are called cryptoperthites, microperthites or perthites. The crystalline phases in a typical perthite, from a granite, are microscopic to megascopic in size and consist of parallel to subparallel strings or blebs of low-albite set in a host crystal of orthoclase or microcline.

The principal minerals in the granitic rocks are the feldspars (orthoclase, orthoclase microperthite, microcline, microcline microperthite or perthite, and plagioclase), quartz, and one or more of the mafic minerals, such as biotite, hornblende, muscovite or even an orthopyroxene. The accessory minerals typical of the granitic rocks are apatite, zircon and Fe-Ti oxides (ilmenite and magnetite). Other less abundant accessory minerals include allanite, sphene, pyrite, tourmaline, fluorite, monazite and xenotime. Some granites contain so-called metamorphic minerals, such as almandine garnet, andalusite, cordierite and sillimanite. These minerals are often regarded as xenocrysts that were incorporated into a silicic magma during the dismemberment and partial assimilation of pre-existing crustal rocks. In the Canberra-Yass area of eastern Australia, Wyborn (1981) has described both rhyolites and granites that contain such xenocrysts. As one would expect the peralkaline granitic rocks usually contain alkalic mafic minerals like aegirine, aegirine-augite, arfvedsonite and riebeckite.

The charnockitic rocks (Le Maitre *et al.*, 1989, p. 13) include granitic members. They are usually called alkali charnockites (equivalent of alkali feldspar granites), charnockites (granites), opdalites (granodiorites) and enderbites (tonalites). These special rocks are usually recognised by the presence of orthopyroxene (often called hypersthene), or fayalite plus quartz. Dark coloured mesoperthite is another characteristic feature of many of these rocks. Mesoperthite is a variety of perthitic feldspar that contains about equal parts of potassium feldspar and plagioclase. Many charnockitic rocks display signs of metamorphic overprinting. Such rocks typically occur in close association with norites and anorthosites in granulite facies metamorphic terranes. They are usually regarded as a special group of igneous-looking plutonic rocks that exist outside the main sequence of igneous rocks.

9.6   Simple and complex textures

Most granitic rocks are medium to coarse grained with a hypidiomorphic-granular texture. Other granites appear to have porphyritic textures. Yet Pitcher (1993, pp. 60–77) devoted a whole chapter of his book on granites to evaluating the evolution of granitic textures. He regards these textures as especially important, and has insisted (p. 60) that 'to understand the evolution of granitic textures is to begin to understand the genesis of granites'. Augustithis (1973) has gone further and produced a copiously illustrated compendium on textural patterns in granites, gneisses and associated rocks. The granitic rocks with the most igneous-looking textures are usually found in subvolcanic settings, yet Bonin (1986, pp. 49–51) has shown that even these rocks often have textures that reveal a complex cooling history. This is because most granitic rocks have long cooling histories and their textures frequently reveal primary magmatic features strongly modified by later metasomatic reactions. Granites usually have the appearance of rocks that have attempted to maintain mechanical and chemical equilibrium over an extended period.

Minerals such as alkali feldspar, amphibole, mica and even zircon, often reveal complex zoning that suggests an interrupted crystallisation history. Small forests have been cut down to supply the paper used to describe the origin of the potassium feldspar megacrysts that occur in granitic rocks and their adjoining thermal aureoles. Some petrologists call these potassium feldspar crystals phenocrysts. Others consider them to be porphyroblasts. The mafic minerals are often enigmatic because they frequently cluster into aggregates or appear as indistinct blotches in the host granite.

The complex rapakivi texture is found in granitic rocks from all the continents. According to Rämö and Haapala (1995, p. 129) most of these rocks are Proterozoic in age (mainly one to 1.7 giga-years) and they make up 'the most voluminous continental silicic intraplate magmatism on Earth'. The type area is the Wilborg batholith adjoining the Gulf of Finland. In this area the granite contains large rounded ovoids of alkali feldspar rimmed by a plagioclase overgrowth. The alkali feldspar ovoids may consist of a single crystal or complex intergrowths of several crystals. As many igneous complexes that include rapakivi granites contain coeval silicic and mafic rocks it is likely that some rapakivi textured rocks are products of magma mixing. Many petrologists have considered this process (Hibbard,1981; Sparks and Marshall, 1986; Frost and Mahood, 1987). Some support the view that these rocks were produced by changing feldspar stabilities in a complex hybrid magma. Others consider this complex texture to result from the gravitative settling of alkali feldspar megacrysts in a compositionally layered magma chamber. To conclude this section on the textures of granitic rocks it is apposite to quote Luth (1976, p. 336) who states that most granitic rocks exhibit textural and mineralogical features that are less related to their ultimate magmatic origin and early history than to their subsequent subsolidus recrystallisation.

9.7 Geochemistry and normative compositions

There is normally a progressive decrease in titanium, aluminium, magnesium and calcium, and a concomitant increase in silica and potassium in the series tonalite $\rightarrow$ granodiorite $\rightarrow$ granite $\rightarrow$ alkali feldspar granite. These changes are also reflected in the normative and modal compositions of the rocks in this series as there is a continuous increase in quartz and potassium feldspar and a concomitant decrease in the normative amounts of anorthite, pyroxene, magnetite and ilmenite, and in the abundance of actual ferromagnesian minerals (see Tables 9.1 and 9.2). This also means that the indices of differentiation discussed in Section 2.8 increase from tonalite, via granodiorite to granite and alkali feldspar granite.

When evaluating the composition of granitic rocks one always has to consider the possibility that some of its phases, or the cores of some of its phases, are derived from older rocks. Many of these rocks appear to have crystallised from magmas strewn with an assortment of xenoliths and xenocrysts. For example, some granites contain zircons that exhibit a succession of overgrowths. Research has confirmed that the cores of these complex crystals may be older than the emplacement ages of their host gran-

ites. Much useful and interesting information has been gathered from dating overgrowths in zircons collected from ancient gneissic granites that are typical of many cratonic areas. For example, material from the core of a zircon from the Yilgarn Block of Western Australia has yielded a crystallisation age of over four giga-years. Since these ancient zircons probably crystallised from silicic magmas it is most likely that the first granitic rocks crystallised in the Hadean Era (see Table 1.2). Granitic gneisses of probable Hadean age are also reported from the Slave Province of the Canadian Shield (Bowring *et al.*, 1989, p. 971).

Geochemical anomalies that probably result from variable dissolution rates in accessory phases have been reported from many granitic rocks of different ages. The Himalayan leucogranites are examples of rocks that appear to have crystallised from magmas that did not remain in the source region long enough to equilibrate fully with all the phases present. Rushmer (1995, p. 131) has shown that the slow dissolution rates of accessory minerals, like apatite, sphene and zircon, can result in the generation of a magma that has a different trace element signature to the one predicted by standard partition coefficients.

9.8 Abundance and distribution in space and time

Granitic rocks are extremely abundant, with extensive spatial and temporal distributions in the continental crust of the Earth (see Figs 9.2 and 9.3). A typical section through the continental crust would contain an 8 km thick layer of granitic rocks and migmatites. According to Brown (1979, p. 106) 'the average composition of the continents is virtually tonalite'. Granitic rocks are particularly common in the Precambrian shield areas. Most of these ancient rocks are sodium-rich tonalites or trondhjemites. The granitic rocks of the Phanerozoic Era usually occur in cordillera-type batholiths that were emplaced near convergent plate boundaries. For example, the Coast Plutonic Complex of British Columbia crops out over an area of about $100\,000\,km^2$ (Roddick and Hutchinson, 1974). In southeastern Alaska, there is an extraordinary tonalite sill that is 5 km wide and about 900 km long (Drinkwater *et al.*, 1990).

Other Phanerozoic granitic bodies occur in non-subduction settings. Such rocks usually belong to the tholeiitic, midalkaline or peralkaline rock associations. Their genesis and emplacement is frequently

Fig. 9.2 Leucogranite to the south of Rotenfels beacon in the central part of the Bremen granite–syenite complex of southern Namibia.

Fig. 9.3 Fresh glaciated granites of the Lyngdal area of southern Norway.

Fig. 9.4 Sharp contact between leucogranite (light) and porphyritic syenite (dark) in the Richtersveld Ring Complex, northwestern Namaqualand, South Africa.

associated with rifting, doming and the ascent of mantle plumes. Many midalkaline and peralkaline rocks were emplaced into subvolcanic ring complexes (see Fig. 9.4), similar to those of the Oslo Magmatic Province and the Younger Granite Province of northern Nigeria (Jacobson *et al.*, 1958).

Cordilleran-type batholiths normally contain many separate intrusive bodies. For example the Lima segment of the Coastal Batholith of Peru accommodates a wide range of intrusive units that have been grouped into nine superunits. Many grani-

tic bodies are also found on the western side of the South Pacific in the Paleozoic Lachlan Fold Belt of eastern Australia. These granites are part of a much more extensive igneous province that includes rocks that now crop out in Antarctica and South Island, New Zealand (Chappell and Stephens, 1988, pp. 71–86). If the pieces of this former province are re-assembled they reveal a belt of granitic rocks that extends for over 3500 km. Chappell *et al.* (1991) have published a map of the granitic and related rocks of eastern Australia and it depicts about 880 discrete bodies and over 20 superunits. This statement is incomplete because over 300 plutonic bodies have yet to be assigned to superunits. In southern New South Wales these Paleozoic granitic rocks have been divided into eastern and western subprovinces.

Alphabet granites and composite batholiths

Chappell and White (1974) and other investigators have attempted to relate the compositions of the igneous rocks that belong to these subprovinces with the composition of the crust that lies beneath them. Chappell and Stephens (1988) claim that the granitic rocks of the eastern subprovince, or I-zone, tend to contain components released during the partial melting of pre-existing igneous crustal rocks that have not experienced much chemical weathering (see Table 9.2). The granitic rocks of the western subprovince, or S-zone, are more likely to contain components released during the partial melting of metasedimentary source rocks (White and Chappell, 1988, pp. 169–181). This research shows that granitic rocks may contain distinctive petrographic and geochemical signatures that can supply worthwhile information about the underlying continental crust.

The following characteristics are typical of composite batholiths, elongate shapes aligned parallel to associated convergent plate boundaries, bulk compositions that are usually in the granodiorite to tonalite range, comagmatic volcanic rocks that are often less silicic than the plutonic rocks, regular geochemical variations across the composite body and the bulk of the rocks belong to the calc-alkaline association. Pankhurst *et al.* (1988, p. 123) examined the isotope geochemistry of the subduction-system rocks of the Andean region of South America and of the neighbouring Antarctic Peninsula. They reported a wide range of initial strontium and neodymium values and identified mixing between mantle-derived magma and the continental crust. In each area the crustal end-member was shown to reflect the age and isotopic composition of the local crustal basement.

Pitcher (1979, p. 3) divided the granitic rocks of convergent plate margins into two broad rock associations. His compositionally restricted association contains approximately 80 per cent granite (sensu stricto), while his compositionally expanded association contains 18 per cent gabbro, 50 per cent tonalite and granodiorite, and only 35 per cent granite (sensu

stricto). Rocks of the compositionally restricted associations are usually peraluminous in composition and are thus likely to be products of anatexis. Rocks of the compositionally expanded association usually contain less evidence of crustal assimilation and their evolution is usually linked to the generation and differentiation of calc-alkaline magmas in active subduction systems (Barbarin, 1990, p. 231).

9.9 Aplites and granitic pegmatites

Aplites and granitic pegmatites are often found together. Occasionally both crystallise in the same intrusive body. Their petrographic modes are often similar but they have entirely different grain sizes. Aplites typically have simple textures, and simple modal compositions (von Leonhard, 1823, p. 51). The term is in fact derived from the Greek word *haploos* that means simple. Brögger (1898) used the term in its modern manner when he employed it to describe various leucocratic fine- to medium-grained dykes, or border facies rocks, with distinctive saccharoidal textures. When examined in detail some aplites appear to have cataclastic textures that have been partly masked by later recrystallisation (Emmons, 1953, p. 73).

The essential minerals in an aplite are normally alkali feldspar and quartz. Some aplites contain over 50 per cent quartz and grade imperceptibly into quartz veins or quartzolite. They occasionally contain metasomatic or pneumatolitic minerals such as beryl, tourmaline, topaz or fluorite and are thus likely to contain somewhat high concentrations of elements like beryllium, boron, chlorine, fluorine and lithium. These chemical components are more frequently concentrated in pegmatites. Some pegmatites also contain a variety of rare minerals rich in elements, like those previously mentioned, and others such as phosphorus, manganese, gallium, rubidium, zirconium, niobium, molybdenum, tin, caesium, the lanthanoids, hafnium, tantalum, tungsten and uranium.

Pegmatites are exceptionally coarse-grained, yet the grain-size within individual pegmatitic bodies may be extremely variable (Martin and Cerny, 1992). The term is derived from the Greek word *pegma* that means a framework. It was first used to describe graphic granites but the meaning has gradually changed. It is now often used to describe any exceptionally coarse-grained igneous-looking rock.

Most pegmatites are granitic in composition. Some petrologists would restrict the use of the term to granitic and syenitic rocks and use the term pegmatitic facies to describe other exceptionally coarse grained plutonic rocks. Individual crystals in pegmatites usually have mean diameters that are greater than 3 cm. In some large granitic pegmatites, such as the Harding Mine Pegmatite in Taos County, New Mexico, individual minerals, for example spodumene, are over 2 m in length. This pegmatite is about 370 m long, with a maximum thickness of 25 m. It displays extraordinary internal zoning. Many pegmatites possess similar anisotropic fabrics or they display graphic textures or comb structures. Most granitic pegmatites occur in dykes, lenses, segregations or veins in the marginal zones, particularly the exocontact zones, in the apices or cupolas of granitic plutons.

Most granitic pegmatites appear to have evolved from crystal-free magmas that accumulated at the tops of large plutonic bodies. These magmas normally have high silica contents, and are also enriched in components like lithium, boron, fluorine, phosphorus and water. Compared to normal granitic magmas the magmas that produce pegmatites have lower solidus temperatures (about 450°C), retarded rates of crystal nucleation and enhanced rates of element diffusion. London (1995, p. 89) has suggested that the extremely coarse-grained pegmatitic texture results from the crystallisation of an undercooled magma in which the rate of crystal nucleation has been suppressed. Retardation in crystal nucleation without a concurrent slowing down in crystal growth can easily explain the grain-size differences between normal aplitic and granitic rocks and pegmatites.

Several elaborate classifications have been proposed for the granitic pegmatites (Cerny, 1982). In the present discussion they will be divided into simple and complex types. The simple pegmatites have simple textures and compositions, whereas the complex pegmatites normally display internal zones with different textures and compositions. They frequently contain high concentrations of rare minerals and rare elements. The shapes of the granitic pegmatites vary greatly because they are mainly controlled by the competency of the enclosing rocks, their depth of emplacement and their initial tectonic and metamorphic setting. Geochemists, mineralogists and economic geologists are fascinated by the complex pegmatites because they contain unusual minerals and elements such as beryllium, niobium, tin, caesium, the rare earth elements and tantalum. Industrial minerals such as beryl, ceramic amblygonite, ceramic and dental feldspar, fluorite, optical quartz, petalite, refractory spodumene, sheet and crushed micas are quarried from pegmatites. These minerals are greatly prized because of their high purity and large size.

Retrograde boiling

Traditionally it has been postulated that the aplites and more particularly granitic pegmatites crystallise from hydrous residual magmas that evolve from crystallising silicic magmas. The water is unable to crystallise at the prevailing temperature. It becomes a supercritical fluid that exsolves from the crystallising rock. This process of retrograde boiling normally generates a fluid enriched in the incompatible elements that easily enters any available fractures and produces aplitic or pegmatitic veins and dykes. London (1992, p. 499) has described granitic pegmatites as 'products of disequilibrium fractional crystallisation through liquidus undercooling'. As we have already noted he supports the idea that the degree of liquidus undercooling and the concentration of incompatible components controls the rate of crystal nucleation and strongly influences the textures of the crystallising rocks. The study of granitic pegmatites has recently been shown to be particularly important because it provides invaluable data on the processes that operate during the evolution of compositionally zoned silicic magma chambers.

If the complex granitic pegmatites evolve from ordinary silicic magmas one would expect to find some rhyolites with similar chemical compositions. Such rocks have been discovered and the Honeycomb Hills rhyolite from Utah in the USA is an excellent example (Congdon and Nash, 1991). These rhyolites are mainly composed of quartz, sanidine and albite in a fine-grained to glassy groundmass. The phases that make these rocks special are rare accessory minerals that contain exceptionally high concentrations of elements like lithium, beryllium, boron, fluorine, phosphorus, chlorine, rubidium, yttrium, zirconium, niobium, tin, caesium, the lanthanoids, hafnium, tantalum, thorium and uranium (Congdon and Nash, 1991, p. 1269). It is probable that the magma responsible for producing these rare-element enriched rhyolites evolved in a cupola above a large body of cooling silicic magma.

9.10 Appinite puzzle

Many composite granitic batholiths and their cara-pace of country rocks contain a spectrum of dark-coloured, hydrous mafic rocks. These rocks can be grouped together as members of the appinite suite. This term was introduced by Bailey (in Bailey and Maufe, 1916) as a group name for a collection of dark-coloured rocks of medium to coarse grain size that appear as pipe-like bodies, inclusions, or mar-ginal facies to granitic plutons. Rocks of the appinite suite were first described from the Ballachulish Gran-ite, near Appin southwest of Fort William in Scot-land. Similar rocks crop out in Donegal, Ireland (Pitcher and Berger, 1972, p. 149), and in many other parts of the world. Particular attention has been focused on the appinitic rocks of western and central Europe. They have variously been called dur-bachites (after Durbach in the Black Forest, Ger-many), nadeldiorites (diorite containing needle-shaped amphibole), redwitzites (after Redwitz in the Fichtelgebirge, Germany) and vaugnerites (after Vaugneray, near Lyon, France). Wilkins and Born-horst (1993) have even described appinites of Archean age from the Northern Complex in the Superior Province of the Canadian Shield. Many Proterozoic appinitic rocks have been described in the petrological literature.

Outcrops of rocks of the appinite suite normally vary in size from about 100 m in diameter to large bodies measuring about 1500 m by 500 m. Most bodies occur as steeply inclined pipes, surrounded by narrow (approximately 20 m) contact meta-morphic aureoles. Individual rocks normally contain large crystals of primary amphibole in a groundmass of plagioclase and alkali feldspar. Some appinitic rocks contain quartz, or quartz ocelli, but the basic members of the suite are more likely to contain nor-mative feldspathoids (see Table 9.3). Apatite, pyrite, rutile, sphene and primary carbonate phases are com-mon accessory minerals. In a standard modal classi-fication most appinitic rocks would be found to belong to one of the following rock types: amphibole melagabbro, amphibole gabbro, amphibole monzo-gabbro, amphibole gabbroic diorite or amphibole diorite. This shows that they could have been just as readily included with the dioritic, or even the lam-prophyric, groups of rocks. The decision to include them with the granitic rocks was taken: (1) because in the past the two groups have usually been included together, and (2) to remind investigators of granitic

Table 9.3 Major element compositions and CIPW norms of the appinite suite.

Rock type	Mean appinite Wright & Bowes, 1979	Mean kentallenite Wright & Bowes, 1979	Vaugnerite Sabatier, 1980
SiO_2	48.9	48.3	46.7
TiO_2	0.6	1	1.7
Al_2O_3	12.5	12.7	17.4
Fe_2O_3*	9.2	9.9	7.9
MgO	14.8	11.6	9.3
CaO	7.2	8.4	8.4
Na_2O	2.6	2.4	1.3
K_2O	2.2	2.1	4.5
P_2O_5	0.3	0.3	1.7
CO_2	0.4	1.3	0
Total	98.7	98	98.9
CIPW norms			
Orthoclase	12.89	12.29	26.87
Albite	19.26	20.01	7.98
Anorthite	15.8	17.52	28.66
Nepheline	1.38	0.06	1.7
Diopside	14.15	23.79	6.02
Olivine	30.25	23.79	20.36
Magnetite	2.65	2.85	3.01
Ilmenite	1.13	1.89	3.27
Apatite	0.69	0.60	2.12

*Total iron as Fe_2O_3

rocks to be always on the lookout for unfamiliar rock types, particularly those of the appinite suite.

Bailey originally suggested that the appinitic rocks were plutonic equivalents of the ordinary or calc-alkaline lamprophyres, spessartite and vogesite (see Section 11.8). The durbachites, redwitzites and vaugnerites have often been linked to other lampro-phyres, such as the minettes and kersantites (Rock, 1991, 115; Sabatier, 1991, pp. 75–76). Many petrolo-gists support the idea of a link between the appinitic rocks and the ordinary lamprophyres (Ayrton, 1991, p. 473; Bowes and McArthur, 1976; Rock, 1977; Rock, 1991). Rock (1977, pp. 140–141) perceptively reminded petrologists that the various origins sug-gested for appinites are remarkably similar to those postulated for the ordinary lamprophyres. O'Connor (1974) asserts that the appinitic and lamprophyric intrusive rocks of East Carlow in Ireland are contem-poraneous and comagmatic, and are all spatially and temporally related to the granitic rocks of the area. The appinites and ordinary lamprophyres only differ in having dissimilar grain sizes and textures. This is probably related to differences in their respective cooling histories. There is much persuasive evidence supporting a close relationship between the appinites, the ordinary lamprophyres, and the mafic microgranular enclaves found in many granitic rocks (Ayrton, 1991, p. 475). Field studies have revealed that some appinites were produced by interaction between a potassium and volatile rich basic (lamprophyric) magma, and a partly, or mostly, crystallised body of silicic magma (see Section 11.8).

9.11   Mineral deposits and silicic magmas

The close spatial and temporal nexus between some mineral deposits and the granitic rocks has long been used as evidence confirming a genetic link between them. According to Clarke (1992, p. 152) the less differentiated, low-potassium, water-undersaturated granitic rocks usually lack economic deposits. Such deposits are much more likely to occur in the highly differentiated, high-potassium, water-oversaturated granitic rocks. Müller and Groves (1995) have recently written a book on the link between gold-copper mineralisation and high-potassium igneous rocks. The mineral deposits associated with silicic rocks usually contain one or more of the following metals: copper, zinc, niobium, molybdenum, silver, tin, tantalum, tungsten, gold, lead, thorium and uranium. Many petrologists have emphasised the link between the enormous volumes of hydrous fluids released during the cooling and crystallisation of granitic batholiths and the production of ore deposits in their exocontact zones and in the apices of parental granitic bodies. Stemprok (1990) has undertaken a comprehensive review of the association of molybdenum, tin and tungsten deposits with granitic rocks. He concluded that most economic deposits of these metals are associated with biotite granites that typically crop out as small stocks that are less than $25 \, \text{km}^2$ in area. As has been noted in Section 9.9 some complex granitic pegmatites produce a range of industrially important minerals and elements. Most of these rare elements are also found associated with the alkaline rocks and carbonatites, thus the granitic pegmatites can be regarded as an important connection between the rocks of the alkaline suite and the granitic rocks.

In areas of active or recently dormant silicic volcanism, such as the Yellowstone Caldera of Wyoming and the Taupo Volcanic Zone of North Island,

New Zealand, surface heat flow values are usually anomalously high and the landscape is strewn with clusters of hot springs and geysers. Geothermal areas of this type are probably activated by heat emanating from silicic magmas impounded beneath them. In some geothermal areas in the Taupo Volcanic Zone superheated water and steam is used to drive turbines that generate electrical power. In 1983 routine maintenance on the production wells in the Ohaaki-Broadlands geothermal field revealed that some pipes that carried high pressure and temperature water and steam were lined with a scale enriched in copper, zinc, silver, gold and lead. Subsequent experiments revealed that these metals are released when superheated water ($260°C$) flashes to steam. Several dissolved gases, like hydrogen sulphide, are normally partitioned into the steam phase, when water flashes to steam the sulphide complexes break down and release the metals they contain (Brown, 1986; Yardley, 1991). Studies of precious metal zoning in the Ohaaki-Broadlands geothermal system have revealed that gold, arsenic, antimony and thallium are enriched at shallow levels of between 200 and 400 m, while silver, selenium, tellurium, bismuth, lead, zinc and copper are enriched at greater depths (Simmons and Browne, 1991, p. 16).

Waiotapu is the largest area (about $18 \, \text{km}^2$) of surface geothermal activity in the Taupo Volcanic Zone (Hedenquist and Browne, 1989; Houghton, 1982). Two dacitic domes lie to the north of this geothermal area, whilst another lies to the west. Most of the surface rocks are felsic ignimbrites, tuffs and volcanic breccias. Waiotapu is characterised by a large area of steaming ground, fumarolic activity related to collapse craters, hot mud pools and widespread rock alteration due to the action of 'acid-sulphate fluids' that arise above the water table. The area also contains about 20 steam-blast eruption craters and their surrounding pyroclastic deposits.

Springs that emerge at the level of the water table usually discharge chloride waters that precipitate siliceous sinter. Champagne Pool is the only large source of undiluted chloride waters in the Waiotapu geothermal area. It lies within an eruption crater and it is about 60 m deep. Bubbles of carbon dioxide constantly rise and break the surface of the pool (see Plate 9.1). Because of rapid convection most of the water in the pool remains at a constant temperature of $74°C$. Orange coloured precipitates are normally observed near the edge of the pool. They contain amorphous arsenic and antimony-sulphur compounds. These compounds contain up to 80 ppm gold, 175 ppm silver, 320 ppm thallium and 170 ppm mercury. Hedenquist and Henley (1985) have suggested that Champagne Pool is the surface manifestation of an active subsurface ore-depositing system.

Drilling in the Waiotapu area has revealed local disseminations of sphalerite (ZnS), galena (PbS) and chalcopyrite (CuFeS$_2$). In Champagne Pool the rising carbon dioxide bubbles maintain a sufficiently low pH to form amorphous arsenic and antimony-sulphur compounds that can absorb gold and silver from fluids strongly undersaturated in these constituents. Outflow from Champagne Pool has built up the Primrose Terraces. They are the largest (1.2 hectares) actively forming siliceous sinter terrace-set in New Zealand. On Lihir Island, east of New Ireland in Papua New Guinea, a major gold deposit has been discovered in a hydrothermally active, collapsed caldera that is similar to the caldera of the Taupo Volcanic Zone.

Hosking (1988) has reviewed the metallogeny of the major tin deposits. He asserts that most primary deposits are spatially and probably genetically related to granitic or rhyolitic rocks. His other major finding was that the abundance and economic value of these deposits has gradually increased with time. Hutchison (1988, p. 229) has re-evaluated this suggestion. He links the generation of economic tin deposits to the evolution of the continental crust and he suggests that the concentration of this metal in the continental crust is the result of the reworking of the crustal rocks by tectonic action, magmatic activity, anatexis, metamorphism and metasomatism. According to him tin is likely to become progressively more concentrated in those belts of the Earth continental crust that have experienced the most tectonic and magmatic reworking.

9.12   Extraterrestrial granites and the nature of KREEP

Before reviewing the granitic rocks found on the Moon it is instructive to try to discover why and how the Moon developed a highland crust that is so unlike the Earth's continental crust. Initially the Moon may have been up to 50 per cent molten (Taylor, 1992, p. 273). As this magma-lava ocean cooled and crystallised plagioclase crystals floated and accumulated in this essentially anhydrous picrobasaltic magma. Convection currents within the magma-

lava ocean would sweep rafts of plagioclase together to form the first crust. A crust and a compositionally zoned mantle containing ferromagnesian cumulates developed by about 4.4 giga-years ago. As the common silicate minerals crystallised the incompatible elements excluded from their lattices concentrated in the residual fluid.

Current knowledge of this residual fluid has come from the study of the lunar highlands alkali suite (Snyder *et al.*, 1995). These rocks contain varying proportions of the primeval residual fluid. Geochemically this fluid contained high concentrations of phosphorus, potassium, zirconium, niobium, the lanthanoids, hafnium, thorium and uranium. Lunar rocks that contain this incompatible element-enriched component are usually called 'KREEPy rocks', with a high KREEP-component. The acronym KREEP (potassium, rare earth elements and phosphorus) was originally given to fragments of glass enriched in the incompatible elements that was discovered among the materials returned by the Apollo Twelve Mission. Later missions discovered a variety of rocks that contained the KREEP-component. These rocks have an intriguing geochemistry because they all have relatively constant ratios between the individual incompatible elements. This suggests that they all formed from magmas that contained varying amounts of an exceptionally similar fluid enriched in the incompatible elements (the primary KREEPy fluid).

An important mineralogical difference between the rocks of the lunar highlands and those from the continental areas of Earth is that most of the lunar rocks do not contain silica minerals (cristobalite, quartz and tridymite) or potassium feldspars. According to Snyder *et al.* (1995, p. 1193) true granitic rocks are rare on the Moon as only six samples have been found in the various collections of Moon rocks. Rhyolitic materials are more common. They have usually been called felsites, or lunar felsites. An excellent example of a rare medium to coarse grained lunar leucogranite is a clast discovered in Apollo Fourteen breccia 14321. It is circa 4.1 giga-years old and it contains about 60 per cent potassium feldspar, 40 per cent quartz, and accessory amounts of fayalitic olivine, high-calcium pyroxene, ilmenite and an iron-nickel metal phase (Warren *et al.*, 1983). This rock contains 74.2 per cent SiO$_2$ and 8.6 per cent K$_2$O. It also has remarkably high concentrations of rubidium (210 ppm), thorium (65 ppm) and uranium (23.4 ppm). Micas and amphiboles are absent in the lunar granitic rocks. This is as one would expect on a planet deficient in water and hydrous phases. Most lunar granites typically have graphic textures. As a group the lunar felsites and granites contain the highest silica and potassium abundances of all the lunar rocks.

The presence of granites and rhyolites among the many picrobasaltic, basaltic and gabbroic and anorthositic rocks of the Moon is somewhat perplexing because there is an apparent lack of intermediate rocks of the type one would expect if the silicic rocks had evolved by normal fractional crystallisation. Snyder *et al.* (1995, p. 1186) argue that the various rocks that have been grouped together as quartz monzodiorites are the supposedly missing group of intermediate rocks. A critical examination of lunar granitic and rhyolitic rocks reveals that even on this planet one is confronted by the enigma of 'granites and granites'.

The evidence currently available suggests that many lunar silicic rocks were produced by liquid immiscibility. This process is likely to occur when a picrobasaltic or basaltic magma cools in an oxygen-depleted environment. Limited fractional crystallisation produces a magma that is slightly enriched in silica and iron. With further cooling the newly evolved magma separates into two liquids and one of them is rhyolitic in composition (Taylor *et al.*, 1980; and see Section 6.2). This petrogenetic model is supported by experimental investigations (Neal and Taylor, 1989, p. 89). The silicic rocks that did not develop by liquid immiscibility are the most evolved members of the highlands alkali suite. It is likely that between 4.35 and 4.0 giga-years ago several different alkalic magmas evolved within the Moon as the result of the commingling of primary picrobasaltic magmas with varying amounts of KREEP-rich fluid. Some of these hybrid magmas became strongly differentiated resulting in the evolution of a silicic magma.

Because of its small size the Moon cooled quickly thus it had a condensed igneous history. The lunar rocks are noteworthy because they are generally old but so little changed by weathering, or destructive tectonic processes. They might be regarded as a substantial part of an archive that contains a comprehensive record of the rise and fall of a magmatic system on a terrestrial-style planet. Petrological data from the Moon are invaluable because the processes operating on the surface and within the Earth have destroyed most records of its early, or Hadean Era, development. The Earth has sustained a profound loss of memory concerning its early history.

Lunar rocks are also important because they contain data on the waning stages in the magmatic history of a cooling planet.

More KREEPy fluids

Petrologists are currently trying to optimise their use of lunar data to unravel the early history of the Earth's crust. They also have to decide whether KREEP-like fluids were important in the Earth's past, are important now, or are likely to become significant in its future? It seems likely that at sometime in the Late Hadean Era, as most of the Earth's protocrust and mantle crystallised the various incompatible elements excluded from the crystallising minerals would concentrate in residual fluids that were analogous to the KREEPy fluids discovered on the Moon.

The origin of the Earth's continental crust and the evolution of the granitic and rhyolitic rocks is probably closely tied to the release of incompatible elements, particularly potassium, during the movement and recycling its primeval KREEPy fluids. The history of the upper crust is a saga of continuing enrichment in potassium, thorium and uranium. It is apposite to note that the glass inclusions in ultramafic xenoliths from the upper mantle are often strongly enriched in silicon, aluminium and potassium (Schiano and Clocchiatti, 1994, p. 623). When evaluating the crust-forming processes it is important to remember that currently the continental crust occupies only about 0.6 per cent of the volume and 0.4 per cent of the mass of the Earth (Newsom, 1995, p. 163).

KREEP-like fluids are still being generated within the Earth. This process is likely to continue operating for at least another giga-year. In the future, as the Earth cools, its convective cooling system will slow down, become more sporadic and eventually stop altogether. During each of these stages less magma will be produced. As the lithosphere thickens and merges into a solid mantle the little magma that is generated is likely to arise from deeper levels and have an alkaline composition. The final magmas that make their way into the crust probably will be enriched in a KREEP-like component and be more akin to lamproites than rhyolites.

The Earth and the Moon have bimodal hypsographic curves or topographic spectra and the continents are totally separate topographic entities. On Venus the hypsographic curve is unimodal and the highlands appear to grade into the lowlands. Ishtar

Terra is the highland area most often considered comparable to the continents of Earth. It is so lofty that it has been argued that it is supported by a sub-crustal dynamic process (Cattermole, 1994, p. 218). Ishtar Terra probably contains at least some lower than normal density rocks, possibly granitic rocks. This is tentatively confirmed by a few gamma-ray measurements recorded from the surface of Venus that reveal at least one area with somewhat high potassium (4.8 per cent K_2O), thorium (6.5 ppm) and uranium (2.2 ppm) values (Vinogradov et al., 1973). On a smaller scale the surface of Venus is covered by many steep-sided volcanic domes or pancakes (see Section 4.5). These volcanic landforms were probably constructed from a viscous silicic lava. Beneath the surface the volcanic domes may contain subvolcanic granitic rocks.

9.13 Granitic rocks on a slowly evolving planet

With normal fractional crystallisation tholeiitic, calc-alkaline and midalkaline picrobasaltic and basaltic magmas follow their various liquid lines of descent and finally converge in the rhyolite or granite field. As silicic rocks from all these suites are found in convergent plate settings it is not surprising that granitic rocks are abundant in these areas. The calc-alkaline suite is particularly abundant in this setting. Tonalites and granodiorites dominate in the zones associated with the volcanic arc, whereas the granodiorites and granites are more characteristic of the back-arc or hinterland areas behind the volcanic arc zone. The leucogranites are typical of continent–continent collision zones.

If one considers the origin of granites in a planetary context one must recognise that the Earth has a unique subduction system that transports hydrous supracrustal materials into the upper mantle. Once subducted these hydrous materials trigger the generation of magmas, particularly calc-alkaline magma. After lengthy differentiation these magmas evolve to form an abundant group of granitic and rhyolitic rocks that are probably unique to planets where Earth-style subduction systems operate. Some petrologists (Fyfe, 1993) claim that the influence of magma underplating on continental processes is currently underestimated. Fyfe (1993, p. 911) advocates the use of more real-time studies in deep probing seismic tomography. He estimates that currently about 'one

third of all continental crust is undergoing active modification at its base'.

As the mid-oceanic ridges are mainly composed of tholeiitic rocks it is not surprising to discover that the typical granitic rocks that evolve in this setting are leucocratic tonalites, or plagiogranites. Many similar leucocratic tonalites were generated in the Archean Era. According to Kroner and Layer (1992) the early Archean Era was an epoch of extensive continental crust development, with most of the ancient granitic rocks evolving from magmas generated by heat supplied by magmatic underplating of the primordial crust. This process is analogous with the one currently operating beneath Iceland. The silicic rocks of the midalkaline suite frequently crop out in intraplate settings and the granitic rocks of this suite are usually alkali feldspar granites, particularly peralkaline types.

Experimental studies have confirmed that silicic magmas are readily generated by the partial melting of a wide range of crustal materials at the temperatures and pressures likely to be found in the continental crust (Winkler and Von Platen, 1957; to Vielzeuf and Montel, 1994). In the early water-saturated experiments partial melting was found to take place between 650 and 700°C. Later fluid-absent investigations of the partial melting of pelitic metasediments by Vielzeuf and Holloway (1988) revealed that large quantities of silicic melt could be produced from the breakdown of biotite at a temperature of 850°C and pressures of 1 GPa. Vielzeuf and Montel (1994) studied the partial melting of calcium-poor metagreywackes and found that they yielded silicic melts at commonly attainable crustal temperatures (850°C), below pressures of 0.7 GPa. The partial melting of these metasediments may occur at pressures and temperatures just above those characteristic of regional metamorphism. Higher temperatures, such as those induced by the introduction of a batch of picrobasaltic magma, are required to produce widespread partial melting.

Thermal reworking of the crust

If one assumes that in the past the distribution of mantle plumes has been essentially random in the continental areas then it is likely that most cratonic areas have passed over several mantle plumes (Crough, 1981). This means that most older areas of continental crust have experienced several epochs

Fig. 9.5 The road traces the contact between the light coloured and jointed Cape Granite and the younger overlying shales and sandstones of the Table Mountain Series. This photograph was taken to the south of Hout Bay in the Cape Peninsula, South Africa.

of thermal reworking, probably resulting in regional metamorphism and the formation of silicic rocks. If large sheets of picrobasaltic magma crystallise slowly at the mantle–crust boundary this process is likely to result in the flotation of plagioclase and the sinking of minerals like olivine and pyroxene. Repeated injection of new batches of picrobasaltic magma may result in the formation of thick bodies of anorthosite that adhere to and augment the volume of the overlying crust (Fyfe, 1993, p. 911). According to Ashwal, (1993, pp. 208–210) the massif-type anorthosites are likely to form in this way and some of them may be later emplaced into the crust by diapirism.

In subduction, collision and post-collision tectonic settings the processes of magmatic differentiation of basic magmas and the partial melting of crustal rocks are likely to merge. This happens because the partial melting of crustal phases is normally triggered by the introduction of hot, basic magma from the upper mantle. Upon assimilating phases that contain lithophile elements a new hybrid magma forms and it has

the capacity to differentiate and produce large volumes of silicic magma (see Fig. 9.5). If basic magma continues to be emplaced into a segment of crust, after the crust has begun to melt it becomes progressively more difficult for the denser basic magma to penetrate the lower density layer of partial melting. This type of underplating may result in a runaway process of crustal melting and silicic magma generation.

The autochthonous granitic rocks occur in thermal harmony with, and appear to grade into, the metamorphosed country rocks that surround them. These granites appear to have formed in place. Some seem to have partly separated from their cocoon of migmatites. They illustrate one way in which the granitic rocks can form in heated segments of continental crust. Such rocks are sometimes called parautochthonous granites. Remarkably well exposed examples of both autochthonous and parautochthonous granites crop out in the northern part of the Bhutan Himalaya. Ganser (1983, p. 99) has stated that he has observed fully mobilised medium- to fine-grained granites with discordant contacts that are still surrounded by migmatitic gneisses. The younger leucogranites of the Bhutan Himalaya are excellent examples of granites produced by anatexis during continent–continent collision.

Most of the granitic rocks of the Phanerozoic orogenic belts such as those found in the cordilleran-type batholiths of the Americas belong to a compositionally expanded calc-alkaline rock association. The primary magmas responsible for the evolution of these rocks were generated in the mantle-wedge of an active subduction system. Regional patterns of changing chemical composition occur in these rocks and they often mimic similar patterns observed in the volcanic rocks from the same tectonic setting. For example, both potassium and silica are likely to increase as one proceeds away from the oceanic trench and across an island arc or continental margin arc.

In 1988 Fyfe presented a review paper on the origin of granites. He asserts that large volumes of granite and associated extrusive rocks are likely to form whenever the continental crust is heated by rising hot mantle materials or it is thickened by collision. The complex nature of the silicic rocks is related to the complexity of the continental crust and the complexity of the processes that lead to thermal perturbations in the crust. Fyfe also emphasises that the present light continental crust acts as a density filter that screens out heavy mantle magmas. This results in complex underplating and the mixing of magmas of dissimilarity composition. Fyfe (p. 341) asserts that the pivotal argument in current debates on the origin of the granitic rocks is concerned with determining the amount of material derived from the mantle as compared to the amount acquired from the crust.

Much has yet to be uncovered about the planets of the Solar System before petrologists can write the definitive history of their crustal rocks. The lunar crust has probably always been different from the Earth's crust, because the Moon seems to have expelled large amounts of its atmophile elements during its initial planet-forming stage. Perhaps one should be guided by Smith's (1980, p. 452) statement that, 'although planetary crusts are so different from each other that direct comparison is difficult, simultaneous study of all planetary crusts provides a tremendous stimulation to the imagination; indeed, only by assembling together data and ideas on the entire Solar System can progress be made and incorrect ideas discarded'.

References

Ashwal, L.D., 1993. *Anorthosites*. Springer-Verlag, Berlin, 422 pp.

Augustithis, S.S., 1973. *Atlas of textural patterns of granites, gneisses and associated rocks*. Elsevier Scientific Publishing Company, Amsterdam, 378 pp.

Ayrton, S.N., 1991. Appinites, lamprophyres and mafic microgranular enclaves: the related products of interaction between acid and basic magmas. In Didier, J. and Barbarin, B. (editors), *Enclaves and Granite Petrology*, pp. 465–476, Elsevier Scientific Publishing Company, Amsterdam, 625 pp.

Bailey, E.B. and Maufe, H.B., 1916. The geology of Ben Nevis and Glen Coe and the surrounding country. *Memoir of the Geological Survey of Scotland*, Sheet 53, Edinburgh, 247 pp.

Barbarin, B., 1990. Granitoids; main petrologic classifications in relation to origin and tectonic setting. *Geological Journal*, 25, pp. 227–238.

Bonin, B., 1986. *Ring Complex Granites and Anorogenic Magmatism*. (A translation from French). North Oxford Academic Publishers Limited, London, 188 pp.

Bowes, D.R. and McArthur, A.C., 1976. Nature and genesis of the appinite suite. *Kristallinikum*, 12, pp. 31–46.

Bowring, S.A., Williams, I.S. and Compston, W., 1989. 3.96 Ga gneisses from the Slave province, Northwest Territories, Canada. *Geology*, 17, pp. 971–975.

Brögger, W.C., 1898. Die Eruptivgesteine des Kristianiagebietes 3, Das Ganggefolge des Laurdalits. Videnskabsselskabet i Kristiania, 1, Matematisk-naturvidenskapelig, Klasse 6, pp. 1–377.

Brown, G.C., 1979. The changing pattern of batholith emplacement during Earth history. In Atherton, M.P. and Tarney, J. (editors) 1979. *Origin of Granite Batholiths: Geochemical Evidence*, pp. 106–115, Shiva Publishing Limited, Orpington, UK, 148 pp.

Brown, K.L., 1986. Gold deposition from geothermal discharges in New Zealand. *Economic Geology*, 81, pp. 979–983.

Brown, P.E., Clarkson, E.N.K., Crampin, S. *et al.*, (editors), 1988. Symposium on The Origin of Granites. *Transactions of the Royal Society of Edinburgh, Earth Sciences*, 79 (2 and 3), pp. 71–346.

Brun, J.P., Gapais, D., Cogne, J.P., Ledru, P. and Vigneresse, J.L., 1990. The Flamanville Granite (Northwest France): an unequivocal example of a syntectonically expanding pluton. *Geological Journal*, 25, pp. 271–286.

Cattermole, P., 1994. *Venus: The Geological Story*. UCL Press, London, 250 pp.

Cerny, P., 1982. *Granitic Pegmatites in Science and Industry*. Mineralogical Association of Canada, Short Course Handbook, 8, 555 pp.

Chappell, B.W. and Stephens, W.E., 1988. Origin of infracrustal (I-type) granite magmas. In Symposium on The Origin of Granites. *Transactions of the Royal Society of Edinburgh, Earth Sciences*, 79 (2 and 3), pp. 71–86.

Chappell, B.W. and White, A.J.R., 1974. Two contrasting granite types. *Pacific Geology*, 8, pp. 173–174.

Chappell, B.W., White, A.J.R. and Wyborn, D., 1987. The importance of residual source material (restite) in granite petrogenesis. *Journal of Petrology*, 28, pp. 1111–1138.

Chappell, B.W., English, P.M., King, P.L., White, A.J.R. and Wyborn, D., 1991. *Map of the Granites and related rocks of the Lachlan Fold Belt. Scale 1:1250 000*. Bureau of Mineral Resources, Geology and Geophysics, Canberra, Australia.

Clarke, D.B., 1992. *Granitoid rocks*. Chapman and Hall, London, 283 pp.

Cloos, H., 1923. *Das Batholithenproblem*. Fortschritte der Geologie u Palaontologie, Heft 1, Berlin.

Congdon, R.D. and Nash, W.P., 1991. Eruptive pegmatite magma: rhyolite of the Honeycomb Hills, Utah. *American Mineralogist*, 76, pp. 1261–1278.

Crough, S.T., 1981. Mesozoic hotspot epeirogeny in eastern North America. *Geology*, 9, pp. 2–6.

Delesse, A., 1861. Etudes sur le Metamorphisme des Roches. *Bulletin Societé Geologiques Francaise*, Serie 218, 541 pp.

Drinkwater, J.L., Brew, D.A. and Ford, A.B., 1990. Petrographic and chemical data for the large Mesozoic and Cenozoic plutonic sills east of Juneau, southeastern Alaska, U.S. *Geological Survey*, Bulletin 1918, Washington, DC, 47 pp.

Emmons, R.C., 1953. Petrogeny of the syenites and nepheline syenites of Central Wisconsin. *Geological Society of America*, Memoir Number 52, pp. 71–87.

Frost, T.P. and Mahood, G.A., 1987. Field, chemical, and physical constraints on mafic–felsic magma interaction in the Lamarck granodiorite, Sierra Nevada, California. *Geological Society of America Bulletin*, 99, pp. 272–291.

Fyfe, W.S., 1988. Granites and a wet convecting ultramafic planet. Symposium on The Origin of Granites. *Transactions of the Royal Society of Edinburgh, Earth Sciences*, 79 (2 and 3), pp. 339–346.

Fyfe, W.S., 1993. Hot spots, magma underplating, and modification of continental crust. *Canadian Journal of Earth Sciences*, 30, pp. 908–912.

Ganser, A., 1983. *Geology of the Bhutan Himalaya*, Denkschriften der Schweizerischen Naturforschenden Gessellschaft, 96, Birkhauser Verlag, Basel, Switzerland, 181 pp.

Green, A.H., 1882. *Geology: Part 1, Physical Geology*. Rivingtons, London.

Guillot, S. and Le Fort, P., 1995. Geochemical constraints on the bimodal origin of High Himalayan leucogranites. *Lithos*, 35, pp. 221–234.

Hedenquist, J.W. and Browne, P.R.L., 1989. The evolution of the Waiotapu geothermal system, New Zealand, based on the chemical and isotopic composition of its fluids, minerals and rocks. *Geochimica et Cosmochimica Acta*, 53, pp. 2235–2257.

Hedenquist, J.W. and Henley, R.W., 1985. Hydrothermal eruptions in the Waiotapu geothermal system, New Zealand; their origin, associated breccias and relation to precious metal deposition. *Economic Geology*, 80, pp. 1640–1668.

Hibbard, M.J., 1981. The magma mixing origin of mantled feldspars. *Contributions to Mineralogy and Petrology*, 76, pp. 158–170.

Holmquist, P.J., 1909. Nagra Jamforelsepunktev emellan nordamerikarsk och fennoskardisk prekambrisk Geologi. *Geologiska Foreningens i Stockholm Forhandlingar*, 31, pp. 25–31.

Holtedahl, O., 1963. Studies of the igneous rock complex of the Oslo region. 19, Charles Lyell's visit to Norway 1837 with remarks on the history of the 'Granite Problem' in the Oslo region, Skrifter utgitt av det Norske Videnskaps-Akademi i Oslo, Matematisk-naturvidskapelig, Ny Serie, Klasse 12, pp. 1–24.

Hosking, 1988. The world's major types of tin deposit. In Hutchison, C.S. (editor) *Geology of Tin Deposits in Asia and the Pacific*, pp. 3–49, Springer-Verlag, Berlin, 718 pp.

Houghton, B.E., 1982. *Geyserland: a guide to the volcanoes and geothermal areas of Rotorua*. Geological Society of New Zealand, Guidebook Number 4, 48 pp.

Hutchison, C.S., 1988. The tin metallogenic province of S.E. Asia and China: a Gondwanaland inheritance. In Hutchison, C.S. (editor) *Geology of Tin Deposits in Asia and the Pacific*, pp. 225–234, Springer-Verlag, Berlin, 718 pp.

Jacobson R.R.E., Macleod, W.N., and Black, R., 1958. Ring-complexes in the Younger Granite Province of Northern Nigeria. *Memoir of the Geological Society of London*, 1, 72 pp.

Kroner, A. and Layer, P.W., 1992. Crust formation and plate motion in the early Archean. *Science*, 256, pp. 1405–1411.

Le Maitre, R.W., 1976. The chemical variability of some common igneous rocks. *Journal of Petrology*, 17(4), pp. 589–637.

Le Maitre, R.W., Bateman, P., Dudek, A., Keller, J., Lameyre, J., Le Bas, M.J., Sabine, P.A., Schmidt, R., Sørensen, H., Streckeisen, A., Woolley, A.R. and Zanettin, B., 1989. *A Classification of Igneous Rocks and Glossary of Terms: Recommendations of the International Union of Geological Sciences Subcommission on the Systematics of Igneous Rocks.* Blackwell Scientific Publications, Oxford, 193 pp.

London, D., 1992. The application of experimental petrology to the genesis and crystallisation of granitic pegmatites. *The Canadian Mineralogist*, 30 (3), pp. 499–540.

London, D., 1995. Pegmatites. In *The Origin of Granites and Related Rocks*, pp. 89–90, Third Hutton Symposium, Abstracts, United States Survey Circular, 1129, 170 pp.

Luth, W.C., 1976. Granitic rocks. In Bailey, D.K. and MacDonald, R. (editors), *The Evolution of the Crystalline Rocks*, pp. 335–417, Academic Press, London, 484 pp.

Martin, R.F. and Cerny, P. (editors), 1992. Granitic Pegmatites. *The Canadian Mineralogist*, 30(3), pp. 497–954.

Mehnert, K.R., 1968. *Migmatites and the Origin of Granitic Rocks.* Elsevier, Amsterdam, 393 pp.

Müller, D. and Groves, D.I., 1995. *Potassium Igneous Rocks and Associated Gold-Copper Mineralization.* Springer-Verlag, Berlin, 210 pp.

Neal, C.R. and Taylor, L.A., 1989. Lunar granite petrogenesis and the process of silicate liquid immiscibility: the barium problem. In Taylor, G.J. and Warren, P.H. (editors), *Workshop on Moon in Transition: Apollo 14, KREEP, and Evolved Lunar Rocks*, pp. 89-93, Lunar and Planetary Institute, Technical Report Number 89–03, 156 pp.

Newsom, H.E., 1995. Composition of the Solar System, Planets, Meteorites, and major terrestrial reservoirs. In Ahrens, T.J. (editor), *Global Earth Physics: A Handbook of Physical Constants*, pp. 159–189, American Geophysical Union Reference Shelf 1, 376 pp.

Nockolds, S.R., 1954. Average chemical compositions of some igneous rocks. *Bulletin of the Geological Society of America*, 65, pp. 1007–1032.

O'Connor, P.J., 1974. Some Caledonian appinitic intrusions in East Carlow, Ireland. *Irish Naturalist Journal*, 18, pp. 103–108.

Pankhurst, R.J., Hole, M.J. and Brook, M., 1988. Isotope evidence for the origin of Andean granites, in Symposium on The Origin of Granites. *Transactions of the Royal Society of Edinburgh, Earth Sciences*, 79 (2 and 3), pp. 123–134.

Pitcher, W.S., 1979. Comments on the geological environments of granites. In Atherton, M.P. and Tarney, J. (editors), *Origin of Granite Batholiths: Geochemical Evidence*, pp. 1–8, Shiva Publishing Limited, Orpington, UK, 148 pp.

Pitcher, W.S., 1993. *The Nature and Origin of Granite.* Blackie Academic and Professional, Glasgow, 321 pp.

Pitcher, W.S. and Berger, A.R., 1972. *The Geology of Donegal: A Study of Granite Emplacement and Unroofing.* Wiley-Interscience, New York, 435 pp.

Ramberg, H., 1970. Model studies in relation to intrusion of plutonic bodies. In Newall, G. and Rast, N. (editors), *Mechanisms of Igneous Intrusions*, pp. 261–286, Geological Journal Special Issue, 2, 380 pp.

Rämö, O.T. and Haapala, I., 1995. One hundred years of Rapakivi Granite. *Mineralogy and Petrology*, 52 pp. 129–185.

Read, H.H., 1957. *The Granite Controversy.* Thomas Murby and Company, London, 430 pp.

Rock, N.M.S., 1977. The nature and origin of lamprophyres: some definitions, distinctions, and derivations. *Earth-Science Reviews*, 13, pp. 123–169.

Rock, N.M.S., 1991. *Lamprophyres.* Blackie and Son, Glasgow, 285 pp.

Roddick, J.A. and Hutchinson, W.W., 1974. Setting of the Coastal Plutonic Complex, British Columbia. *Pacific Geology*, 8, pp. 91–108.

Rushmer, T., 1995. Application of Rock Deformation Experiments to Melt Segregation in the Lower Crust, pp. 130–131 in The Origin of Granites and Related Rocks, Third Hutton Symposium, Abstracts, United States Survey Circular, 1129, 170 pp.

Sabatier, H., 1991. Vaugnerites: special lamprophyre derived mafic enclaves in some Hercynian granites from Western and Central Europe. In Didier, J. and Barbarin, B. (editors), *Enclaves and Granite Petrology*, pp. 63– 81, Elsevier Scientific Publishing Company, Amsterdam, 625 pp.

Schiano, P. and Clocchiatti, R., 1994. Worldwide occurrence of silica-rich melts in sub-continental and sub-oceanic mantle minerals. *Nature*, 368, pp. 621–624.

Sederholm, J.J., 1907. Om granit och gneis. *Bulletin de la Commission Geologique de Finlande*, Helsingfors, 23, pp. 1–90.

Shand, S.J., 1947. *Eruptive Rocks, Their Genesis, Composition, Classification, and Their Relation to Ore Deposits With a Chapter on Meteorites*. Third Edition, Thomas Murby, 488 pp.

Shimizu, M., 1986. *The Tokuwa Batholith, Central Japan: An Example of Occurrence of Ilmenite-Series and Magnetite-Series Granitoids in a Batholith*. University of Tokyo Press, Tokyo, 146 pp.

Simmons, S.F. and Browne, P.R.L., 1991. Active Geothermal Systems of the North Island, New Zealand. *Geological Society of New Zealand Miscellaneous Publication* 57, 70 pp.

Smith, J.V., 1980. Planetary crusts: a comparative review. In Papike, J.J. and Merrill, R.B., (editors) in Proceedings of the Conference on the Lunar Highlands Crust, pp. 441–456, *Geochimica et Cosmochimica Acta*, Supplement 12, Pergamon Press, New York, 505 pp.

Smith, R.B. and Braile, L.W., 1984. Crustal structure and evolution of an explosive silicic volcanic system at Yellowstone National Park. In *Explosive Volcanism: Inception, Evolution, and Hazards*, pp. 96–109, Studies in Geophysics, Geophysics Study Committee, National Academy Press, Washington, 176 pp.

Snyder, G.A., Taylor, L.A. and Halliday, A.N., 1995. Chronology and petrogenesis of the lunar highlands alkali suite: cumulates from KREEP basalt crystallisation. *Geochimica et Cosmochimica Acta*, 59 (6), pp. 1185–1203.

Sparks, R.S.J. and Marshall, L.A., 1986. Thermal and mechanical constraints on mixing between mafic and silicic magmas. *Journal of Volcanology and Geothermal Research*, 29, pp. 99–124.

Stemprok, M., 1990. Intrusion sequences within ore-bearing granitoid plutons. *Geological Journal*, 25, pp. 413–417.

Taylor G.J., Warner, R.D., Keil, K., Ma, M.-S. and Schmitt, R.A., 1980. Silicate liquid immiscibility, evolved lunar rocks and the formation of KREEP. In Papike, J.J. and Merrill, R.B., (editors) in Proceedings of the Conference on the Lunar Highlands Crust, pp. 339–352, *Geochimica et Cosmochimica Acta*, Supplement 12, Pergamon Press, New York, 505 pp.

Taylor, S.R., 1992. *Solar System Evolution: A New Perspective*. Cambridge University Press, Cambridge, 307 pp.

Termier, P., 1904. Les schistes cristallins de Alpes occidentales, Ninth International Geological Congress, Vienna, p. 571.

Tuttle, O.F. and Bowen, N.L., 1958. Origin of granite in the light of experimental studies in the system $NaAlSi_3O_8$–$KAlSi_3O_8$–SiO_2–H_2O. *Memoir of the Geological Society of America*, 74, 153 pp.

Vielzeuf, D. and Holloway, J.R., 1988. Experimental determination of the fluid-absent melting relations in the pelitic system; consequences for crustal differentiation. *Contributions to Mineralogy and Petrology*, 98, pp. 257–276.

Vielzeuf, D. and Montel, J.M., 1994. Partial melting of metagreywackes. Part 1, Fluid-absent experiments and phase relationships. *Contributions to Mineralogy and Petrology*, 177, pp. 375–393.

Vigneresse, J.L., 1990. Use and misuse of geophysical data to determine the shape at depth of granitic intrusions. *Geological Journal*, 25, pp. 249–260.

Vinogradov, A.P., Surkov, Y.A. and Kirnozov, F.F., 1973. The contents of uranium, thorium and potassium in the rocks of Venus as measured by Venera 8. *Icarus*, 20, pp. 253–259.

von Leonhard, K.C., 1823. *Charakteristik der Felsarten*. Vol. 1, Engelmann, Heidelberg. 230 pp.

Warren, P.H., Taylor, G.J., Keil, K., Shirley, D.N. and Wasson, J.T., 1983. Petrology and chemistry of two 'large' granite clasts from the Moon. *Earth and Planetary Science Letters*, 64, pp. 175–185.

Wegmann, C.E., 1935. Zur Beutung der Migmatite. *Geologische Rundschau*, 26, pp. 305–350.

Weinberg, R.F., 1995. Diapirism and Decompression Melting. In *The Origin of Granites and Related Rocks*, pp. 162–163, Third Hutton Symposium, Abstracts, United States Survey Circular, 1129, 170 pp.

Whalen J.B., Currie, K.L. and Chappell, B.W., 1987. A-type granites: chemical characteristics, discrimination and petrogenesis. *Contributions to Mineralogy and Petrology*, 95, pp. 407–419.

White, A.J.R. and Chappell, B.W., 1988. Some supracrustal (S-type) granites of the Lachlan Fold Belt, in Symposium on The Origin of Granites. *Transactions of the Royal Society of Edinburgh, Earth Sciences*, 79 (2 and 3), pp. 169–181.

Wilkins, R.T. and Bornhorst, T.J., 1993. Archean appinites from the Northern Complex, Michigan. *The Journal of Geology*, 101, pp. 107–114.

Winkler, H.G.F. and Von Platen, H., 1957. Experimentelle Gesteinsmetamorphose. *Geochimica et Cosmochimica Acta*, 13, pp. 42–59.

Wright, A.E. and Bowes, D.R., 1979. Geochemistry of the appenite suite. In Harris, A.L., Holland C.H. and Leake, B.E. (editors), *Caledonian of the British Isles —*

Reviewed, pp. 699–704, Geological Society of London, Special Publication Number 8, 768 pp.

Wyborn, D., Chappell, B.W. and Johnston, R.M., 1981. Three S-type volcanic suites from the Lachlan Fold Belt, Southeast Australia. *Journal of Geophysical Research*, 86 (B11), pp. 10 335–10 348.

Wyllie, P.J., 1983. Experimental studies of biotite- and muscovite-granites and some crustal magmatic sources. In Atherton, M.P. and Gribble, C.D. (editors), *Migmatites, Melting and Metamorphism*, pp. 12–26, Shiva Publishing Limited, Nantwich, U.K., 326 pp.

Yardley, B., 1991. The successful alchemist. *New Scientist*, 131, Number 1781, pp. 20–24.

CHAPTER 10 — Magmas and rocks of the midalkaline suite

> In the typical basalt-trachyte association the differentiation may proceed still further in the alkaline direction and give phonolitic trachyte....

Bowen, 1928, p. 239.

10.1 Survey of the midalkaline suite

Rocks of the midalkaline suite occupy the transitional field between the common rocks of the subalkaline suite and the rare rocks of the alkaline suite. Their sigma values normally vary between 2.5 and 10 (see Section 2.7). The volcanic rocks and magmas of this suite typically range from basalt, via trachybasalt, basaltic trachyandesite and trachyandesite to trachyte or trachydacite (see Table 10.1). Some trachydacitic magmas evolve further and become rhyolites, whereas a few silica undersaturated trachytic magmas evolve into phonolites. Trachyte is the pivotal magma in the midalkaline lineage. The principal plutonic rocks of this suite are monzogabbro, monzodiorite, monzonite, syenite and quartz monzonite, as revealed in Section 2.7.

In 1813 Haüy (in Brongniart, 1813, p. 43) introduced the name trachyte to describe the distinctive rocks of the Drachenfels volcanic dome on the Rhine river southwest of Bonn in Germany. This dome is about 450 m in diameter. It is mostly composed of a porphyritic rock with phenocrysts of sanidine, plagioclase (oligoclase-andesine), biotite and quartz, set in a fine-grained groundmass. The groundmass contains aligned laths of feldspar. This gives the groundmass a trachytic texture. In 1802 von Buch described and named a similar rock from the Puy-de-Dome in the Chaine des Puys of the Massif Central in France. He called the rock domite. Trachytes commonly occur both as lavas and as pyroclastic rocks. Most of the lavas are porphyritic with many rod and lath shaped minerals arranged in subparallel fluidal patterns. The phenocrysts are typically sanidine or anorthoclase, with lesser amounts of the ferromagnesian minerals, Fe-rich augite, hornblende and biotite. Eutaxitic struc-

tures are abundant in many pyroclastic trachytes. The peralkaline trachytes usually contain phases such as aegirine, aegirine-augite, aenigmatite ($Na_2Fe_5Ti-Si_6O_{20}$), arfvedsonite or riebeckite. Foid-bearing trachytes are intermediate in composition between the trachytes and phonolites and they normally carry less than 10 per cent total feldspathoids. The trachyte from the Drachenfels volcanic dome carries 75 per cent sanidine, 11 per cent plagioclase, and 10 per cent augite, with the rest of the rock being taken up by sphene, fluorapatite, iron-titanium oxides and glass.

10.2 Abundance and distribution

A perusal of the geological record reveals that a high proportion of all known midalkaline volcanic rocks were extruded during the Cenozoic Era, but even in this era they have a low abundance. They do have an extensive distribution on the continents and on oceanic islands and seamounts. Most trachytes crop out in a volcanic centre, or field, that contains larger volumes of comagmatic basaltic, particularly hawaiitic rocks. The largest known outcrop of trachyte is located in the main rift valley of Kenya and northern Tanzania (Baker and Mitchell, 1976, p. 482). These extensive (1770 km³), flat-lying flows of flood trachyte are usually peralkaline with high abundances of sodium (about 6 per cent Na_2O), fluorine and chlorine. Their singular chemical composition is believed to have greatly reduced their viscosity. This attribute and their high eruption rate enabled them to spread out and cover large areas. In the South Turkana region of the northern Kenya Rift there are a series of large (50 km in diameter), low-angle (slopes averaging about 5°) shield-shaped trachytic volcanoes.

Table 10.1 Major element compositions and CIPW norms of the midalkaline suite. After Le Maitre, 1984, pp. 252–253.

	Hawaiite	Potassic trachybasalt	Mugearite	Shoshonite	Benmoreite	Latite	Sodic trachyte	Potassic trachyte
SiO_2	48.71	48.88	52.61	53.06	58	58.23	63.88	64.22
TiO_2	2.63	2.54	1.96	1.45	1.21	1.11	0.53	0.6
Al_2O_3	16.33	15.42	16.94	16.85	17.28	17.17	16.51	16.9
Fe_2O_3	4.47	5.09	4.19	4.21	3.81	3.64	3.18	2.88
FeO	7.3	6.67	5.85	4.96	3.78	3.29	2.21	1.75
MnO	0.19	0.18	0.16	0.17	0.18	0.13	0.17	0.11
MgO	5.53	6.04	3.95	4.65	2.07	2.6	0.61	0.73
CaO	8.62	8.88	7.2	7.46	4.9	5.33	1.69	1.9
Na_2O	4.04	3.12	4.61	3.43	5.64	3.98	6.96	5.13
K_2O	1.55	2.5	1.95	3.21	2.71	4.12	4.13	5.62
P_2O_5	0.63	0.68	0.58	0.55	0.42	0.4	0.13	0.16
Total	100	100	100	100	100	100	100	100
CIPW norms								
Quartz	0	0	0	0.5	3.12	6	4.32	8.54
Orthoclase	9.16	14.76	11.54	18.97	16.01	24.31	24.4	33.23
Albite	31.93	26.41	38.96	29.03	47.73	33.7	58.84	43.4
Anorthite	21.83	20.68	19.76	21.07	13.82	16.82	1.64	6.48
Nepheline	1.21	0	0	0	0	0	0	0
Diopside	13.55	15.11	9.79	9.87	6.24	5.61	4.79	1.47
Hypersthene	0	2.17	6.97	10.41	4.28	5.25	0.08	1.17
Olivine	9.35	7.05	1.8	0	0	0	0	0
Magnetite	6.49	7.38	6.07	6.1	5.52	5.27	4.62	4.18
Ilmenite	5	4.83	3.73	2.75	2.29	2.11	1	1.14
Apatite	1.48	1.61	1.38	1.3	0.99	0.93	0.31	0.39
Total	100	100	100	100	100	100	100	100

These shields are surmounted by structurally complex central zones that contain a cluster of vents, dykes and domes (Webb and Weaver, 1975, p.18).

Rocks of the midalkaline suite are usually divided into normal and peralkaline types. Both groups may be further subdivided into sodic or potassic suites (see Table 10.1). The sodic suite is the more abundant of the two. It usually consists of midalkaline basalt → hawaiite → mugearite → benmoreite → sodic trachyte → trachydacite. Rocks of the rare potassic suite consist of midalkaline basalt → potassic trachybasalt → shoshonite → latite → potassic trachyte. Rocks of the sodic series may be separated from those of the potassic series by using the formula $(Na_2O - 2.0 >/< K_2O)$. Peralkaline trachytes are also quite common. They usually contain about 3 per cent normative nepheline and 1.5 per cent normative aegirine (acmite).

10.3   Daly Gap

Daly (1925) was the first to record that on many oceanic islands there was a noteworthy lack of midalkaline suite rocks of intermediate composition. This paucity of intermediate rocks is now known as the Daly Gap. Chayes (1977) and Booth *et al.* (1978) have shown that this gap is real for the extrusive rocks of the Canary Islands and the Azores respectively. A similar paucity of intermediate rocks has been reported from various continental areas where there are outcrops of midalkaline rocks. An explanation of the Daly Gap is presented in Section 10.5 on the flood trachytes of the East African Rift.

10.4   Petrogeny's residua system

Trachytes occupy a paradoxical position in petrogeny's residua system (quartz-nepheline-kalsilite: Bowen, 1937; Schairer, 1950) because they lie astride a thermal ridge between two thermal troughs (see Fig. 10.1). The compositions of the materials in one trough are rhyolitic, whereas those in the other are phonolitic. Petrogeny's residua system is important because it illustrates that at low pressures there is a

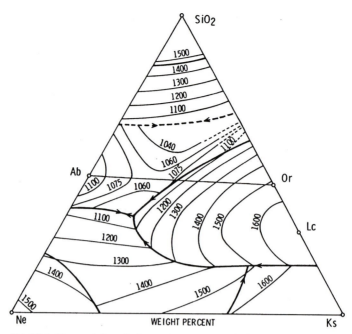

Fig. 10.1 Phase relations in the system quartz-nepheline-kalsilite, or petrogeny's residua system. Materials of trachytic composition plot on the thermal ridge between two thermal troughs. The low-temperature troughs are occupied by materials of either rhyolitic or phonolitic composition. Adapted from Bowen (1937) and Schairer (1950).

thermal barrier that normally induces magmas that fall on either side of it to evolve towards different minimum points. As we have already seen suites of rocks that contain trachytes do not always terminate in the trachyte field. Some silica oversaturated trachytic magmas terminate in the rhyolite field whereas foid-normative trachytic magmas may terminate in the phonolite field. Peralkaline rhyolites frequently evolve from peralkaline trachytic magmas. This differentiation process usually consists of fractional crystallisation assisted by volatile transfer. It often takes place in a thin buoyant upper layer that develops in long-lived, compositionally zoned magma chambers. The petrology of the trachytic and midalkaline rocks is best understood by considering actual examples of these rocks.

10.5   Trachytes of the East African Rift

The upper Cenozoic alkaline rocks of Kenya and northern Tanzania evolved concurrently with the development of the eastern branch of the East African rift system. This petrographic province contains three main suites of magmatic rocks, They are the melanephelinite-nephelinite suite, the normal alkaline suite and the midalkaline suite. Some volcanoes contain rocks from more than one of these suites. This implies that magmas with different degrees of alkalinity were available at much the same time. Magmatic activity began in the northern Turkana area about 30 million years ago. Since then the magmatic focus has moved progressively south. In the main rift zone there is a general decrease in alkalinity with time. There is also a general tendency for the least alkaline magmas to erupt within the central part of the rift valley. These relationships probably relate to a progressive ascent of a zone of melting under the rift valley. Most of the trachytes are less than seven million years old and they evolved from a midalkaline magma that was emplaced as the graben developed. Although there are a few extensive flows of porphyritic benmoreites the midalkaline rocks usually display a marked Daly Gap between 48 per cent and 58 per cent SiO_2.

Small volumes of alkali rhyolite are usually also present.

Extensive floods of peralkaline trachyte cover much of the floor of the central and southern part of the rift and trachytic pyroclastic rocks are strewn over an even wider area. The latter materials were probably derived from caldera complexes that now lie buried beneath the younger materials in the central part of the rift. Locally both the trachytes and the rhyolites occur as plug domes. A chain of large shield volcanoes traces the medial line of the rift. These volcanoes are mainly composed of basaltic and rhyolitic rocks, but some, like Menengai, contain compositionally zoned ash-flow tuffs.

A variety of fluid dynamic processes operate within magma chambers. They can influence the course of magmatic differentiation and the characteristics of the successive eruptions. Buoyant residual liquids are likely to rise along the walls of magma chambers and accumulate at the top of the chamber at a rate that can account for the volumes of salic material that erupt from the midalkaline shield volcanoes. Generally the compositions of the materials extruded during a particular eruption depend on a balance between the rate of boundary-layer segregation and the volume of material extruded. Magma densities help explain why the rocks of the midalkaline suite often lack intermediate members. The initial basalt-trachybasalt magma has a density of about $2710 \, kg/m^3$, its density rises to $2790 \, kg/m^3$ in the ferrobasalt-ferrotrachybasalt field, and then decreases rapidly through the intermediate field to trachyte. The latter has a density of $2450 \, kg/m^3$. Iron enrichment followed by iron depletion is mainly responsible for these changes in density.

It is also known that magmas of intermediate composition only exist for a small temperature interval whereas both the basaltic-trachybasaltic and the trachytic magmas are usually present in a magma chamber over a much wider range in temperatures. When new, hot, basaltic-trachybasaltic magma is being added to a magma chamber, differentiation is not likely to go beyond the ferrobasalt-ferrotrachybasalt field. Once the replenishment process stops, the residual magma cools and moves rapidly through the intermediate field and into the trachyte field. Magma chambers are usually thermally stratified and a gravitationally stable upper zone of silicic magma develops soon after the process of magma replenishment stops.

10.6 Peralkaline tendency

Usually as fractional crystallisation proceeds in the midalkaline suite the $Al_2O_3/(Na_2O + K_2O)$ ratio of the residual magma decreases and they become more peralkaline (Shand, 1950, p. 229; see Section 2.9). In the Hawaiian midalkaline suite (Macdonald, 1968, p. 502) this ratio decreases from 3.7 in the basalts to 2.8 in the hawaiites, 2.3 in the mugearites, 2.0 in the benmoreites and 1.6 in the sodic trachytes. The main reason for this peralkaline tendency is that the feldspars that precipitate from the various magmas that evolve in this suite have a consistently higher $Al_2O_3/(Na_2O + K_2O)$ ratio than the magmas in which they occur (see Section 2.9). For most practical purposes rocks are peralkaline if their Al_2O_3 contents in weight percent are less than ($1.6 \, Na_2O$ $+ 1.1 \, K_2O$) in weight percent.

10.7 Syenites

Alkali feldspar is the only essential mineral in the syenites. The rock also may contain small amounts of plagioclase, one or more mafic minerals, and either quartz or a foid. They are usually medium- to coarse-grained with granular, hypidiomorphic granular, or trachytoid textures. Many syenites contain a single perthitic feldspar. The common mafic minerals found in these rocks are hornblende, biotite, augite (usually iron-rich), arfvedsonite, riebeckite, aegirine and aegirine-augite. In some syenites the abundance of iron-titanium oxides exceeds five per cent. Most syenites occur in small subvolcanic intrusions, or form local igneous facies within larger compound intrusive bodies, such as the Fongen-Hyllingen layered complex of central Norway (Sørensen and Wilson, 1995). Many syenites are interpreted as having formed as layers that evolved within compositionally layered magma chambers.

10.8 Cenozoic volcanic rocks of eastern Australia

The Eastern Australian Volcanic Province is an example of an extensive volcanic belt that contains large volumes of midalkaline rocks. It crops out discontinuously for some 4400 km, from the Murray Islands in the Torres Strait, north of Queensland, to Tasmania in the south (Johnson, 1989). The province also extends westwards across Victoria into South Australia. It contains rocks that range in age from about 70 million years (Late Cretaceous) to 4600 years before the present. The youngest rocks occur in the extreme north and south of the province. In north Queensland the youngest rocks are about 13 000 years old, whereas in South Australia the youngest rocks are 4600 years old (Duggan et al., 1993). The least differentiated rocks such as the alkali basalts, basanites and melilitites contain a wide range of xenoliths and xenocrysts derived from the upper mantle (Johnson, 1989). Spinel lherzolite xenoliths are particularly common.

The on-shore volcanoes of the Eastern Australian Volcanic Province developed along the western border of the Tasman Sea. This sea and its northward extension the Coral Sea opened between 95 and 53 million years ago. During the last 33 million years, as Australia and the Tasman Sea have moved northwards, chains of volcanoes have developed both on land and in the sea (see Fig. 10.2). Most of the submarine volcanoes belong to either the Tasmantid or the Lord Howe seamount chains. The Tasmantid Seamounts form a prominent 1400 km long north–south volcanic chain that transects the Tasman abyssal plain between the east coast of Australia and the Lord Howe Rise. Over the summits of these seamounts the depth of water increases progressively northwards. Thermal modelling, augmented by radiometric dating, all show that these linear submarine volcanic chains probably developed as the result of the movement of the Australian plate over a relatively fixed mantle plume (McDougall and Duncan, 1988; Vogt and Conolly, 1971).

The Derwent-Hunter Seamount is a typical example of the volcanoes of the Tasmantid seamount chain. It is approximately 15 to 16 million years old. Rock samples collected from its surface reveal that it contains an abbreviated differentiation sequence that ranges in composition from midalkaline basalt to mugearite. On a titanium-yttrium-zirconium discriminant diagram (Pearce and Cann, 1973) the basic rocks all plot in the within-plate field (Middlemost, 1989, p. 315).

Lava fields, central volcanoes and volcanic plugs

Rocks of the Eastern Australian Volcanic Province belong to three main groups: the lava fields, the central volcanoes and the leucite basanites of the western volcanic chain. The latter are discussed in Section 11.6. Most rocks from the lava fields are basalts, basanites or hawaiites. They frequently appear as thin flows that cover extensive areas. In a few areas, such as the Liverpool Range in central New South Wales, the piles of lava are more than 1000 m thick. The younger lava fields of western Victoria, southeastern South Australia and northern Queensland frequently contain many small scoria cones and maars (see Figs 10.3 and 10.4). Some lava fields contain major lava-tube systems, such as those associated with Undarra Volcano southwest of Cairns in north Queensland (Atkinson et al., 1975). Lava fields are typical of the Eastern Highlands of Australia and they were mainly extruded between 55 and 30 million years ago. A second major period of this type of volcanism began about 5 million years ago.

The volcanic products of the central volcanoes of Eastern Australia are mainly hawaiites with lesser volumes of the other members of the midalkaline lineage. Small volumes of rhyolite and phonolite are also found. Like the seamounts of the Tasman Sea the central volcanoes show a progressive decrease in age from north to south. Cape Hillsborough in north Queensland is about 35 million years old, whereas Mount Macedon in central Victoria is only 6 million years old.

The various Miocene central volcanoes of New South Wales have been subjected to different amounts of erosion. If one uses field data gleaned from them all, one can piece together a composite volcanic history. This model midalkaline volcano has a low-angle shield form, surmounted by a complex centralised system of vents (see Fig. 10.5). The central zone contains clusters of domes, dykes, plugs and small cones, scattered amidst a variety of pyroclastic materials. Originally much of the surface of the central and proximal zones of the volcano is likely to have been covered by trachytic pyroclastic materials, while the distal zone was mainly covered by flows of hawaiitic composition. Beneath a thin veneer of pyroclastic rocks the proximal zone is likely to contain large volumes of hawaiitic lava and more differentiated midalkaline rocks. The upper part of the central zone is probably mainly composed of trachytic, rhyolitic and even phonolitic materials. At depth one would expect to find a range of compositions including rocks of intermediate composition, such as mugearites and benmoreites. The large Tweed Volcano on the New South Wales–Queensland border contains coarse-grained rocks at its

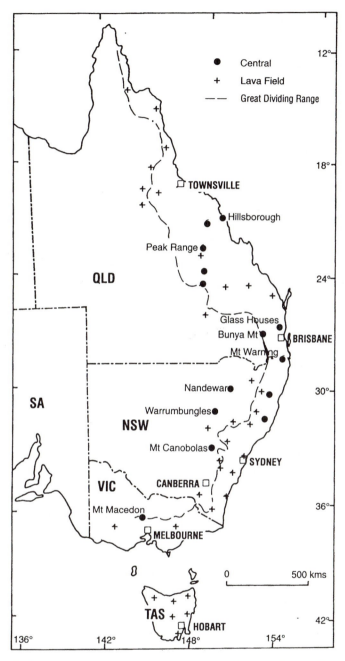

Fig. 10.2 Sketch map of the Eastern Australian Volcanic Province that developed during the last 33 million years as Australia moved northwards.

centre. They probably crystallised in a large zoned magma chamber that lay beneath, and partly intruded, the surficial volcanic edifice. It has been suggested that when the various central volcanoes of the Eastern Australian Volcanic Province were active they contained extensive high-level magma chambers that limited the height, or more particularly the weight, of the volcanic construction that

Fig. 10.3 Maars in the Victorian segment of the Eastern Australian Volcanic Province.

could be placed upon them. This constraint probably controlled the morphology of the central zone of these volcanoes (Middlemost, 1985, p. 55).

The morphology of these central volcanoes is similar to that of the Plio-Pleistocene volcanoes of the South Turkana region in the northern part of the Kenya Rift and the volcanoes of the Rainbow Range of British Columbia in Canada. Webb and Weaver (1975, p. 18) regarded these volcanoes with their low-angled, stratified flanks surrounding a complex centralised system of vents as a distinct volcanic landform that they called the Turkana-type.

10.9   Oslo Magmatic Province

Extrusive and intrusive rocks of the Oslo (originally called Kristiania) Magmatic Province of southern Norway crop out over an area of $6500\,km^2$ (see Figs 10.6 and 10.7). The magmatic activity can be related to the evolution of the Oslo paleorift. This tectonic unit is composed of three distinct grabens that are designated the Akershus graben in the

north, the Vestfold graben in the centre and the Skagerrak graben to the south (Ro *et al.*, 1990b). Most of the igneous rocks crop out in the central part of the Akershus and Vestfold grabens. Magmatic activity started in the south, circa 300 million years ago (Carboniferous/Permian) and then shifted northwards at a rate of 1–2 cm per year (Sundvoll *et al.*, 1990, p. 67). The total life-span of the paleorift was about 60 million years. Geophysical exploration has shown that beneath the entire paleorift the Moho has been elevated by 3–5 km (Ro *et al.*, 1990a, p. 16). There is also a gravity high that coincides with the paleorift. In the Akershus and Vestfold grabens there are strong magnetic anomalies that can be directly related to both the gravity anomalies and the surface outcrop patterns of the major igneous rock bodies.

The Oslo Magmatic Province has become well known because much excellent research has been carried out on these diverse midalkaline rocks by many pioneers of petrology, like von Buch (1810), Kjerulf (1855–1880), Brögger (1890–1933), Goldschmidt (1911–1928) and Barth (1945–1954). In the Permian, the Oslo region was occupied by a rift valley that was

Fig. 10.4 Pyroclastic beds in Tower Hill Volcano in southern Victoria, Australia.

probably as impressive topographically as the current rift valleys of eastern Africa (see Sections 10.5 and 11.4). The volcanic and subvolcanic rocks of both areas are similar. Both contain rocks of the alkaline and midalkaline suites.

In Oslo post-Permian erosion has removed several thousand metres of material thus exposing rocks that were emplaced deep within the paleorift. The large Permian calderas are a feature of the present topography and geology of the city of Oslo and its region. According to Oftedahl (1978a, p. 160) the region contains 15 calderas, 3 probable calderas, and 7 ring complexes (see Figs 10.6 and 10.7). Some volcanic rocks that subsided into these calderas have been preserved, thus plutonic and supracrustal rocks have been juxtaposed and are now preserved in the same region. The igneous rocks range in composition from ferromagnesian cumulates to syenites and granites (see Table 10.2). At the present level of erosion most of the rocks are monzodiorites, monzonites, syenites and quartz monzonites, and their volcanic equivalents.

The Oslo Magmatic Province contains approximately 500 km^3 of volcanic rocks but in the Permian

Fig. 10.5 Trachytic domes, plugs and dykes in the central part of the Warrumbungle Volcano in central New South Wales, Australia.

Fig. 10.6 Rhomb porphyry from Kolsas in the Oslo Magmatic Province of southern Norway.

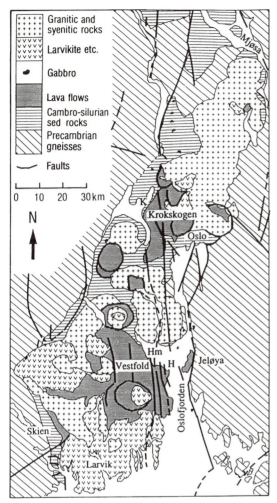

Legend:
- Granitic and syenitic rocks
- Larvikite etc.
- Gabbro
- Lava flows
- Cambro-silurian sed rocks
- Precambrian gneisses
- Faults

0   10   20   30 km

N

Fig. 10.7 Sketch map showing the main groups of magmatic rocks of the Oslo Magmatic Province of southern Norway. After Neumann, Sundvoll and Øverli, 1990, p. 90.

10.2). Basaltic rocks (sensu lato) dominate the lower part of the sequence while the more evolved types of lava gradually become more abundant at higher levels within the sequence. They dominate the uppermost part of the sequence. The earliest group of lavas is designated B1. In different parts of the province B1 is extremely variable as regards the number of flows, their aggregate thickness and their composition. The oldest B1 lavas crop out at Skein in the southwest corner of the province. They are usually melanocratic basanites (ankaramites). In the central part of the province the B1 lavas are normal basanites. Quartz-normative midalkaline basaltic rocks such as the lavas of the Krokskogen Plateau (Weigand, 1975, p. 23) crop out in the north. According to Oftedahl (1978a, p. 163) the B1 lavas erupted through fissures linked to a string of sill-shaped magma chambers. Some lavas lie over Early Permian sediments, whereas in other areas they cover older sedimentary rocks of Cambrian to Devonian age. The extrusion of the B1 lavas was followed by a brief period of sedimentation that was marked by the laying down of a bed of basaltic detritus. This distinctive layer is followed by a thin sandstone unit.

This pause in volcanic activity ended with the eruption of a series of rhomb porphyries. This term was used by von Buch (1810) in his pioneering book, *Reise durch Norwegen und Lappland.* These rocks are porphyritic latites. They were called rhomb porphyries because they usually contain complex feldspar phenocrysts that display rhomboidal outlines. Many large phenocrysts consist of cores of plagioclase with mantles and patches of alkali feldspar. Individual flows are readily mapped because the textures of the rocks in the various flows are thoroughly different. This is because the phenocrysts vary in abundance and display distinctive shapes and sizes. Most of the complex phenocrysts originally crystallised as anorthoclase and subsequently became mantled; after that the cores unmixed to form an antiperthitic intergrowth. It is postulated that these phenocrysts initially crystallised in a latitic magma at moderately high pressures and temperatures and the exsolution took place while the crystals were in transit to the surface. The groundmass of the rhomb porphyries usually contains clouded alkali feldspar and clinopyroxene, typically aegirine-augite. Minor amounts of chlorite, sericite, apatite, carbonate and iron-titanium oxides also may be present. Lacunae between cooling units are often marked by layers of sedimentary rocks. Modern rhomb-porphyries are not common, but they do occur in the volcanic materials extruded

they were probably much more voluminous. Confirmation of the original size of the province was obtained from: (1) the distribution of the many dykes of appropriate age and composition that crop out outside the present grabens; (2) the presence along the Oslo Fjord Fault scarp of a fanglomerate unit that contains some of the missing volcanic material; and (3) fragments of volcanic rock found in explosion breccia pipes that lie outside the present grabens. The lavas of the Oslo Magmatic Province form an interfingering sequence. Traditionally they have been called basalts (B), rhomb porphyries (RP), trachytes (T) and rhyolites (R) (see Table

Table 10.2 Major element composition of some igneous rocks of the Oslo Magmatic Province.

Rock type	Alkali basalt	K-trachybasalt	Basalt	Hawaiite	Shoshonite	Latite	Latite	Potassic trachyte	Potassic trachyte	Rhyolite
Local name	Px basalt	All basalts	Aphyric basalt	Plag. basalt	Rhomb porphyry	Latite tuff	Larvikite	Trachyte	Trachyte tuff	Ignimbrite
Reference	Weigand, 1975	Weigand, 1975	Weigand, 1975	Weigand, 1975	Brögger, 1933	Holtedahl, 1943	Neumann, 1980	Holtedahl, 1960	Holtedahl, 1943	Holtedahl, 1943
SiO_2	44.6	48.2	48.9	49.3	53.51	57.33	57.77	63.27	63.56	75.44
TiO_2	3.59	2.9	2.7	2.98	1.25	0.52	1.18	0.59	1.08	0.07
Al_2O_3	12.3	14.1	14.5	16.2	18.38	17.96	18.65	16.88	15.04	12.33
Fe_2O_3	7.13	6.91	6.85	7.76	7.92	2.79	4.42	1.47	6.75	0.49
FeO	6.34	5.23	6.19	4.32	0.97	4.07	1.3	2.63	0.39	1
MnO	0.2	0.2	0.2	0.19	0.28	0.04	0.11	0.03	0.13	0.11
MgO	7.96	5.7	5.72	4.7	1.34	1.55	1.6	1.84	0.6	0.52
CaO	11.4	8.48	8.29	7.58	5.07	3.12	4.47	2.98	0.7	0.01
Na_2O	2.95	3.71	3.25	3.79	4.68	4.7	5.16	2.68	3.3	2.38
K_2O	1.86	2.6	1.23	1.7	4.29	4.36	3.66	6.8	6.73	7.13
P_2O_5	0.73	0.44	0.41	0.41	0.71	0.7	0.49	0.22	0.7	0.31
Total	99.06	98.47	98.24	98.93	98.4	97.14	98.81	99.39	98.98	99.79
CIPW norms										
Quartz	0	0	0	0	0	4.76	2.96	11.98	17.94	33.1
Orthoclase	11.19	15.72	7.51	10.26	26.01	26.51	21.94	40.42	39.97	41.41
Albite	11.55	23.39	28.48	32.72	38.74	40.94	44.33	22.85	28.04	19.83
Anorthite	15.08	14.47	22.09	22.61	16.89	11.21	17.2	13.45	0	0
Nepheline	7.51	4.75	0	0	1.05	0	0	0	0	0
Corundum	0	0	0	0	0	1.53	0	0.2	2.7	1.36
Diopside	30.39	20.89	14.49	10.69	3.5	7.67	1.64	0	0	0
Hypersthene	0	0	14.72	5.83	0	0	4.91	7.32	2.35	2.76
Olivine	11.17	10.08	2.02	7.24	5.59	0	0	0	0	0
Magnetite	4.42	3.98	4.36	3.89	4.08	4.67	3.56	2.15	5.28	0.7
Ilmenite	6.95	5.66	5.33	5.79	2.44	1.03	2.29	1.12	2.07	0.13
Apatite	1.74	1.06	1	0.97	1.7	1.68	1.17	0.51	1.65	0.71

from Mount Kilimanjaro in northern Tanzania and Mount Erebus on Ross Island off Victoria Land in Antarctica.

After the extrusion of the first group of rhomb porphyries (i.e. RP1 to 10) the composition of the lavas and the nature of the eruptions changed. This produced a second period of basic eruptions (B2). Currently such lavas are only found in three areas, that is, in the Glitrevann caldera, in the area around Krokskogen and in the Nittedal area. In the Nittedal area B2 attains a thickness of 200 m. It has many small flows and near surface intrusions. These eruptions were followed by the extrusion of two more rhomb porphyries.

The thick B3 basic flows are now only recorded from inside the calderas. As most calderas lack suitable vents for the transport and extrusion of this lava Oftedahl (1978a) has proposed that the B3 lava was erupted outside the calderas and flowed passively into them. This hypothesis is supported by the idea that it would be difficult for magma of the density of basalt to travel through a large body of less dense felsic material that was probably partly molten. The Oslo calderas are circular collapse structures and they exhibit evidence of vertical subsidence of 4000–5000 m. At the present level of erosion they are mainly filled by large bodies of monzonitic, syenitic and granitic rock. When mobile these materials appear to have stoped their way up into cylindrical cavities created by ring-fracturing and cauldron subsidence. There is no evidence that large basaltic volcanoes ever existed above these calderas. The only volcanism associated with the calderas was highly explosive in character and felsic in composition. Oftedahl has estimated that the volume of pyroclastic rock ejected from the Baerum caldera was large enough to account for its subsidence. It is postulated that all the young rhomb porphyries from RP13 onwards flowed into caldera basins that had already subsided.

Basic lavas B4 and B5 are considered to have erupted from fissures. B5 is approximately 2000 m thick and it contains all the different types of basic lavas found in the Oslo Magmatic Province. Currently most of the plutonic rocks crop out in the northern part of the province. They belong to two broad groups: (1) the subvolcanic microgabbroic plugs of alkali basaltic or basanitic composition that have traditionally been called Oslo essexites; and (2) the larger plutonic bodies that include monzonites (larvikites), quartz monzonites, syenites (nordmarkites and tonsbergite), and granites (eker-

ites). An Oslo essexite is essentially composed of augite and plagioclase (andesine-labradorite), with biotite-phlogopite, potassium feldspar and olivine usually being present in smaller amounts. Common accessory minerals include iron-titanium oxides and apatite. Nepheline is sometimes present, but it invariably occurs as a normative phase.

The more differentiated plutonic rocks have traditionally been called alkaline rocks because they contain one or more of the minerals aegirine augite, aegirine, arfvedsonite and riebeckite. Most of the larvikites are monzonites and are chemically akin to the rhomb porphyries. Their essential minerals are perthitic alkali feldspar, oligoclase and titanaugite. Common minor and accessory minerals include biotite, quartz or nepheline, iron-titanium oxides, apatite and sphene. Dykes of diverse composition are abundant. They are particularly common in the country rocks adjoining the plutonic bodies and include leucocratic microsyenites (bostonite), lamprophyres (camptonite) and arfvedsonite-bearing felsic rocks (lindöite).

The main difficulty encountered when trying to understand the origin of the Oslo Magmatic Province is the presence of such large volumes of monzonitic and syenitic rocks and their volcanic equivalents. Strontium, neodymium and lead isotopic studies have established that the more primitive rocks of the province were generated by partial melting of a 'mildly depleted source' in the upper mantle (Neumann et al., 1990). Geochemical data on the monzonitic and syenitic rocks attests that they were produced from magmas that had been contaminated by upper crustal continental materials (Neumann et al., 1988). It is suggested that extensive rifting triggered local decompression melting in the upper mantle, resulting in the generation of basanitic to alkali basaltic magma. It rose into the lower crust but halted when its density matched that of the surrounding crustal rocks. Cooling facilitated magmatic differentiation that resulted in the generation and upward movement of batches of less dense residual magma. This process eventually left a large mass of dense cumulates and layered gabbroic rocks in the lower crust (Neumann et al., 1986) and a smaller body of latitic magma in the upper crust. Additional magmatic differentiation, intermittent magmatic replenishment and density layering induced by gravity all assisted in the evolution of the diverse magmatic rocks currently exposed in the Oslo Magmatic Province.

10.10   Cenozoic midalkaline lava fields of western Saudi Arabia

In the Cenozoic, several large lava fields, or harrats, were extruded in the western part of the Arabian Peninsula. Rocks belonging to this petrographic province now cover an area of about $180\,000\,km^2$. The lavas of the large Harrat Rahat extend from Al Madinah al Munawwarah in the north to the outskirts of Jiddah in the south (Camp and Roobol, 1989). This lava field forms a volcanic plateau that is a major topographic feature in western Saudi Arabia. The main mass of volcanic rocks measures about $310\,km$ north–south and $75\,km$ east–west. Tongues of lava occupy the ravines that cut through the Red Sea escarpment. Individual flows may extend southwestwards away from the main mass of lavas for distances of up to $100\,km$. Extrusion began in the south but gradually the volcanic focus shifted northwards actuating a $310\,km$ long linear vent system. These vents are part of an even more extensive system of vents that trends northwards. It has been called the Makkah-Madinah-Nafud volcanic line. The more differentiated lavas, such as the trachytes and comendites, are among the youngest rocks. They typically crop out along the Makkah-Madinah-Nafud volcanic line (Camp *et al.*, 1991). This volcanism was contemporaneous with oceanic tholeiitic volcanism in the Red Sea and alkaline volcanism (tephrite to phonolite) in some other lava fields of western Saudi Arabia.

The total volume of volcanic rocks in the Harrat Rahat plateau is about $2000\,km^3$ (Camp and Roobol, 1991, p. 15). Over 80 per cent of these lavas are alkali basalts that contain various amounts of normative nepheline as shown in Table 10.3. Camp and Roobol (1991, p. 19) used a TAS diagram to divide the basalts into what they called olivine transitional basalts (75 per cent) and alkali olivine basalts (25 per cent). The alkali olivine basalts contain more total alkalis and more normative nepheline than the more abundant olivine transitional basalts. Most of the olivine transitional basalts belong to the oldest stratigraphic unit, whereas alkali olivine basalt appears to occur in all stratigraphic units. The more evolved lavas only occur in the younger units. These lavas typically belong to the sodic midalkaline series and they range from hawaiites via mugearites and benmoreites to trachytes (see Table 10.3).

It has been proposed that the primary magma that evolved to generate the midalkaline suite of rocks of the Harrat Rahat plateau was produced by about 15 per cent partial melting of a garnet lherzolite source rock at a depth of about $100\,km$. After segregating from its source rocks the primary magma rose and accumulated in a magma chamber at the crust mantle boundary. While resident in this magma chamber fractional crystallisation produced the alkali olivine basaltic magma that rose towards the surface. Some erupted, but much of it became trapped in high-level magma chambers. Continued cooling and magmatic differentiation resulted in the development of a series of compositionally zoned magma chambers. These chambers were periodically replenished by new primitive magma and became the source for the diverse magmas required to generate the midalkaline rocks of the Harrat Rahat plateau (Camp and Roobol, 1991, p. 32).

10.11   Petrogenesis of the midalkaline suite

In any unembellished discussion of the origin and evolution of igneous rocks from the main sequence it is important to begin by sorting them into their principal lineages. The midalkaline rocks comprise the most intriguing of these lineages because they fill the verge between the abundant subalkaline lineage and the much less abundant, but petrologically important, alkaline lineage (see Fig. 10.8). If a differentiating magma experiences slight changes in composition it may either be nudged into, or out of, the midalkaline field. For example, if an alkaline magma digests minor amounts of normal, silica oversaturated, crustal material the composition of the new hybrid magma is likely to be propelled towards, or even into, the midalkaline field. Many petrographic provinces that are principally composed of midalkaline rocks also contain some rocks that are either undersaturated or oversaturated in silica. To account for this apparent anomaly it is often suggested that the various magmatic rocks of the province are products of assimilation coupled with fractional crystallisation (AFC).

Most evolved midalkaline rocks, particularly those derived from parental magmas known to have experienced limited crustal assimilation, are regarded as products of low pressure, fractional crystallisation of a more basic magma. This fractionation process usually takes place in a compositionally layered subvolcanic magma chamber and the level of silica

Table 10.3 Mean chemical composition and CIPW norms of the midalkaline suite from Harrat Rahat, Kingdom of Saudi Arabia. After Camp and Roobol, 1991, p. 20.

Major elements	Transitional basalt	Alkali basalt	Hawaiite	Mugearite	Benmoreite	Trachyte
SiO_2	48.03	47.12	47.84	52.2	56.61	60.93
TiO_2	1.53	2.48	2.54	1.88	0.96	0.36
Al_2O_3	16.04	16.08	16.16	16.77	17.14	17.58
FeO*	10.81	11.85	11.88	11.22	9.78	6.24
MnO	0.17	0.19	0.2	0.26	0.28	0.21
MgO	9.02	7.67	6.37	2.87	1.19	0.4
CaO	10.88	9.78	8.58	6.01	4.2	2.32
Na_2O	2.9	3.59	4.38	5.73	6.23	7.75
K_2O	0.43	0.8	1.24	2.08	3.04	4.4
P_2O_5	0.2	0.5	0.77	1.19	0.57	0.15
Total	100.01	100.06	99.96	100.21	100	100.34
CIPW norms						
Orthoclase	2.54	4.73	7.33	12.28	17.97	26
Albite	22.73	23.08	26.27	41.21	48.33	50.47
Anorthite	29.48	25.4	20.77	13.89	9.83	0.19
Nepheline	0.94	3.9	5.78	3.9	2.38	8.15
Diopside	18.92	16.29	13.84	6.91	6.32	9.08
Olivine	20.33	18.92	17.46	13.88	10.55	4.42
Magnetite	1.7	1.87	1.91	1.81	1.48	1
Ilmenite	2.91	4.71	4.82	3.57	1.82	0.68
Apatite	0.46	1.16	1.78	2.76	1.32	0.35
Trace elements				Parts per million		
Sc	28	23	20	14	15	5
V	224	228	183	59	15	12
Cr	319	215	144	2	<1	2
Ni	178	122	84	4	3	13
Cu	110	72	56	23	23	13
Zn	80	86	94	116	133	172
Ga	18	18	21	22	27	41
Rb	5	7	11	19	27	51
Sr	367	550	652	590	535	186
Y	21	25	29	44	53	76
Zr	111	197	283	428	669	1142
Nb	13	23	35	48	68	116
Ba	87	108	187	316	565	243
La	6	17	29	50	62	93
Ce	30	50	69	105	125	175

saturation of the more evolved magmas is usually determined by the initial level silica saturation of the parental magma.

Mass balance mixing calculations are often used to test the feasibility of fractional crystallisation or assimilation coupled with fractional crystallisation to generate the successive magmas in the midalkaline lineage. One might try to calculate the proportions of the phases that would have to be added or subtracted to produce magmas that belong to the sequence hawaiite, mugearite, benmoreite and trachyte. If one uses average chemical compositions, one discovers that it is arithmetically feasible to generate a mugearite from a hawaiite by removing about 12 per cent augite, 2 per cent olivine and 2 per cent iron-titanium oxides. Benmoreite can be generated from mugearite by removing 13 per cent plagioclase, 7 per cent augite, 3 per cent iron-titanium oxides and 2 per cent olivine. A trachyte can be generated from a benmoreite with the removal of about 25 per cent plagioclase, 4.5 per cent augite and 4.5 per cent iron-titanium oxides. Alternatively, if alkalis can migrate upwards and collect near the top of a compositionally layered magma chamber, any trachytic magma that forms

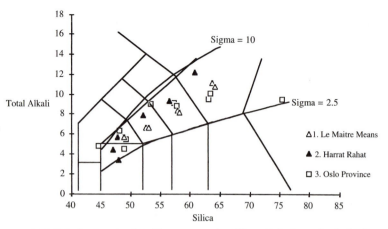

Fig. 10.8 TAS diagram showing: (1) average rocks of the midalkaline suite (Le Maitre, 1984, pp. 252–253); (2) midalkaline rocks from Harrat Rahat, Kingdom of Saudi Arabia (Camp and Roobol, 1991, p. 20); and (3) midalkaline rocks from the Oslo Magmatic Province (see Table 10.2).

in the roof-zone is likely to require the extraction of a much smaller proportion of solid phases.

Rare trachytic rocks, such as those from the Jos Plateau of Nigeria, contain lherzolitic xenoliths of probable upper mantle origin (Wright, 1969, p. 285). These evolved magmas were probably generated at greater depths and at pressures that are higher than normal for magmas of this composition. Another possible explanation is that the mantle-derived xenoliths had a complex emplacement history and they were only entrained by the trachytic magma at a late stage in this process. An extraordinary origin has been proposed for a bizarre group of trachytic rocks that occur in some melanephelinite-carbonatite complexes. These trachytes are regarded as mobilised fenites and their origin will be reviewed in Chapters 11 and 12.

References

Atkinson, A., Griffin, T.J. and Stephenson, P.J., 1975. A major lava tube system from Undarra Volcano, north Queensland. *Bulletin Volcanologique*, 39, pp. 1–28.

Baker, B. H. and Mitchell, J.G., 1976. Volcanic stratigraphy and geochronology of the Kedony-Olorgesailie area and the evolution of the South Kenya rift valley. *Journal of the Geological Society of London*, 132, pp. 467– 484.

Booth, B., Croasdale, R. and Walker, G.P.L., 1978. A quantitative study of five thousand years of volcanism

on Sao Miguel, Azores. *Philosophical Transactions of the Royal Society of London*, 288, pp. 271–319.

Bowen, N.L., 1928. *The Evolution of Igneous Rocks.* Princeton University Press, Princeton, New Jersey, 334 pp.

Bowen, N.L., 1937. Recent high-temperature research on silicates and its significance in igneous geology. *American Journal of Science*, 33, pp. 1–21.

Brögger, W.C., 1933. Die Eruptivgesteine des Oslogebietes VII. *Die chemische Zusammenstzung der Eruptivgesteine des Oslogebietes*, Skrifter utgitt av det Norske Videnskaps-Akademi i Oslo, Matematisk-naturvidenskapelig, Klasse 1933, No. 1, 147 pp.

Brongniart, A., 1913. Essai d'une classification mineralogique des roches melangees. *Journal des Mines*, Paris, 34, pp. 5–48.

Camp, V.E. and Roobol, M.J., 1989. The Arabian continental alkali basalt province: Part 1; Evolution of Harrat Rahat, Kingdom of Saudi Arabia. *Geological Society of America Bulletin*, 101, pp. 71–95.

Camp, V.E. and Roobol, M.J., 1991. *Explanatory notes to the geological map of the Cenozoic Lava Field of Harrat Rahat, Kingdom of Saudi Arabia.* GM-123, Ministry of Petroleum and Mineral Resources, Directorate General of Mineral Resources, Jiddah, Saudi Arabia, 37 pp.

Camp, V.E., Roobol, M.J. and Hooper, P.R., 1991. The Arabian continental alkali basalt province: Part 2; Evolution of Harrats Khaybar, Ithnayn, and Kura, Kingdom of Saudi Arabia. *Geological Society of America Bulletin*, 103, pp. 363–391.

Chayes, F. 1977. The oceanic basalt-trachyte relation in general and in the Canary Islands. *American Mineralogist*, 62, pp. 666–671.

Daly, R.A., 1925. The geology of Ascension Island. *Proceedings of the American Academy of Arts and Sciences* (Philadelphia), 60, pp. 3–124.

Duggan, M.B., Knutson, J. and Ewart, A., 1993. *Warrumbungle, Nandewar and Tweed Volcanic Complexes*. IAVCEI, Excursion Guide C4, Australian Geological Survey Organisation, Canberra, Australia, 40 pp.

Holterdahl, O., 1943. Some structural features of the district near Oslo. *Studies of the Igneous Rock Complex of the Oslo Region 1*. Skrifter utgitt av det Norske Videnskaps-Akademi i Oslo, Matematisk-naturvidenskapelig, Klasse 1943, No. 2, 71 pp.

Holterdahl, O. (editor), 1960. Geology of Norway. *Norske Geologisk Undersøkelse, Oslo*, 208, 540 pp.

Johnson, R.W. (editor), 1989. *Intraplate Volcanism in Eastern Australia and New Zealand*. Cambridge University Press, Cambridge, 408 pp.

Le Maitre, R.W., 1984. A proposal by the IUGS Subcommission on the Systematics of Igneous Rocks for a chemical classification of volcanic rocks based on the total alkali silica (TAS) diagram. *Australian Journal of Earth Sciences*, 31 (2), pp. 243–255.

Macdonald, G.A., 1968. Composition and origin of Hawaiian lavas. In Coats, R.R., Hay, R.L. and Anderson, C.A. (editors), *Studies in Volcanology: A Memoir in Honour of Howel Williams*, pp. 477–522, The Geological Society of America, Boulder, Memoir 116, 678 pp.

McDougall, I. and Duncan, R.A., 1988. Age progressive volcanism in the Tasmantid Seamount Chain, a hotspot trace. *Abstracts, Geological Society of Australia*, 21, p. 262.

Middlemost, E.A.K., 1985. *Miocene Shield Volcanoes of New South Wales*. Geological Society of Australia, New South Wales Division, Publication Number 1, pp. 49–58.

Middlemost, E.A.K., 1989. The Derwent-Hunter submarine volcano: the product of a long defunct subduction zone. *Journal of Volcanology and Geothermal Research*, 37, pp. 311–323.

Neumann, E.-R., 1980. Petrogenesis of the Oslo region larvikites and associated rocks. *Journal of Petrology*, 21 (3), pp. 498–531.

Neumann, E.-R., Pallesen, S. and Andresen, P., 1986. Mass estimates of cumulates and residues after anatexis in the Oslo Graben. *Journal of Geophysical Research*, 91 (11), pp. 11 629–11 640.

Neumann, E.-R., Tilton, G.R. and Tuen, E., 1988. Sr, Nd and Pb isotope geochemistry of the Oslo Rift Igneous Province, Southeast Norway. *Geochimica et Cosmochimica Acta*, 52, pp. 1997–2007.

Neumann, E.-R., Sundvoll, B. and Overli, P.E., 1990. A mildly depleted upper mantle beneath southeast Norway: evidence from basalts in the Permo-Carboniferous Oslo Rift. *Tectonophysics*, 178, pp. 98–107.

Oftedahl, C., 1978a. Main Geologic Features of the Oslo Graben. In Ramberg, I.B. and Neumann, E.-R. (editors) *Tectonics and Geophysics of Continental Rifts*, pp. 149–165, D. Reidel Publishing Company, Dordrecht, The Netherlands, 444 pp.

Oftedahl, C., 1978b. Cauldrons of the Permian Oslo Rift. *Journal of Volcanology and Geothermal Research*, 3, pp. 343–371.

Pearce, J.A. and Cann, J.R., 1973. Tectonic setting of basic volcanic rocks determined using trace element analyses. *Earth and Planetary Science Letters*, 19, pp. 290–300.

Ro, H.E., Larsen, F.R., Kinck, J.J. and Husebye, E.S., 1990a. The Oslo Rift – its evolution on the basis of geological and geophysical observations. *Tectonophysics*, 178, pp. 11–28.

Ro, H.E., Stuevold, L.M., Faleide, J.I. and Myhre, A.M., 1990b. Skagerrak Graben, the offshore continuation of the Oslo Graben. *Tectonophysics*, 178, pp. 1–10.

Schairer, J.F., 1950. The alkali feldspar join in the system $NaAlSiO_4$–$KAlSiO_4$–SiO_2. *Journal of Geology*, 58, pp. 512–517.

Sørensen, H.S. and Wilson, J.R., 1995. A strontium and neodymium isotopic investigation of the Fongen-Hyllingen Layered Intrusion, Norway. *Journal of Petrology*, 36 (1), pp. 161–187.

Sundvoll, B., Neumann, E.-R., Larsen, B.T. and Tuen, E., 1990. Age relations among Oslo Rift magmatic rocks: implications for tectonic and magmatic modelling. *Tectonophysics*, 178: 67–87.

Vogt, P.R. and Conolly, J.R., 1971. Tasmantid guyots, the age of the Tasman Basin, the motions between the Australian plate and the mantle. *Bulletin of the Geological Society of America*, 82, pp. 2577–2584.

Webb, P.K. and Weaver, S.D., 1975. Trachyte shield volcanoes; a new volcanic form from south Turkana, Kenya. *Bulletin Volcanologique*, 39 (2), pp. 294–312.

Weigand, P.W., 1975. Studies of the igneous rock complex of the Oslo region, 24, Geochemistry of the Oslo basaltic rocks. *Skrifter utgitt av det Norske Videnskaps-Akademi i Oslo*, 1, pp. 1–38.

Wright, J.B., 1969. Olivine nodules in trachyte from the Jos Plateau, Nigeria. *Nature*, 223. pp. 285–286.

CHAPTER 11

Alkaline magmatism

Perhaps the outstanding feature of the more notably alkaline types is that they contain one or more members of the group of minerals known as feldspathoids.

Bowen, 1928, p. 234.

11.1 Survey of the alkaline rocks

In contemporary particle physics more attention is given to exotic particles that are absent from ordinary matter than to the quarks and electrons that occur in ordinary matter. According to Weinberg (1993, p. 48) this happens because the fundamental character of matter will be better understood by studying exotic particles. Similar reasoning can be used to explain why so much research effort is currently concentrated on studying the rare alkaline rocks. Woolley (1987) in his introduction to *Alkaline Rocks and Carbonatites of the World* states that as the alkaline rocks have 'the most extreme compositions of all igneous rocks, so an understanding of their genesis is essential if we are to understand fully the workings of the solid Earth'.

The primary magmas responsible for the alkaline rocks normally equilibrate at greater depths than the magmas that yield the more common subalkaline rocks so the alkaline rocks and their xenoliths are more likely to provide novel data about mantle materials and mantle processes. In summing up a recent symposium on mantle metasomatism and alkaline magmatism it was asserted that: (1) alkaline magmatism is present in nearly every tectonic setting; (2) mantle metasomatism is important in the generation of most alkaline magmas; and (3) the crust is seldom involved in the genesis of alkaline rocks whatever their tectonic setting (Morris and Pasteris, 1987, p. 4).

The expression alkaline igneous rocks is ambiguous, because it is defined in several different ways (Sørensen, 1974b). It is used to describe the particular rocks in a group of igneous rocks that contain greater than normal amounts of alkali metals. In addition it is used to describe igneous rocks with high alkali metal abundances that have developed a special mineral assemblage. Rocks of the second group typically contain feldspathoids, melilites, sodium-bearing pyroxenes, such as aegirine and aegirine-augite, or sodium-bearing amphiboles, such as arfvedsonite, kaersutite and riebeckite. Consequently, there are at least eight different groups of alkaline rocks: (1) peralkaline silica-oversaturated rocks; (2) peralkaline silica-saturated rocks; (3) foid-bearing rocks in which feldspathoids make up less than 10 per cent of the felsic minerals; (4) feldspar-foid rocks in which between 10 per cent and 60 per cent of the felsic minerals are feldspathoids; (5) foidites and foidolites in which over 60 per cent of the felsic minerals are feldspathoids; (6) alkaline rocks that contain essential amounts of a primary carbonate mineral; (7) melilitic rocks; and (8) the alkaline members of the extended lamprophyre clan that includes the extraordinary lamproites and kimberlites.

The silica-oversaturated and the silica-saturated peralkaline rocks are usually the more differentiated members of the midalkaline suite (peralkaline trachytes and peralkaline trachydacites) that were discussed in Chapter 10. For practical purposes the rocks of the basanite → tephrite → phonotephrite → tephriphonolite → phonolite suite can be regarded as the normal suite of alkaline rocks (see Table 11.1). They are easily defined on the TAS classification diagram and they form an important, albeit minor, part of the main sequence of igneous rocks. The most challenging alkaline rocks are those rocks that evolved outside the main sequence of igneous rocks. These rare rocks are the ones likely to reveal the most about mantle materials and processes. They include the kimberlites, lamproites, ultrapotassic rocks and melilitites (Mitchell, 1986; Mitchell and Bergman, 1991; Rock, 1991).

Table 11.1 Major element and CIPW normative compositions of the normal alkaline suite.

	Basanite Le Maitre, 1984 p. 254	Tephrite Le Maitre, 1984 p. 254	Phonotephrite Le Maitre, 1984 p. 254	Tephriphonolite Le Maitre, 1984 p. 254	Phonolite Le Maitre, 1984 p. 255
SiO_2	44.06	44.74	49.03	53.64	57.34
TiO_2	2.86	3.4	1.89	1.37	0.62
Al_2O_3	13.69	15.07	17.29	18.83	19.87
Fe_2O_3	3.86	5.68	3.9	3.56	2.5
FeO	8.48	7.02	5.62	3.13	1.92
MnO	0.19	0.18	0.16	0.15	0.16
MgO	10.53	6.19	4.35	2.32	0.77
CaO	10.67	10.86	8.12	4.96	2.02
Na_2O	3.28	3.76	4.62	5.88	8.3
K_2O	1.67	2.31	4.21	5.69	6.35
P_2O_5	0.71	0.79	0.81	0.47	0.15
Total	100	100	100	100	100
CIPW normative minerals					
Orthoclase	9.87	13.64	24.87	33.64	37.54
Albite	8.91	12.41	15.05	23.25	26.82
Anorthite	17.7	17.41	14.01	8.17	0
Nepheline	10.2	10.51	13	14.35	21.68
Aegirine	0	0	0	0	3.02
Diopside	24.49	24.77	16.87	10.59	7.12
Wollastonite	0	0	0	0	0.16
Olivine	16.15	4.69	5.03	1.16	0
Magnetite	5.6	8.24	5.66	5.15	2.12
Ilmenite	5.42	6.46	3.59	2.59	1.18
Apatite	1.66	1.87	1.92	1.1	0.36

11.2   Tectonic settings

Alkaline rocks erupt in most tectonic settings. They are particularly abundant in: (1) continental rifts; (2) intraplate areas perturbed by mantle plumes; (3) outer volcanic arcs, or back-arc zones, in areas of active subduction; or (4) plate boundary zones that have experienced continent–continent collision. Bailey (1974, p. 150) made the sage observation that alkaline rocks are not exclusively restricted to zones of continental rifting, yet in all areas where such rifting occurs alkaline magmatism is strongly in evidence. A survey of Cenozoic rift zones reveals that they tend to form on the crests of crustal arches, domes or up-warps. According to Le Bas (1980, p. 33) 'crustal swelling and uplift has long been associated with alkaline igneous activity'. He has identified three different scales of uplift and doming that relate to the processes of alkaline magmatism. They are lithospheric doming, crustal doming and sub-volcanic doming.

A well preserved lithospheric dome extends over most of East Africa. Its long-axis is approximately 1500 km in length and extends from lakes Albert and Turkana in the north to Lake Malawi in the south. This huge dome is centred on Lake Victoria. As it evolved and sundered a variety of potassium-rich alkaline rocks were emplaced into the western rift system and sodium-rich alkaline rocks in the eastern rift system. The dome evolved over tens of millions of years but was most active towards the end of the Cretaceous Period (65–80 million years). Le Bas (1980, p. 34) has attempted to link the evolution of lithospheric domes with the expansion of the rocks at the base of thick sections of continental lithosphere. Lithospheric domes are not unique to Earth. They occur in the Tharsis and Elysium volcanic provinces of Mars and are broadly similar to the regional topographic rises on Venus (Cattermole, 1994).

Well preserved examples of crustal domes include the Kenya Dome centred on the Nairobi area, the Rhine Dome of western Europe and the Rungwe Dome of southern Tanzania. These domes are usually between 200 and 500 km in diameter, with a maximum uplift of approximately 1 km. Rifting generally develops along their crests. The alkaline rocks associated with rifts include foid-bearing basalts and picrobasalts, tephrites, phonolitic tephrites, tephritic phonolites, phonolites and olivine nephelinites, and a range of other rare and often bizarre rocks discussed in the next two sections. Large volumes of trachyte and rhyolite are usually extruded after the main graben-forming stage.

11.3 Sodic rocks

Most alkaline rocks are sodium-rich and belong to the basanite → tephrite → phonotephrite → tephriphonolite → phonolite suite, or if they are coarser grained the foid gabbro → foid monzodiorite → foid monzosyenite → foid syenite suite. Traditionally these rocks have received a plethora of different names. Larouziere (1989, p. 187) has for example prepared a list of 109 different terms that have been used to name the foid-bearing syenites. It is interesting that the first of these names is agpaite, a term that is still used to name a special type of peralkaline nepheline syenite that contains complex titanium and zirconium silicate minerals. A typical agpaitic nepheline syenite crops out at Lujavr-Urt in the Lovozero Alkaline Complex of the Kola Peninsula, northwestern Russia. These rocks are also called lujavrites. Besides alkali feldspar and nepheline they normally contain aegirine, arfvedsonite, eudialyte and accessory minerals enriched in incompatible elements. An unusual type of leucocratic nepheline syenite is called mariupolite. It was originally described from Mariupol in the Oktyabr'skii Massif that lies to the north of the Sea of Azov in the Ukraine (Kogarko et al., 1995, pp. 54–55). Mariupolite is essentially composed of nepheline, albite and aegirine. It is unique because it normally contains over 11 per cent Na_2O.

In the past there has also been a trend to regard all the rocks of the alkaline suite as particular types of either tephrites (foid gabbros) or phonolites (foid syenites) and to ignore or group together the other members of the suite. For example, Sørensen (1974c, p. 32) has called a rock that contains 47.0 per cent SiO_2 and 8.3 per cent $(Na_2O + K_2O)$ a mafic phonolite. It would be a basanite in the TAS classification. Nephelinite is another term that is often used in a broad sense to include foiditic rocks, tephrites and phonotephrites (Le Bas, 1989). It is recommended that the term nephelinite should be used as a name for feldspar-free, foiditic rocks that are essentially composed

of nepheline and clinopyroxene (Le Maitre et al., 1989, p. 96).

When perusing the literature on silica-undersaturated alkaline plutonic rocks one often discovers an apparent lack of foid monzodiorites and foid monzosyenites. Table 11.2 contains chemical data on a selection of alkaline plutonic rocks from the Gardar Igneous Province of South Greenland. It shows that the precise petrographic nature of the plutonic rocks that are interposed between the foid gabbros and foid syenites is camouflaged by the loosely defined term syenogabbro. In other areas, such as the volcanic islands off the coast of Brazil, the term gauteite has been misused to describe both foid monzodiorites and foid monzosyenites.

Tephrite, phonotephrite, tephriphonolite and phonolite

The tephrites are ultrabasic to basic volcanic rocks that are essentially composed of calcic plagioclase, clinopyroxene and one or more feldspathoids. As many tephrites have vitrophyric textures it is apposite to classify them on the TAS diagram, where they occupy a field where they are surrounded by the foidites, picrobasalts, basalts, trachybasalts and phonotephrites. Their plutonic equivalents are the foid gabbros or theralites. A typical foid gabbro contains titanaugite, plagioclase (labradorite) and nepheline, with minor amounts of olivine, iron-titanium oxides, apatite and possibly some phlogopite. The basanites are tephrites that contain more than 10 per cent modal, or normative, olivine (see Table 11.1). In the past many basanites have been called ankaramites or limburgites. Limburgite is essentially a glassy basanite. The plutonic equivalent of basanite is olivine-foid gabbro. Basanites and olivine-foid gabbros are petrologically important because many suites of comagmatic alkaline rocks that terminate in phonolite field have parental magmas of basanitic composition.

The phonotephrites are basic alkaline volcanic rocks that are essentially composed of plagioclase, feldspathoids, augite, alkali feldspar and minor amounts of olivine, iron-titanium oxides and apatite (Rittmann, 1973, p. 134). They can be precisely defined on the TAS diagram (Le Maitre et al., 1989, p. 28). The equivalent plutonic rock is the foid monzodiorite.

Tephriphonolites are basic to intermediate alkaline volcanic rocks that are essentially composed of alkali

Table 11.2 Major element and CIPW normative compositions of alkaline rocks originally called syenogabbro from the Gardar Igneous Province, South Greenland.

	Foid gabbro (tephrite) Watt, 1966, p. 54, No 86-61158	Foid monzodiorite (phonotephrite) Watt, 1966, p. 50, No 69-61162	Foid monzosyenite (tephriphonolite) Watt, 1966, p. 44, No 45-61133	Foid monzosyenite (tephriphonolite) Watt, 1966, p. 44, No 45-61267
SiO_2	48.2	53.69	55.15	57.06
TiO_2	3.21	1.91	1.58	0.99
Al_2O_3	14.49	15.56	15.47	15.43
Fe_2O_3	2.94	2.03	1.84	2.35
FeO	10.4	7.86	7.64	5.36
MnO	0.2	0.17	0.19	0.16
MgO	2.97	1.48	1.2	0.54
CaO	6.93	4.67	4.06	2.38
Na_2O	4.8	5.95	5.9	6
K_2O	2.9	4.2	4.65	5.8
P_2O_5	1.43	0.7	0.5	0.2
H_2O^+	0.9	1.04	1	0.54
H_2O^-	0	0	0	0
CO_2	0	0	0	2.5
Total	99.37	99.26	99.18	99.31
CIPW normative minerals				
Orthoclase	17.4	25.2	28	35.5
Albite	30.5	36.3	37.8	38.2
Anorthite	9.6	3.4	2	0
Nepheline	5.8	8.1	7.1	5.8
Aegirine	0	0	0	3.4
Diopside	13.3	13.3	13	9.5
Olivine	9.5	5.3	5.1	3.2
Magnetite	4.3	3	2.7	1.9
Ilmenite	6.2	3.7	3.1	2
Apatite	3.4	1.7	1.2	0.5

feldspars, sodic plagioclase, feldspathoids and augite, with minor amounts of olivine, iron-titanium oxides and apatite (Rittmann, 1973, p. 134). They are also readily defined on the TAS diagram (Le Maitre *et al.*, 1989, p. 28). The mean major element compositions and CIPW norms of both the phonotephrites and tephriphonolites are given in Table 11.1. The plutonic rock equivalent to the tephriphonolite is the foid monzosyenite.

Most phonolites are alkaline intermediate volcanic rocks that contain unusually high concentrations of Al_2O_3 and alkali metals. A typical nepheline phonolite contains over 60 per cent normative alkali feldspar and about 20 per cent normative nepheline (see Tables 11.1–11.3). The chemical and normative compositions of the singular flood phonolites of the Eastern Rift of Kenya are presented in Tables 11.4 and 11.5. Many phonolites are peralkaline and contain normative, and often modal, aegirine.

A most interesting active phonolitic volcano is Mount Erebus on Ross Island, Antarctica (Kyle, 1995). Its summit crater contains an actively convecting and continuously degassing lava lake (Kyle, 1994). Each day the volcano produces several small Strombolian eruptions that emanate from either the lava lake or an adjacent vent. The highly vesicular volcanic bombs ejected by these explosive eruptions contain abundant large anorthoclase feldspar megacrysts that may be up to 9 cm in length. Similar phonolites have erupted from Mount Kenya and Mount Kilimanjaro in East Africa and these African anorthoclase phonolites have been called kenytes (Gregory, 1900, p. 211). The rocks of Mount Erebus belong to a volcanic province that ranges in composition from basanite to phonolite. According to Kyle (1981, p. 495) the more differentiated rocks of this sodic alkaline suite all evolved by fractional crystallisation from a basanitic parental magma.

Table 11.3 Mean major element compositions and CIPW norms of melteigites, ijolites, urtite and nephelinite.

	Melteigite Le Bas, 1977, p. 301	Melteigite Nockolds, 1954, p. 1028	Ijolite Le Bas, 1977, p. 307	Ijolite Nockolds, 1954, p. 1028	Urtite Nockolds, 1954, p. 1028	Nephelinite Le Maitre, 1976, p. 629
SiO_2	36.73	41.9	40	42.58	42.59	40.6
TiO_2	3.51	2.21	2.38	1.41	0.35	2.66
Al_2O_3	7.8	12.2	15.25	18.46	27.42	14.33
Fe_2O_3	6.51	6.41	6.61	4.01	2.49	5.48
FeO	6.69	4.32	3.39	4.19	1.89	6.17
MnO	0.22	0.22	0.16	0.2	0.09	0.26
MgO	9.28	5.45	3.5	3.22	0.69	6.39
CaO	18.19	16.6	16.36	11.38	4.38	11.89
Na_2O	2.29	5.1	7.1	9.55	14.12	4.79
K_2O	1.79	2.66	3.1	2.55	3.82	3.46
P_2O_5	1.13	1.24	0.88	1.52	0.44	1.07
H_2O^+	3.05	0.87	0.64	0.55	0.42	1.65
H_2O^-	1.15	0	0.07	0	0	0.54
Co_2	0	0.82	0.56	0.38	1.3	0.6
Total	98.34	100	100	100	100	99.89
CIPW norms						
Orthoclase		6.8		10.05	8.34	3.24
Anorthite	6.07	2.54	0.59		0.6	7.57
Leucite	8.81	7.51	14.35	3.92	10.96	13.9
Nepheline	11.15	23.66	35.52	43.92	64.82	22.48
Diopside	37.18	42.13	18.77	29.05	6.94	33.15
Wollastonite	0	0	8.41	0	0	0
Larnite (Cs)	10.65	0	9.44	0	0	0
Olivine	6.19	0	0	0	0	2.38
Magnetite	10.05	8.22	4.56	5.83	3.72	8.14
Hematite	0	0	3.48	0	0	0
Ilmenite	7.1	4.37	4.53	2.71	0.6	5.17
Apatite	2.8	2.94	2.05	3.62	1	2.57
Calcite		1.83	1.3	0.9	3.02	1.4

Foidites and foidolites

After the modal compositions of the foidites and foidolites have been adjusted to remove the M, or mafic and related mineral component, they are formally defined as rocks that contain more than 60 per cent feldspathoids. These rocks fall outside the main sequence of igneous rocks, and include many feldspathoid cumulate rocks that usually seem to have evolved as the result of the flotation of feldspathoids in a tephritic or melilititic parental magma (see Section 11.5). Nephelinites and sodalitites are typical of the sodic foidites and nephelinolites and sodalite foidolite are their plutonic equivalents.

According to Le Bas (1989, p. 1310) the best way to separate the nephelinites from the melanephelinites and basanites is to use their normative nepheline content. In his classification the nephelinites contain more than 20 per cent normative nepheline,

Table 11.4 Chemical and CIPW normative composition of a typical meimechite. After Arndt *et al.*, 1995, p. 48, No. 24–4.

SiO_2	40.2	Anorthite	4.28
TiO_2	1.89	Leucite	4.04
Al_2O_3	2.54	Nepheline	0.09
FeO	12.87	Diopside	13.12
MnO	0.2	Larnite (Cs)	0.99
MgO	35.9	Olivine	70.69
CaO	5.18	Magnetite	2.53
Na_2O	0.02	Ilmenite	3.6
K_2O	0.87	Apatite	0.58
P_2O_5	0.25		99.92
Total	99.92		
	ppm		ppm
V	187	Sr	301
Cr	3052	Y	12
Co	130	Zr	152
Ni	1923	Nb	30
Cu	81	Ba	194
Zn	97	Pb	3
Ga	7	Th	2
Rb	47		

Table 11.5 Chemical and CIPW normative composition of a primitive sodic basanite from Saint Helena. After Weaver *et al.*, 1987, p. 256.

Major oxides	per cent	CIPW norms		Trace elements	ppm	Trace elements	ppm
SiO_2	42.94	Orthoclase	4.91	V	240	Ba	335
TiO_2	2.65	Albite	11.83	Cr	494	La	38.7
Al_2O_3	12.19	Anorthite	19.81	Co	64.2	Ce	82
Fe_2O_3*	13.37	Nepheline	6.1	Ni	241	Nd	38
MnO	0.19	Diopside	25.77	Zn	108	Eu	2.44
MgO	11.36	Olivine	21.68	Rb	19	Tb	0.83
CaO	10.79	Magnetite	3.33	Sr	753	Hf	5.2
Na_2O	2.63	Ilmenite	5.22	Y	23.5	Ta	3.81
K_2O	0.8	Apatite	1.35	Zr	233	Th	4.45
P_2O_5	0.56			Nb	55	U	1.16
L.O.I.	2.3			Cs	0.13		
Total	99.78						

while the basanites and melanephelinites contain less than 20 per cent of this normative mineral. The value of this classification is illustrated by Tables 11.3 and 11.5. Nephelinite is shown to contain 22.5 per cent normative nepheline, whereas sodic basanite only contains 6.1 per cent of this mineral. Le Bas (1989, p. 1310) also provides a way of separating the sodic basanites from the melanephelinites. The sodic basanites usually contain more than 5 per cent normative albite whereas the melanephelinites contain less albite.

The term melanephelinite was originally used to describe volcanic rocks that contained abundant pyroxene and some nepheline. More recently the term has been used to describe either olivine-rich or pyroxene-rich varieties of nephelinite that contain less than 20 per cent normative nepheline (Le Bas, 1989, p. 1304). Most olivine melanephelinites contain more than 10 per cent MgO, whereas the pyroxene melanephelinites contain less. Pyroxene melanephelinite lavas are the dominant type of rock that occurs interbedded with carbonatites in the Eastern Branch of the African Rift System (see Section 11.4 and Chapter 12). The problems associated with the classification of the leucite-bearing basanites and tephrites will be investigated in Section 11.6.

Modal mafic mineral content is usually used to subdivide the nephelinolites into urtites, ijolites and melteigites. The urtites are the leucocratic nephelinolites, the ijolites are the normal nephelinolites, whereas the melteigites are melanephelinolites. Table 11.3 displays the chemical and normative compositions of these rocks. The abundance of normative feldspathoids increases from about 25 per cent in the melteigites to 75 per cent in the urtites. Some melteigites contain melilite. The presence of this mineral in some melteigites and ijolites is revealed by the appearance of larnite (Cs) in their CIPW norms. It is likely that many melteigites are pyroxene cumulate rocks. They complement the urtites as these nepheline-rich rocks are likely to be flotation cumulates (see Section 6.13). As the pyroxenes found in the melteigites are likely to contain over 50 per cent silica, it is probable that a few melteigites have compositions that protrude into the foid gabbro field on the TAS diagram. Because the sodic foidolites are frequently comagmatic with carbonatites they will all be examined in more detail in Chapter 12.

Foiditic magmas are unusual because of the huge range in densities of the phases that crystallise in them. This frequently results in a separation of the contemporaneous feldspathoidal and ferromagnesian minerals. Most olivines and pyroxenes have densities that are greater than $3.2\,g/cm^3$. Melilite (2.9–3.1 g/cm^3) and phlogopite (2.75–2.95 g/cm^3) are less dense. Minerals in the plagioclase group have densities that range from 2.62 to $2.76\,g/cm^3$. Nepheline, kalsilite and the alkali feldspars have similar densities that vary between 2.55 and $2.67\,g/cm^3$. Haüyne, leucite, nosean and sodalite have densities that are less than $2.5\,g/cm^3$.

Somma-Vesuvio revisited

It is interesting to take a closer look at the alkaline lavas and tephra that erupted from Somma-Vesuvio in the period 1631 to 1944 (Villemant *et al.*, 1993).

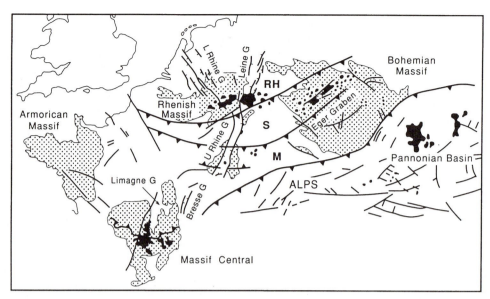

Fig. 11.1 Sketch map of the Tertiary-Quaternary Alkaline Province of central and western Europe showing that these rocks are associated with several grabens, such as the Limagne, Bresse, Rhine, Ruhr, Leine and Eger grabens and horsts. After Wilson and Downes, 1991.

Many of these rocks are highly porphyritic and it is often difficult to separate the effects of fractional crystallisation from crystal accumulation. The process of crystal accumulation was particularly important in the eruptions of 1906 and 1944. Geochemical modelling shows that these rocks were produced by the mechanical mixing of a slightly differentiated magma with varying proportions of large crystals available in the high-level magma chamber beneath the volcano. According to Villemant *et al.* (1993, p. 303) crystal accumulation was controlled by density differences between the magma and the resident minerals. Minerals that are denser than the magma, like clinopyroxene, tended to accumulate at the bottom of the magma chamber, whereas the minerals like leucite that were lighter than the magma float to the top. When at the top the buoyant crystals cluster together and form a layer that is usually also enriched in buoyant volatile components. If the build-up of volatiles results in an explosive eruption the buoyant feldspathoidal crystals are likely to become entrained in the extruding lava or tephra.

Central and western Europe

Wilson and Downes (1991) have reviewed the geochemistry and petrology of the more primitive vol-

canic rocks of the Tertiary-Quaternary Alkaline Province of central and western Europe. In this province extension and magmatism are closely related. This has resulted in the province being divided into several major structural units. Located in the south are the Limagne and Bresse grabens. They are associated with the Massif Central of France. The Rhine, Ruhr and Leine grabens are further north, whilst in the east the Eger graben cuts the Bohemian Massif (see Fig. 11.1). The main phase of volcanic activity in this petrographic province occurred in the Miocene and Pliocene. Later activity was restricted to a few scattered volcanic fields. Most of the volcanic activity was located on horst blocks, on the shoulders of rift structures, or on or near major structural dislocation and accommodation zones associated with the grabens (Fig. 11.2).

A notable example of a subprovince within this petrographic province is in the Massif Central of France (Downes, 1987; Maury and Varet, 1980). It is an area of regional uplift and rifting. The Limagne graben formed during the Oligocene. Prior to this event, between about 65 and 35 million years ago, the only visible magmatic activity consisted of the eruption of small volumes of silica undersaturated lava. Volcanic activity decreased during the main phase of graben formation, but resumed about 20 million years ago and continued to about 3450

Fig. 11.2 A quarry in phonolitic pyroclastic beds from the crater ring of the Laacher See Volcano in the East Eifel District of Germany.

years before the present. The main volcanic fields within the Massif Central are Cantal (11 to 3 million years), Velay (10 to 1 million years), Aubrac (8 to 6 million years), Deves (5.4 to 0.8 million years), Mount Dore (4.3 to 0.4 million years) and the Chaine des Puys (50 000 to about 3500 years). Two principal magma series are present. One has produced a mid-alkaline suite of rocks (see Chapter 10), whereas the other contains rocks of the alkaline basanite → tephrite → phonotephrite → tephriphonolite → phonolite suite. A few nephelinites with variable K_2O/Na_2O ratios are also found. Their compositions range from melilite nephelinite to leucite melanephelinite. Many lavas from this subprovince that have in the past been called basalts (sensu lato), are now known to be basanites or trachybasalts. For example, the rock that Maury and Varet (1980, p. 147) call their average basalt contains 43.92 per cent SiO_2, 4.91 per cent $(Na_2O + K_2O)$ and 10.7 per cent normative nepheline.

According to Wilson and Downes (1991, p. 842) the various primary magmas that generated the rocks of the Tertiary-Quaternary Alkaline Province of cen-

tral and western Europe contain a mixture of chemical components from the lithosphere and a mantle source, possibly a mantle plume. The lithospheric component included amphibole and phlogopite bearing metasomatised mantle materials. A broad understanding of the composition of the lithospheric component can be gleaned from the study of the interstitial glasses generated by partial melting (incongruent melting) of amphibole and phlogopite bearing xenoliths. Such xenoliths have been collected from the Massif Central and the northern Hessian Depression. The compositions of these glasses are extremely variable (Wilson and Downes, 1991, p. 838). Six are silica oversaturated, whereas the other five are silica undersaturated. The samples with the most extreme compositions are sample Bt39 that contains 13 per cent normative quartz and sample VP8379 that contains 19 per cent normative nepheline. Wilson and Downes (1991, p. 813) suggest that the geochemical differences between the various volcanic fields of western and central Europe are mainly caused by differences in the petrography of the particular lithospheric blocks beneath these volcanic fields. The lithosphere in central and western Europe is composed of a mosaic of abutting blocks.

Wilson *et al.* (1995, p. 182) have suggested that the main non-lithospheric component in the various alkaline primary magmas that evolved to produce this petrographic province was melilititic in composition. Furthermore they suggest that the melilititic magma was probably generated within the boundary layer at the base of the lithosphere beneath central and western Europe (see Section 11.5 on the melilitic rocks). The generation of an extensive body of melilititic magma is probably contingent on the docking of a large mantle plume beneath central and western Europe. Crustal magmatic activity and volcanism was then focused by structural features that constrained magmatic pathways. A less likely way of explaining the origin of these alkaline rocks would be to suggest that during the past 65 million years a cluster of small mantle plumes has impinged on the base of the lithosphere beneath central and western Europe.

Lovozero and other alkaline complexes

The Lovozero and Khibina alkaline complexes adjoin one another in the central part of the Kola Peninsula in northwestern Russia (Kogarko *et al.*, 1995; Kukharenko *et al.*, 1965). They are large (650 and 1327 km^2), layered, silica-undersaturated, peralkaline complexes. Their evolution has been related to the formation of the Central Kola Rift System. The Lovozero Alkaline Complex is especially well known as the type area for the rocks lujavrite and urtite, and because it contains a range of rare minerals including loparite [(Ce,Na,Ca)$_2$(Ti,Nb)$_2$O$_6$]. Loparite is mined at Lovozero for its niobium. The complex is also interesting because it contains both extrusive and intrusive silica-undersaturated sodic alkaline rocks. They all appear to have erupted at the same time, circa 395 million years ago. The compositions of the volcanic rocks range from basanite to phonolite.

Geophysical investigations have revealed that the exposed intrusive rocks are part of a laccolith that continues downwards for more than 7 km. The intrusion comprises two units: an upper layered intrusion that passes downwards into a stock-shaped body. Where exposed the contacts between the intrusive rocks and the older surrounding granitic gneiss are very steep. The intrusive sequence is readily divided

into four units. Unit one contains the oldest rocks. They are mainly nepheline syenites and nepheline-nosean syenites and they only crop out along the margins of the laccolith. The contact between this unit and unit two dips inwards towards the centre of the complex at about 25°. Layered rocks from unit two fill most of the outcrop area. They display rhythmic layering, with the individual layers being of lujavrite (melanocratic nepheline syenite), foyaite (nepheline syenite with a trachytoid texture), juvite (coarse-grained nepheline syenite) and urtite (leucocratic nephelinolite). These layers dip inwards towards the centre of the complex and individually their thickness ranges from a few centimetres to hundreds of metres. This layering is particularly distinct in the main body of the intrusion but it seems to disappear towards the margins where the rock becomes coarser-grained, even pegmatitic.

Unit three overlies and appears to intrude unit two. Dips along the contact between these units increase progressively from the margins to the centre giving unit three a funnel-shape. The rocks in this upper unit form the summits of the mountains in the Lovozero Massif. They are mainly coarse-grained eudialyte [Na$_4$(Ca,Fe^{2+})$_2$ZrSi$_6$O$_{17}$(OH,Cl)$_2$] lujavrites, but the unit also contains layers of foyaite and urtite. The final episode in the evolution of the complex was the emplacement of a series of lamprophyric dykes belonging to the camptonite-monchiquite-sannaite family (see Section 11.8).

Kogarko *et al.* (1995) have recently reviewed the petrology of the many alkaline complexes that crop out in Russia and the Commonwealth of Independent States. The largest of these complexes is the 240-million-year-old Guli, or Gulinskii, complex. It crops out between the Maimecha (or Meimecha) and Kotui rivers in north-central Siberia. Included among its country rocks are large volumes of lava of meimechitic composition (SiO$_2$ < 53 per cent, Na$_2$O + K$_2$O < 1 per cent, MgO > 18 per cent, and TiO$_2$ > 1 per cent: see Table 11.4). The Guli Alkaline Complex occupies an area of about 1550 km^2. Its evolutionary history is readily divided into six intrusive episodes. Rocks intruded during the first episode cover about 60 per cent of the total area of the complex. They are dunites, peridotites and pyroxenites and they contain a variety of minor and accessory minerals including nepheline, phlogopite, apatite, perovskite, calcite, iron-titanium oxides and chromite.

The characteristic rock emplaced during the second intrusive episode is an olivine melilitolite, or kugdite. These rocks are cut by the alkali peridotites, alkali pyroxenites (jacupirangites), melteigites, shonkinites and malignites of the third intrusive episode. During the fourth intrusive episode the complex was intruded by a cluster of ijolitic rocks. After their emplacement the site of intrusive activity migrated to the centre of the complex where a cluster of peralkaline, nepheline-bearing syenites was intruded. The final intrusive episode attests to a remarkable change in magma composition, with the injection of phoscorites, calcite-carbonatites and dolomite-carbonatites (see Section 12.1). Phoscorite is a rare and unusual rock. Its name is a mnemonic that describes a phosphate-rich rock that evolves in the core of a carbonatite complex (Russell *et al.*, 1955, p. 199).

The meimechites of the Maimecha-Kotui region of northern Siberia are a most interesting group of magnesian alkaline lavas. Their high magnesian and nickel contents and their forsteritic olivines meet the criteria usually employed to identify primary mantle magmas, yet they crop out in close association with potassic picrobasalts, basanites and foidites (Arndt *et al.*, 1995, pp. 48–49). Table 11.4 presents the chemical composition of a typical meimechite, with low silica (40.2 per cent) and high MgO (35.9 per cent), TiO_2 (1.89 per cent) and K_2O (0.87 per cent) values. This rock also contains just over 70 per cent normative olivine and 4 per cent normative leucite. The meimechitic lavas are approximately the same age as the Siberian continental flood lavas that crop out to their immediate southwest. It is probable that the two magmatic events are linked. The initial meimechitic magma was probably produced by the Siberian mantle plume partially melting the lowermost continental lithosphere. On its passage through the subcontinental lithosphere the magma collected the many olivine xenocrysts now present in the meimechitic lavas.

tephrites, tephriphonolites and phonolites (Borley, 1974, p. 320; Fuster *et al.*, 1968). These particular rocks usually contain the feldspathoid haüyne, with lesser amounts of the other feldspathoids such as nepheline and sodalite (see Fig. 11.5).

Table 11.5 presents the major and trace element compositions of a primitive basanite from Saint Helena Island in the South Atlantic Ocean (Weaver *et al.*, 1987, p. 256). This sample is regarded as typical of the primitive sodium-rich alkaline rocks that erupt on oceanic islands and seamounts. In the past such rocks have often been called primitive oceanic island basalts or OIBs. The volcanic rocks of Tahiti in the Society Islands of the south-central Pacific are particularly interesting as they belong to both the mid-alkaline and alkaline suites. Rocks of the latter suite range from foid picrobasalts, via basanites, tephrites, phonotephrites, tephriphonolites to phonolites. The phonolites contain about 30 per cent normative nepheline (McBirney and Aoki, 1968, p. 534). Tahiti is renowned for its haüyne phonotephrites or tahitites. They were first described by Lacroix (1917, p. 583).

Nephelinites and melanephelinites are also found on oceanic volcanic islands. The melilitic rocks from oceanic islands, such as those from Hawaii, are discussed in Section 11.5. It is rare to find primitive alkaline rocks from oceanic areas that contain more potassium and incompatible elements than the basanites of Saint Helena Island. Examples of these extraordinary potassic rocks occur on the South Atlantic islands of Tristan da Cunha (see Table 11.6) and Gough (Baker *et al.*, 1964; Le Maitre, 1962; Weaver *et al.*, 1987). These rocks are discussed in Section 11.6.

Volcanic islands

Assorted midalkaline and alkaline magmatic rocks crop out on volcanic islands and seamounts in all the oceans (Borley, 1974; Clague, 1987; Mitchell-Tome, 1970). For example, a wide range of these rocks, including carbonatites (Woolley, 1989: see Chapter 12), have erupted on the volcanic islands of the Atlantic Ocean. On the western islands of the Canary Archipelago (Tenerife, Gomera, Heirro and La Palma; see Plate 11.1; Figs 11.3, and 11.4) one finds sodium-rich basanites, tephrites, phono-

11.4 East African Rift System and flood phonolites

The association between continental rifting and alkaline volcanism is nowhere more clearly developed than in the African Rift System. Compared with other Cenozoic rift systems the Kenyan and Ethiopian sections of this system have extruded much larger volumes of alkaline volcanic rocks. For example, the Kenya Rift has produced a volume of volcanic rocks that is 50 times greater than the Rio Grande Rift of New Mexico (Keller *et al.*, 1991). In 1990 the Kenya Rift International Seismic Project (KRISP)

Fig. 11.3 The view to the east from the summit of Teide Volcano on Tenerife in the Canary Islands, showing Montana Blanca and the older bedded alkaline pyroclastic rocks and lavas of the Portillo-Tauce escarpment. This escarpment marks the edge of a large collapse caldera that developed prior to the construction of the Teide volcanic edifice. The dark lavas in the foreground are phonolites.

set out to determine the structure of the uppermost mantle beneath the Kenya Rift (Keller *et al.*, 1994). Refraction and wide-angle reflection seismic data obtained by the project revealed that the lithosphere was thinnest under the Kenya Dome in the Lake Naivasha area. It is likely that beneath this area there is a hot, active mantle plume that has thinned the mantle lithosphere to a greater extent than the crust.

Beneath Lake Turkana in the north, the relative thickness of the crust and lithospheric mantle is consistent with stretching. It is proposed that a pre-rift lithosphere of 100 km has been thinned to 50 km. The Kenya Crustal Dome is about 700 km from north to south, and 500 km from east to west. Beneath this area, but more particularly beneath the rift valley itself, very low upper mantle P-wave velocities extend down to depths of about 165 km. These low seismic velocities probably mean that there is a zone of 3–6 per cent partial melting extending down beneath the present Kenya Rift (Mechie *et al.*, 1994).

The extraordinary feature of East Africa volcanism is the enormous volumes of flood trachyte (see Chapter 10), flood phonolite and nephelinite that have erupted. Gregory (1921) was the first to recognise the vast volumes of phonolitic lava within and surrounding the Kenya Rift Valley. These phonolites have been divided into three main types. The Gwasi-type phonolites crop out as minor intrusions and local flows within, or marginal to, the central nephelinitic volcanoes. They range in age from 22 million years ago to the present. The plateau-type flood phonolites are by far the most abundant. They were mainly erupted between 14 and 11 million years ago (see Table 11.7). The Kenya-type phonolites crop out throughout the intrarift sequence.

According to Lippard (1973, p. 217) the volume of the plateau-type flood phonolites is about $25\,000\,km^3$. This exceeds the total volume of phonolite found elsewhere on Earth by several orders of magnitude. The Miocene phonolites of the Uasin Gishu Plateau on the western shoulder of the

Fig. 11.4 A haüyne phonolite plug at Jedey, San Miguel de la Palma in the Canary Islands.

Kenya Rift erupted as seven separate flows with a combined volume of approximately $600\,km^3$. These lavas were extruded before the main phase of graben development in the area that is now the rift valley. They then flowed westwards for distances of up to 60 km. The voluminous fourth flow covered an area of over $2400\,km^2$ to a thickness of some 100 m.

These porphyritic phonolites usually have a uniform chemical composition (Tables 11.7 and 11.8). They frequently contain phenocrysts of nepheline and alkali feldspar (sanidine and anorthoclase) in roughly equal proportions, with lesser amounts of titanomagnetite, clinopyroxene (augite to aegirine), andesine, amphibole (magnesio-katophorite), biotite-phlogopite and apatite. The groundmass normally contains alkali feldspars, alkali pyroxenes, alkali amphiboles and aenigmatite ($Na_2Fe_5TiSi_6O_{20}$). Goles (1976, p. 50) has estimated that in the Miocene there was probably between 50 000 and 100 000 km^3 of phonolitic magma beneath the area of the present Kenya Rift. In addition this magma was probably derived from a parental magma that was at least

Fig. 11.5 Caldera Taburiente on San Miguel de la Palma in the Canary Islands showing a section through the alkaline lavas and pyroclastic rocks that have erupted to form this island.

Table 11.6 Chemical and CIPW normative composition of a primitive basanite from Tristan da Cunha. After Weaver *et al.*, 1987, p. 256.

Major oxides	per cent	CIPW norms		Trace elements	ppm	Trace elements	ppm
SiO_2	45.18	Orthoclase	18.23	V	235	Ba	979
TiO_2	3.27	Albite	12.98	Cr	2	La	73.4
Al_2O_3	17.67	Anorthite	23.06	Co	27	Ce	156
Fe_2O_3*	11.61	Nepheline	10.07	Ni	6	Nd	72
MnO	0.19	Diopside	13.6	Zn	103	Eu	4.2
MgO	4.44	Olivine	10.31	Rb	82	Tb	1.13
CaO	9.34	Magnetite	2.82	Sr	1461	Hf	7.5
Na_2O	3.68	Ilmenite	6.3	Y	31.4	Ta	6.16
K_2O	3.04	Apatite	2.63	Zr	324	Th	8.9
P_2O_5	1.12			Nb	82	U	2
L.O.I.	0.27			Cs	0.77		
Total	99.81						

five times as voluminous as the differentiated phonolitic magma. Once formed the phonolitic magma acted as a density filter and locally prevented the eruption of more primitive and denser magmas. According to Hay *et al.* (1995, p. 415) the plateau-type flood phonolites are likely to have been generated by the partial melting (14 to 15 weight per cent) of an alkali basalt (7 per cent normative nepheline) that had underplated the crust of central Kenya in the Early to Mid Miocene (23 to 14 million years).

Table 11.7 Major element composition of plateau-type flood phonolites from central Kenya.

	Flood phonolite Hay et al., 1995, p. 412, No 25	Flood phonolite Hay et al., 1995, p. 412, No 136	Flood phonolite Hay et al., 1995, p. 412, No 138	Flood phonolite Hay et al., 1995, p. 412, No 157	Flood phonolite Hay et al., 1995, No 528	Mean flood phonolite	Recalculated mean flood phonolite	Mean flood phonolite CIPW norm	
SiO_2	54.1	54.1	55.3	54.9	55.5	54.78	55.56	Orthoclase	35.44
TiO_2	0.79	0.42	0.34	0.72	0.45	0.54	0.55	Albite	32.06
Al_2O_3	19	18.6	19.6	19.2	20.1	19.3	19.57	Anorthite	1.51
Fe_2O_3*	5.26	6.93	6.25	5.58	5.02	5.81	5.89	Nepheline	18.53
MnO	0.23	0.39	0.23	0.32	0.19	0.27	0.28	Diopside	4.6
MgO	0.93	1.01	0.46	0.9	0.5	0.76	0.77	Olivine	2.15
CaO	1.97	1.93	1.14	1.21	1.34	1.52	1.54	Magnetite	4.36
Na_2O	7.12	6.62	8.35	8.01	7.82	7.58	7.69	Ilmenite	1.06
K_2O	5.6	5.88	5.72	5.7	6.14	5.81	5.89	Apatite	0.29
P_2O_5	0.27	0.09	0.06	0.11	0.08	0.12	0.12		
H_2O^+	3	2.43	1.32	2.12	1.66	2.11	2.14		
Total	98.27	98.4	98.77	98.77	98.8	98.6	100		

Table 11.8 Trace element composition of plateau-type flood phonolites from central Kenya.

Trace elements	Flood phonolite Hay *et al.*, 1995, p. 412, No 25 ppm	Flood phonolite Hay *et al.*, 1995, p. 412, No 136 ppm	Flood phonolite Hay *et al.*, 1995, p. 412, No 138 ppm	Flood phonolite Hay *et al.*, 1995, p. 412, No 157 ppm	Flood phonolite Hay *et al.*, 1995, p. 412, No 528 ppm	Mean flood phonolite ppm
Cr	6	5	5	9	7	6.4
Ni	9	13	21	13	10	13.2
Cu	6	8	7	7	6	6.8
Zn	137	234	170	204	138	177
Ga	24	28	29	24	22	25.4
Rb	123	195	195	194	172	176
Sr	94	249	89	177	47	131
Y	48	62	107	64	33	62.8
Zr	579	865	846	1436	723	890
Nb	188	348	303	461	274	315
Cs	1.43	2.17	2.45	2.89	1.57	2.1
Ba	543	163	46	151	3535	888
La	134	230	255	247	152	204
Ce	234	301	299	348	210	278
Nd	74.04	92.44	104.28	104.53	55.48	86.15
Sm	11.28	13.33	15.43	14.66	7.53	12.45
Eu	2.46	2.84	1.97	3.28	1.81	2.47
Gd	–	11.4	15.4	11.1	6.5	11.1
Tb	1.59	1.89	2.57	1.89	1	1.79
Ho	2.26	2.74	3.95	–	–	2.98
Tm	0.8	1.1	1.67	1.16	0.66	1.08
Yb	4.74	6.67	10.8	7.49	4.62	6.86
Lu	0.74	0.84	1.4	0.97	0.59	0.91
Hf	11.8	16.2	18.9	27.1	13.9	17.6
Ta	9.1	17.3	18	21.8	15.9	16.4
Pb	12.1	26.6	20.7	10	17.1	17.3
Th	18.4	35.1	33.8	45.1	28.7	32.2
U	4.4	11.9	4.74	9.54	2.52	6.62

When considering the petrology of the Kenya Rift the foremost problem is why does this small area of our planet possess such huge volumes of flood phonolite, flood trachyte and nephelinite. It appears likely that the lithosphere beneath the area had been previously enriched in alkalis and incompatible elements by mantle metasomatism. Perhaps the first magma released from the mantle plume that created the Kenya Crustal Dome was a melanephelinite. It differentiated to form the nephelinites and closely related rocks. Further differentiation may have resulted in the generation of phonolitic magma. The removal of huge quantities of magma from the diapir and its concomitant cooling is likely to have triggered rifting in the overlying rocks of the crustal dome. The magma that remained in the apex of the diapir gradually changed in composition and moved into the midalkaline field as it assimilated silica-rich crustal materials, cooled and differentiated. After the extrusion of the flood trachytes the more primitive mela-nephelinite magma could again break through the various layers in the crustal dome.

11.5 Melilitic rocks

Besides the melilite-bearing lamprophyric rocks, the melilitic rocks are defined as including all the other igneous rocks that contain more than 10 per cent modal melilite (Le Maitre *et al.*, 1989, p. 12; Streckeisen, 1978). According to the IUGS Subcommission on the Systematics of Igneous Rocks the minerals of the melilitic group should be regarded as mafic minerals and a component of their M group of mafic and related minerals (Le Maitre *et al.*, 1989, p. 4). In the past some petrologists have considered melilite to be a type of silica undersaturated pyroxene. Others have deemed them to be high-calcium, low-silica analogues of diopside and plagioclase, just as the normal

Table 11.9 Major element composition of selected melilitic rocks.

	Mean olivine melilitite, Sutherland, South Africa, McIver and Ferguson, 1979, p. 119	Mean uncompahgrite, Kisingiri, Western Kenya Le Bas, 1977, p. 304	Olivine melilitite, Homa Mountain, Western Kenya Le Bas, 1977, p. 301, No 116A	Mean melilite nephelinite, Hawaii Macdonald, 1968, p. 502	Olivine melilitite, Canary Islands, Wilson *et al.*, 1995, p. 187, No. LL1415	Olivine melilitite, Balcones Province, Texas, Spencer 1969, p. 281, No 25	Olivine melilitite, Upper Rhine Graben, Wilson *et al.*, 1995, p. 186, No H3
SiO_2	33.52	33.95	36.03	36.6	37.4	38.01	39.05
TiO_2	3.03	4.04	4.01	2.8	3.78	3.1	2.14
Al_2O_3	7.21	6.78	8.03	10.8	9.35	11.53	9.14
Fe_2O_3	1.74	8.13	7.45	5.7	12.4	3.57	11.3
FeO	9.56	6.6	6.32	8.9	0	8.25	0
MnO	0.23	0.14	0.26	0.1	0.2	0.19	0.19
MgO	19.55	8.44	10.94	12.6	14.02	12.4	18.19
CaO	16.76	26.18	19.34	13.6	14.27	15.49	13.74
Na_2O	1.86	2.46	3.1	4.1	2.97	3.11	2.31
K_2O	1.62	1.13	2.5	1	1.41	1.24	1.21
P_2O_5	1.29	0.51	0.76	1.1	1.38	1.16	0.7
H_2O^+	0	0.71	0.58	0	0	0.21	0
H_2O^-	0	0.09	0.95	0	0	1.08	0
CO_2	0	0.39	0	0	0	0.12	0
Total	96.37	99.55	100.27	97.3	97.18	99.46	98.24

feldspathoids are effectively low-silica analogues of the alkali feldspars. In the IUGS Subcommission's classification of the melilite-bearing lamprophyre, alnöite (Le Maitre *et al.*, 1989, p. 11), it is implied that melilite is a type of feldspathoid. Nowadays most petrologists regard melilite as a high-calcium, magnesium-bearing, feldspathoid (Bates and Jackson, 1987, p. 238). Natural igneous melilites usually contain the following end-member components akermanite ($Ca_2MgSi_2O_7$), soda-melilite ($CaNaAlSi_2O_7$), and iron-akermanite ($Ca_2Fe^{2+}Si_2O_7$). A typical igneous melilite is likely to contain 60 per cent akermanite, 30 per cent soda-melilite and 10 per cent iron-akermanite (Le Bas, 1977, p. 56).

Melilitic rocks never contain primary quartz, seldom contain feldspars, but often contain feldspathoids, like nepheline and leucite. Anorthite is found in the olivine melilitites from Saltpetre Kop, Sutherland, South Africa (McIver and Ferguson, 1979, p. 115). A typical olivine melilitite might contain 30.8 per cent olivine, 28.0 per cent augite, 14.3 per cent melilite, 13.0 per cent iron-titanium oxides, 3.4 per cent perovskite and 10.5 per cent 'irresolvable phases' (Shand, 1950, p. 443). In the CIPW norms of these silica-undersaturated melilitic rocks the presence of larnite (Ca_2SiO_4: originally called calcium orthosilicate by Cross *et al.*, 1902) can be taken to signify the presence, or latent existence, of melilite. The olivine melilitite from Homa Mountain in Western Kenya contains over 20 per cent normative larnite, whereas the melilite nephelinite from Hawaii contains only 10 per cent of this normative phase (see Table 11.9).

Le Bas (1973) developed a special alkaline norm for the feldspathoidal and melilitic igneous rocks. It is an interesting innovation because the nepheline composition used in this calculation is $Na_3KA_4Si_4O_{16}$ and not $NaAlSiO_4$. The latter is used in the CIPW norm. Calculations that use this more realistic nepheline composition produce a substantial reduction in the amount of potassium available to form leucite. This yields normative leucite abundances that correspond to modal abundances. The alkaline norm also favours the formation of minerals such as akermanite, andradite, gehlenite, kalsilite, iron-akermanite, perovskite and sphene. Melilite is represented by the end-members akermanite, iron-akermanite and gehlenite. While this calculation does not consider the compositions of real igneous melilites that often contain sodium, it is much better than the CIPW norm for evaluating the composition of igneous rocks that contain less silica than the picrobasalts. Table 11.10 compares a CIPW norm with a Le Bas alkaline norm.

Melilite is normally only found in volcanic and subvolcanic rocks. At greater depths magmas of similar composition crystallise as pyroxenites, or hornblende- or phlogopite-bearing pyroxenites (Yoder, 1979,

Table 11.10 Comparing the CIPW and alkaline norms of an olivine melilitite. After Le Bas, 1973, p. 91.

	Olivine melilitite Homa Mountain, Western Kenya	CIPW normative minerals	CIPW norm weight per cent	Alkaline normative minerals	Alkaline norm weight per cent
SiO_2	36.03	Anorthite	0.61	Kalsilite	3.12
TiO_2	4.01	Leucite	11.58	Nepheline	19.48
Al_2O_3	8.03	Nepheline	14.21	Gehlenite	0.6
Fe_2O_3	7.45	Diopside (wo)	10.56	Akermanite	19.82
FeO	6.32	Diopside (en)	9.13	Iron–akermanite	7.47
MnO	0.26	Diopside (fs)	0	Diopside (wo)	9.05
MgO	10.94	Larnite or cs	20.14	Diopside (en)	5.85
CaO	19.34	Olivine (fo)	12.69	Diopside (fs)	2.6
Na_2O	3.1	Olivine (fa)	0	Olivine (fo)	9.88
K_2O	2.5	Magnetite	9.6	Olivine (fa)	4.83
P_2O_5	0.76	Hematite	0.83	Hematite	7.45
H_2O^+	0.58	Ilmenite	7.62	Perovskite	6.82
H_2O^-	0.95	Apatite	1.76	Apatite	1.76
CO_2	0	Water	1.53	Water	1.53
Total	100.27	Total	100.26	Total	100.26

p. 407). Volcanic melilitic rocks are called melilitites, while the coarser grained subvolcanic rocks are designated melilitolites (Le Maitre *et al.*, 1989, p. 12). A few special varietal names are also used. They include katungite (potassic melanocratic olivine melilitite), kugdite (olivine melilitolite), okaite (haüyne melilitolite), turjaite (nepheline melilitolite) and uncompahgrite (pyroxene melilitolite). The type uncompahgrite from Mount Uncompahgre, Gunnison county, Colorado, contains between 63 and 67 per cent melilite with augite, iron-titanium oxides, perovskite, cancrinite, apatite, calcite, melanite and phlogopite (Johannsen, 1938, pp. 321–322).

Melilite occurs in some foidites such as the melilite nephelinites, melilite melanephelinites and melilite leucitites. It is also found in some chondritic meteorites. The most important melilite-bearing lamprophyric rocks belong to the alnöite-polzenite family (see Section 11.8). Melilite-bearing rocks are frequently found in melanephelinite-carbonatite complexes (Le Bas, 1977). Olivine melilitites have erupted from the active natrocarbonatite volcano Oldoinyo Lengai in northern Tanzania (Peterson and Kjarsgaard, 1995). The natrocarbonatite magmas are considered to have evolved as immiscible liquids from a primary carbonated olivine melilite magma of low to moderate peralkalinity (Kjarsgaard *et al.*, 1995, p. 187).

Some petrologists (Wilson *et al.*, 1995, p. 182) regard the melilitites and melilitolites as the most primitive members of the alkaline suite. Others (Le Maitre *et al.*, 1989, p. 6) consider them to form a special group of rocks that lie outside the main

Table 11.11 Trace element composition of selected olivine melilitites.

	Olivine melilitite Canary Islands Wilson *et al.*, 1995 No. LL1415	Olivine melilitite Upper Rhine Graben Wilson *et al.*, 1995 No. H3
Sc	23.7	27
V	296	264
Cr	373	928
Ni	337	545
Rb	32.7	45
Sr	1335	1039
Y	37	21
Zr	317	229
Nb	90	229
Ba	759	1011
La	101	75.2
Ce	216	136.8
Nd	100	54.5
Sm	17.5	9
Eu	5.26	2.61
Gd	n.d.	7.02
Er	n.d.	1.7
Yb	1.88	1.45
Lu	0.25	n.d.
Pb	4.58	4.13
Th	10.13	14
U	2.52	2.15

sequence of igneous rocks. Perhaps they should be categorised as a distinctive group of melafoidites, or melafoidolites, that usually have remarkably low silica abundances and high contents of magnesium plus calcium (see Tables 11.9 and 11.11). Melilitites normally plot in the foidite field of the TAS diagram because they contain less than 41 per cent silica. They usually also contain more than 10 per cent normative larnite (Le Bas 1989, p. 1300). The genesis of the olivine melilitites is often linked with that of the kimberlites, but according to Mitchell (1986, p. 354) fresh melilite has never been found in an authentic kimberlite.

Although melilitites are rare they are petrologically important. This is because: (1) they crop out in both oceanic and continental settings; (2) they have high chromium, nickel and Mg-values (greater than 70) and thus appear to be primary magmas; and (3) they often contain high-pressure mantle-derived xenoliths and xenocrysts (Brey, 1978; Jackson and Wright, 1970; Wilson *et al.*, 1995). Their low-silica, high-(MgO + CaO) and high-titanium values are reflected in their mineralogy as their modes are usually dominated by melilite, olivine, clinopyroxene and perovskite ($CaTiO_3$). Most melilitites are extremely enriched in incompatible elements, particularly niobium and the light lanthanoids (Wilson *et al.*, 1995, p. 186).

Melilitites on oceanic islands

In the Hawaiian Archipelago olivine melilitites erupt as part of the post-erosional alkaline stage (Clague, 1987, p. 237–238). These post-erosional alkaline lavas contain many spinel lherzolite and dunite xenoliths. In addition, several vents, notably those in Salt Lake Crater to the immediate north of Honolulu International Airport on Oahu and on the small island of Ka'ula in Kauai County, contain garnet-bearing peridotites and pyroxenites, including garnet lherzolites (Jackson and Wright, 1970; Maaloe *et al.*, 1992). Clague (1987, p. 239) has proposed that the post-erosional alkaline magmas probably equilibrated at depths of between 75 to 110 km. These depth estimates are approximately at the base of the oceanic lithosphere that is estimated to be about 90 km thick for the 90 to 100-million-year-old crust beneath the Hawaiian Archipelago. In 1982 Clague and Frey presented a detailed trace element investigation of the post-erosional Honolulu Volcanics on Oahu. They suggested that these lavas, including the melilitites, were generated by the partial melting of 2–11 per cent of a recently enriched garnet lherzolite source. During the partial melting process phlogopite,

amphibole and a titanium-rich phase remained in the residuum, but apatite was completely melted.

Melilitites are also found among the younger rocks of Gran Canaria in the Canary Archipelago (Hoernle and Schmincke, 1993; and Table 11.9). These rocks also contain mantle-derived xenoliths. Further south in the Atlantic Ocean larnite (or akermanite and gehlenite) normative igneous rocks have been described from the islands of Cape Verde, Fernando de Noronha and Trindade (Mitchell-Thome, 1970). Malaita in the Solomon Island Arc in the Southwestern Pacific contains melilite melanephelinites (Allen and Deans, 1965).

Continental melilitites

Le Bas (1977) has described olivine melilitites, melilite nephelinites, melilite melanephelinites, turjaites and uncompahgrites from the Homa Bay carbonatite area of western Kenya. The melilite melanephelinites typically contain about 20 per cent melilite, while the olivine melilitites contain about 45 per cent melilitite and the uncompahgrites contain between 60 and 80 per cent melilite (see Table 11.9). Olivine melilitites that contain between 16 and 21 per cent normative larnite have been reported from In Teria in the Algerian Sahara. These rocks are interesting as they contain harzburgitic and dunitic xenoliths that show that the lithospheric mantle of this region has been subjected to multi-stage metasomatism (Dautria *et al.*, 1992).

Another continental area that is well known for its melilitic rocks is the Balcones Igneous Province of Texas. It contains more than 200 sills, laccoliths, plugs and volcanic remnants that extend for some 300 km from east of Austin to west of Uvalde, with most of the rocks being clustered around Uvalde. Many subvolcanic rocks of this province were discovered during drilling programmes searching for petroleum. According to Barker and Young (1979) petroleum has been produced from at least 60 of the buried igneous centres in this province. Approximately half the rocks in the Balcones Igneous Province are melilite-olivine nephelinites (see Table 11.9), and they are essentially composed of melilite, olivine, titanaugite and nepheline, with accessory amounts of apatite, phlogopite, perovskite and iron-titanium oxides (Spencer, 1969, pp.274–275).

Melilitites as has already been recorded are an important part of the Tertiary-Quaternary Alkaline

Province of western and central Europe (Wilson and Downes, 1991). This magmatic province is related to extension and its most primitive rocks include alkali basalts, alkali picrobasalts, basanites, leucitites, melilitites and nephelinites. Wilson *et al.* (1995) have recently described the melilitites that crop out in the Freiburg, Hegau and Urach volcanic fields of the Upper Rhine Graben (Table 11.9). The Hegau and Urach fields occur on the eastern shoulder of the rift, whereas the Freiburg field lies within the rift and contains the well-known Kaiserstuhl carbonatite volcano. Carbonatites are also found in the Hegau volcanic field. The Urach volcanic field lies to the south of Stuttgart and it contains more than 350 individual diatremes. These diatremes usually contain tuff and volcanic breccia of olivine melilitite and olivine-melilite nephelinite composition. Many mantle-derived xenoliths including some spinel-phlogopite wehrlites (clinopyroxene peridotites) are included in these rocks. These inclusions are particularly interesting because clinopyroxene is the main carrier of calcium and sodium in the upper mantle and the abundance of phlogopite may reach 10 per cent in a few samples. Such xenoliths are probably products of mantle metasomatism.

Most of these melilitic rocks erupted in the mid-Miocene, concurrent with the updoming of the basement rocks in the Vosges and Black Forest areas on either side of the Upper Rhine graben. According to Wilson *et al.* (1995, p. 182) the magma that crystallised to form the melilitites probably represent a near-primary partial melt of the thermal boundary layer at the base of the European lithosphere. The volume of this magma was probably augmented by the infiltration of fluid generated by partial melting in a subjacent mantle plume. Evidence supporting such a plume includes crustal doming and recent seismic tomographic investigations of the upper mantle beneath central and western Europe (Blundell *et al.*, 1992, p. 176).

An extraordinary chain of olivine melilitites crop out over a 400-km wide zone that lies parallel to the coasts of Namibia and western South Africa (Cornelissen and Verwoerd, 1975; McIver and Ferguson, 1979; Taljaard, 1936). In the belt that is closest to the rift-generated South Atlantic coast the olivine melilitites are frequently associated with more differentiated rocks, such as phonolites. Further inland they are more likely to be associated with carbonatites. Even further inland they become less abundant, more undersaturated in silica and have higher Mg-values (Brey, 1978, p. 85). The distribution of these hinterland olivine melilitites overlaps with a suite of kimberlitic rocks. These systematic changes in the petrology of the melilitic rocks are probably related to the increase in lithospheric thickness as one moves inland from the rifted coastline. The most silica-undersaturated of these olivine melilitites differ from authentic kimberlites in having much higher CaO/MgO ratios. According to Brey (1978, p. 85) this chemical difference is probably the result of dolomite and enstatite being stable at the pressures at which melilititic magmas are generated. At the greater depths where kimberlitic magmas equilibrate dolomite and enstatite are usually replaced by magnesite and diopside (Kushiro *et al.*, 1975).

11.6 Potassic and ultrapotassic rocks and mantle metasomatism

The potassic and ultrapotassic volcanic rocks have low abundances, but like the sodic alkaline rocks they have wide distributions in diverse tectonic settings. If one excludes the lamprophyric rocks (see Section 11.8), the following areas contain most of the better-known examples of potassic and ultrapotassic basic and ultrabasic rocks: (1) the Eifel Province of Germany (see Fig. 11.2); (2) the Roman Province of Central Italy; (3) the Southwestern Ugandan Province; (4) the potassic midalkaline rocks of the Mediterranean island arcs, Anatolia and Northwestern Iran; (5) the Western Aldan Alkaline Province of eastern Siberia; (6) the Dongbei Province of Northeastern China; (7) the Eastern Sunda Arc Province of Indonesia; (8) the leucite basanite subprovince of Eastern Australia; (9) the Sapucai Graben Province of Eastern Paraguay; (10) the Serra do Bueno Province of Minas Gerais, Brazil; and (11) the various potassic rocks from oceanic islands, such as the Tristan da Cunha Group, Gough Island, Cape Verde Group, and the Marquesas Group.

Silica-undersaturated, potassic and ultrapotassic volcanic and subvolcanic rocks typically contain leucite, or less commonly kalsilite (Gupta and Yagi, 1980; Peccerillo, 1992). Because leucite is unstable in the plutonic environment (Scarfe *et al.*, 1966), rocks of similar chemical composition that crystallise at greater pressures are likely to be phlogopite or potassian richterite bearing pyroxenites. It is apposite to recall that in Section 2.3 it was shown that the CIPW normative composition of a typical phlogopite has 53 per cent olivine, 27 per cent leucite and 20 per cent kalsilite. Phlogopite pyroxenites and even kalsilite pyroxenites (locally known as yakutites) have been reported from the Malomurunskii lopolith, a part of the Murun Complex in the Western Aldan Alkaline Province in southeastern Siberia (Kogarko et al., 1995, pp. 186–189; Mitchell et al., 1994). Mantle-derived xenoliths that contain both phlogopite and potassian richterite are included in the Wesselton kimberlite from South Africa (Erlank and Finger, 1970; see Section 3.4).

Many foiditic rocks contain some modal or CIPW normative leucite. For example, the rock that Le Maitre (1984, p. 255) called sodic foidite has a mean CIPW normative leucite content of 8.3 per cent and a mean CIPW normative nepheline content of 21.5 per cent. This means that one cannot simply call all leucite-normative rocks potassic rocks. To avoid this and other ambiguities it is often advisable to call all alkaline rocks that contain more K_2O than Na_2O potassium-dominant rocks. The term potassic is normally used in a special way ($Na_2O–2.0 < K_2O$) to define the potassic trachybasalts, shoshonites and latites of the midalkaline suite (see Chapter 10 and Le Maitre et al., 1989, p. 28).

The island of Tristan da Cunha in the South Atlantic Ocean contains a well-known suite of potassic midalkaline rocks (Baker et al., 1964). Table 11.6 provides geochemical and normative data on a primitive basanite from this suite. It contains 3.7 per cent Na_2O, 3.0 per cent K_2O, 10.1 per cent normative nepheline, 18.2 per cent normative orthoclase, but no normative leucite. These data show that it may be difficult to define the potassic basalts and basanites because they frequently contain slightly more Na_2O than K_2O. The consensus among petrologists is that the potassic midalkaline rocks contain higher than normal amounts of potassium. In practice potassic basanites and basalts are usually only separated from their more abundant sodic analogues when they belong to a suite that contains more readily defined potassic members, such as potassic trachybasalt, shoshonite or latite. Table 11.12 contains geochemical data on two potassic trachybasalts from southwestern Uganda. They are readily shown to be potassic as they contain more K_2O than Na_2O. When considering the classification of the rocks of the alkaline suite, and the foiditic and foidolitic rocks, it is usually convenient to follow Le Bas (1989, p. 1299) and use the term potassic to describe all the potassium-dominant rocks.

Ultrapotassic rocks

When petrographers originally attempted to develop a systematic classification of the feldspathoid-bearing rocks many made the erroneous assumption that the leucite-bearing rocks were potassic analogues of the more abundant nepheline-bearing rocks. Once this assumption is made it is difficult to reconcile modal and chemical classifications. This is principally because nepheline only contains 42 per cent silica whereas leucite contains 55 per cent silica. One can readily see the implications of this difference if one compares the chemical composition of the average leucitite that contains 47.1 per cent silica (Nockolds, 1954, p.1031), with its supposed analogue the average nephelinite that contains only 40.6 per cent silica (Le Maitre, 1976, p. 629). Ultrapotassic is another useful term. It is used to define the rare but important potassium-dominant rocks in which the K_2O/Na_2O ratio is greater than three (Bergman, 1987, p. 104; see Section 11.9 on the lamproites).

The leucite basanites of southeastern Australia are interesting because their mean K_2O/Na_2O ratio is three (Cundari, 1973, p. 472) thus they lie astride the boundary between the potassic and ultrapotassic. These leucite-bearing lavas crop out in a narrow (90 km) north–south trending belt that extends for about 640 km from Byrock in central New South Wales to Cosgrove in northern Victoria (see Section 10.8). Their ages vary from 16 million years in the north to 6 million years in the south. This pattern of ages supports the idea that the origin of these rocks is linked to a stationary sublithospheric magmatic focus that triggered their extrusion on to a northward moving Australian Plate. These lavas usually contain about 40 per cent diopside, 24 per cent leucite, 14 per cent olivine, 12 per cent iron-titanium oxides (magnetite, maghemite, ilmenite and perovskite), and lesser amounts of phlogopite and titaniferous potassian richterite, apatite and interstitial sanidine and nepheline (Cundari, 1973, p. 469). The CIPW norms of these rocks all contain leucite and nepheline, while some contain small amounts of aegirine showing that they have a tendency to be peralkaline. According to Johnson (1989, p. 341) these leucite-bearing lavas possess strontium, neodymium and lead isotopic ratios that are distinctly different from the other rocks of the Eastern Australian Volcanic Province (see Section 10.8). These ratios are interpreted as comparable to those measured in other potassic and ultrapotassic rocks that originated

Table 11.12 Chemical compositions of selected potassic and ultrapotassic rocks from southwestern Uganda. After Lloyd *et al.,* 1991, p. 54.

Rock type	Ultrapotassic foidite	Ugandite	Mafurite	Leucite tephrite	Potassic trachybasalt	Potassic trachybasalt
Location and identification	Katwe-Kikorongo, No 14	Kachuba, Bunyaruguru, No 2	Mafuru Crater, Bunyaruguru, No 7	Katwe-Kikorongo, No 12	Bufumbira, No 4	Bufumbira, No 5
Comment	Kalsilite-bearing	Ultrapotassic lava	Ultrapotassic foidite	Leucite-bearing	Leucite-bearing	Leucite-bearing
SiO_2	38.6	40.47	40.7	43.6	50.05	52.94
TiO_2	5.66	3.52	3.38	4.68	1.67	1.55
Al_2O_3	7.56	5.38	7.62	10.24	16.82	18.32
Fe_2O_3	6.05	4.03	4.56	4.98	2.99	2.09
FeO	7.32	6.47	5.26	6.81	7	6.05
MnO	0.2	0.13	0.16	0.18	0.17	0.18
MgO	10.8	24.84	17.43	8.53	4.35	2.3
CaO	14.3	8.06	10.8	11.76	8.63	6.38
Na_2O	1.38	0.68	0.99	2.6	2.88	3.6
K_2O	5.08	3.46	6.56	4.76	4.46	5.36
P_2O_5	0.57	0.29	0.71	0.57	0.73	0.66
H_2O^+	0.64	1.11	1.13	0.95	0.21	0.38
CO_2	0.52	0.36	0.23	0.35	0.05	n.d.
Total	98.68	98.8	99.53	100.01	100.01	99.81
Trace elements	ppm	ppm	ppm	ppm	ppm	ppm
F	n.d.	1000	n.d.	n.d.	1300	700
Cl	n.d.	100	n.d.	n.d.	300	500
Cr	528	9.6	801	438	83	n.d.
Ni	165	1021	515	107	n.d.	n.d.
Rb	102	137	154	123	165	171
Sr	1677	2043	1559	1603	1109	1381
Y	18	n.d.	17	13	n.d.	n.d.
Zr	361	394	202	334	440	451
Nb	203	111	166	196	131	193
Ba	1068	2418	2407	1300	1254	1075
La	130	68	170	127	91	50

from the partial melting of segments of lithospheric mantle enriched in incompatible elements.

A remarkable group of Cretaceous volcanic rocks has been described from the Kamchatka Peninsula in the far east of Russia. They include potassic komatiites, potassic picrites, ultrapotassic picrobasalts and ultrapotassic basanites. The chemical analyses of these rocks provided by Kamenetsky *et al.* (1995, p. 647) are slightly dubious, because they have high values for loss on ignition. If one accepts these data then one has to record that there are some remarkable potassic and ultrapotassic rocks in the Valaginsky Range of Kamchatka. The ultrapotassic basanite (K-139) contains 4.5 per cent K_2O and if its chemical composition is converted into normative values it is found to contain over 20 per cent normative leucite, with additional smaller amounts of normative nepheline and larnite. According to Kamenetsky *et al.* (1995, p. 659) rocks that are similar to the primitive

potassic picrites and potassic komatiites of eastern Kamchatka are also found on Karaginsky Island in the Bering Sea and in the Koryak highland to the north of Kamchatka.

Western Branch of the African Rift System

In Section 11.4 we explored the sodic alkaline rocks of the Kenya Rift. We will now examine the extraordinary potassic and ultrapotassic rocks of the Western Branch of the African Rift System. These rocks crop out in four volcanic fields that lie astride the borders of western Uganda, Rwanda and eastern Zaire. The two southern volcanic fields, called the South Kivu and Virunga-Bufumbira fields, crop out on the shores of Lake Kivu. The central volcanic field is situated between lakes Edward and George. It is

Table 11.13 CIPW norms of selected potassic and ultrapotassic rocks from southwestern Uganda. These are the same samples as those presented in Table 11.12.

CIPW Norms	Ultrapotassic foidite Katwe-Kikorongo, No 14 Kalsilite-bearing	Ugandite Kachuba, Bunyaruguru, No 2 Ultrapotassic lava	Mafurite Mafuru Crater, Bunyaruguru, No 7 Ultrapotassic foidite	Leucite tephrite Katwe-Kikorongo, No 12 Leucite-bearing	Potassic trachybasalt Bufumbira, No 4 Leucite-bearing	Potassic trachybasalt Bufumbira, No 5 Leucite-bearing
Orthoclase	0	0	0	8.9	26.42	31.84
Albite	0	0	0	0	17.24	24.8
Anorthite	0	1.45	0	2.24	19.84	18.09
Leucite	24.15	16.46	30.95	15.37	0	0
Nepheline	5.89	3.21	1.48	12.07	3.89	3.16
Aegirine	0.51	0	5.12	0	0	0
Diopside	28.21	24.37	7.34	41.47	14.97	7.82
Larnite (Cs)	10.1	2	12.5	0	0	0
Olivine	10.17	38.91	30.2	2.24	8.4	6.71
Magnetite	8.07	6.02	4.18	7.33	4.35	3.06
Hematite	0.48	0	0	0	0	0
Ilmenite	11.06	6.88	6.55	9.03	3.19	2.97
Apatite	1.36	0.7	1.68	1.35	1.7	1.55

sometimes called the Bunyaruguru-Katwe-Kikorongo field. The northern volcanic field is in the Fort Portal and Kasekere areas, where it is located to the east of the spectacular Ruwenzori horst. This range of up-faulted rocks is also known as the Mountains of the Moon. It contains ten peaks that are over 4500 m in elevation and the highest peak attains an elevation of 5119 m. Most of the volcanoes in the southern and northern fields are steep-sided ash cones whereas those in the central field are principally maars surrounded by tuff rings (Lloyd *et al.*, 1991, p. 29).

Because of their exotic modal compositions the nomenclature that has been devised to classify these rare potassic and ultrapotassic rocks is extensive, complicated and often awkward to use. Clinopyroxenes, particularly diopside and augite, are the essential phases in most of these rocks. Olivine usually occurs in the groundmass of the Mg-rich rocks. Melilite is characteristic of the calcium-rich and silica-poor rocks of the central volcanic field. Feldspar is confined to the basic and intermediate rocks of the southern volcanic fields. Leucite is the typical feldspathoid of the southern volcanic fields. When leucite occurs as a phenocryst in these highly silica-undersaturated rocks it is normally absent from the groundmass. The groundmass typically contains phases that are more silica deficient, like kalsilite and nepheline. Kalsilite, nepheline and melilite are common in the lavas of the central volcanic field.

Nepheline is usually absent from the ultrapotassic rocks of the Bufumbira area (Bufumbira is the Ugandan part of the Virunga volcanic field). Feldspathoids are not essential phases in the carbonatitic volcanic rocks of the northern volcanic field where primary calcite is an essential phase.

The Western Branch of the African Rift System is renowned in the petrological community because it contains three extraordinary volcanic rocks. These rocks are ugandite, mafurite and katungite (Holmes, 1937, p. 201; Holmes, 1942, p. 199; Holmes and Harwood, 1937, p. 11). They are known collectively as the kamafugitic rocks (KAtungite + MAFurite + UGandite + ITE; Sahama, 1974, p. 96). Ugandites range in composition from melanocratic leucitites to ultrapotassic basanites. They were first described from the Bufumbira area in southwestern Uganda, where the type rock is mainly composed of clinopyroxene, olivine, leucite, and glass. The sample featured in Tables 11.12 and 11.13 is a feldspar-free olivine ugandite from the Bunyaruguru area.

Mafurite is a volcanic rock named after the Mafuru Crater in the Bunyaruguru area. It is an ultrabasic and ultrapotassic volcanic rock that usually contains phenocrysts of olivine and clinopyroxene in a groundmass of diopside, kalsilite, olivine, perovskite and phlogopite. Tables 11.12 and 11.13 contain the major element, trace element and the CIPW normative composition of a typical mafurite. This ultrabasic rock is exceptional as it contains 6.56

per cent K_2O and just over 30 per cent normative leucite.

Tables 11.12 and 11.13 also contain geochemical and normative data on two midalkaline potassic trachybasalts from the Bufumbira area of the southern volcanic field. The parental magma from which the midalkaline lavas of the Western Branch of the African Rift System evolved seems to have formed as the result of assimilation of silicic crustal materials by a primary, or near-primary, ultrapotassic magma (Lloyd *et al.*, 1991, p. 70).

Katungite was named after an isolated volcanic cone called Katunga that lies south of Bunyaruguru in the central volcanic field. Similar rocks crop out in Katwe Crater (Lloyd *et al.*, 1991, pp. 54 and 56). Katungite is an ultrapotassic melanocratic olivine melilitite and it usually also contains subordinate amounts of leucite, kalsilite and nepheline embedded in a glassy groundmass (see Section 11.5). The melilites are essentially akermanites with up to 25 per cent of the soda-melilite end-member component (Lloyd *et al.*, 1991, pp. 51–52). If one computes the CIPW norms of the groundmass glass of this rock one discovers that it contains kalsilite, aegirine and larnite. It is thus not only strongly silica-undersaturated it is also peralkaline. This is unusual in a primary mantle-derived magma and it is reminiscent of the lamproites (see Section 11.9). The notion of peralkaline primary magmas should set one pondering the nature of the phases in the source area where such magmas are generated.

Over the years there has been considerable discussion of the chemical composition of the volcanic rocks and xenoliths from the Western Branch of the African Rift System. This is because they show exceptional enrichment in phosphorus, potassium, titanium, rubidium, strontium, yttrium, zirconium, niobium, barium and the light lanthanoids. The leucite lamproites are the only other major group of rocks that contain these elements at comparable levels of abundance. Bell and Powell (1969) have noted the consistency of the K:Rb, Ca:Sr and Nb:Zr ratios in the Ugandan lavas and concluded that it was likely that their magmas were either generated by a similar process, or were derived from similar source materials. It is also indisputable that these ultrapotassic, ultrabasic magmas cannot be produced by simple partial melting of normal lherzolitic mantle. Because the ultrapotassic lavas of southwestern Uganda frequently contain xenoliths of phlogopite wehrlite, or phlogopite and amphibole bearing clinopyroxenite, it is likely that the lithospheric mantle under this area

has been changed by extensive mantle metasomatism. With a moderate to high degree of partial melting (about 25 per cent), and at a pressure of about 3 GPa (between 100 and 110 km), phlogopite wehrlite is likely to produce a range of ultrapotassic primary magmas capable of yielding the potassic and ultrapotassic rocks of the Western Branch of the African Rift System (Edgar, 1991, p. 81). This petrogenetic model is similar to most other speculations about the origin of ultrapotassic rocks in that it postulates that the source region contains phlogopite, yet phlogopite is only stable to pressures equivalent to 250 km. This means that the maximum depth of the source region for ultrapotassic magmas is likely to be 250 km.

11.7 'Provincia Magmatica Romana'

A particularly well studied area of potassic and ultrapotassic magmatism is the Roman Magmatic Province, or the Roman Comagmatic Region (Washington, 1906) of central Italy. It covers an area of about $6000 \, km^2$, and extends from the Lago di Bolsena Caldera in the north to Somma-Vesuvio in the south. The Vulsini Volcanoes, Mount Vico Volcano, Cimino Volcano and the Mount Sabatini Volcanic Complex are north of Rome; the Colli Albani Volcanic Complex, Ernici Volcano and Roccamonfina Volcano lie between Rome and Napoli; and Campi Flegrei, Ischia and Somma-Vesuvio are located in the Campanian area (Gupta and Yagi, 1980, pp. 66–76; and see Fig. 11.6). According to Peccerillo and Manetti (1985, p. 379) the alkaline rocks of the Roman Magmatic Province belong to four groups: (1) a trachybasalt to trachyte suite of midalkaline potassic rocks that are saturated in silica, and have K_2O/Na_2O ratios of about one; (2) a leucite tephrite to leucite phonolite alkaline suite of high-potassium rocks that are strongly undersaturated in silica and have high potassium contents, high K_2O/Na_2O ratios, and high incompatible element contents; (3) ultrabasic leucite and kalsilite bearing melilitites that are strongly undersaturated in silica and have high K_2O/Na_2O ratios; and (4) a series of lamprophyric rocks that are essentially minettes and are slightly oversaturated in silica but strongly enriched in potassium (see Section 11.8). Peccerillo and Manetti (1985, p. 389) also proposed that all the parental magmas responsible for these suites were produced by the partial melting of phlogopite-bearing lithospheric mantle over a range of pressures.

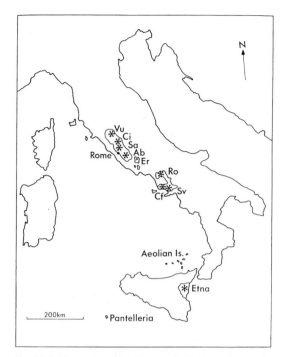

Fig. 11.6 Sketch map showing the distribution of the active volcanoes of Italy, where Ab = Alban Hills, Ci = Cimino, Cf = Campi Flegrei, Er = Ernici, Ro = Roccamonfina, Sa = Sabatini and Sv = Somma-Vesuvio. After Appleton, 1972, p. 427. Figure 4.3 shows the Aeolian Islands in more detail.

The rocks that erupted from Roccamonfina Volcano may be regarded as typical of the Roman Magmatic Province. They include a wide range of materials that belong to at least two different lineages. These lineages can be distinguished by age, geochemistry and petrography (Appleton, 1972, p. 426). Tables 11.14 and 11.15 contain major element, trace element and normative data on a selection of primitive lavas from the volcano (Giannetti and Ellam, 1994). The lavas include alkali basalts, potassic trachybasalts, potassic basanites and an ultrapotassic tephrite. This denotes that this volcano has produced potassium-dominant rocks of both the midalkaline and the alkaline suites. The potassic basanite and the ultrapotassic tephrite are particularly interesting as their respective normative leucite contents are 14.6 and 21.8 per cent (see Table 11.15). Appleton (1972, p. 429) includes geochemical data on the more differentiated rocks of both the midalkaline and alkaline suites. The former lineage ends in potassic trachyte, whereas the latter terminates in potassic phonolite. Whilst the phonolites contain just over 10 per cent feldspathoids, the trachytes are usually silica-saturated and contain over 90 per cent feldspar. Geochemical data show that the differentiated rocks of both series are probably products of low pressure fractional crystallisation. Local variations in composition were probably caused by magma mixing or wall-rock reaction.

Giannetti and Ellam (1994, p. 30) have used strontium isotopic ratios to show that the primary magmas responsible for the two series of rocks that issued from Roccamonfina Volcano were produced by the partial melting of dissimilar source rocks. The primary magma responsible for the midalkaline (or low-potassium) suite probably equilibrated with rocks that contained amphibole and possibly garnet, whilst the primary magma that produced the alkaline (or high-potassium) suite is likely to have formed from the partial melting of a lithospheric mantle source enriched in phlogopite and pyroxene. Rogers (1992, p. 93) has compared the geochemistry of the ultrapotassic primary, or near-primary, magmas from the Roman Magmatic Province with those from Southwestern Uganda. He proposed that the low-titanium, ultrapotassic primary magma of the Roman Magmatic Province is likely to be a product of the partial melting of lithospheric mantle that had been metasomatised by elements emanating from deeply subducted materials. The tectonic history of the Roman Magmatic Province is complex. There is evidence of active subduction beneath the Calabrian Arc, yet there is also cogent evidence of local doming and rifting along the eastern margin of the Tyrrhenian extension zone (see Fig. 11.7).

Another belt of strongly alkaline magmatic rocks crops out to the east of the Roman Magmatic Province in the Umbria-Latium region (Stoppa and Lavecchia, 1992). These Late Pleistocene rocks include kalsilite melilitites, olivine-leucite melilitites and leucite-wollastonite melilitites and a carbonatite diatreme. The latter is a calcite-carbonatite and it crops out just outside the village of Polino. The diatreme is particularly interesting as it contains xenoliths of phlogopite peridotite. Most of the many melilitites and melilitolites from the Umbria-Latium Province contain more than 20 per cent leucite and kalsilite. The geochemistry of the primitive rocks from this province is broadly similar to the geochemistry of the primitive rocks of the Roman Magmatic Province. Both suites are of similar age and evolved in similar tectonic settings in the rifted border area of the Tyrrhenian extensional zone.

Fig. 11.7 Dykes of variable thickness in the wall of the Valle del Bove on the southeastern flank of Etna Volcano, Italy.

11.8   Lamprophyric rocks

According to Rock (1991, p. 149) the 'lamprophyres deserve a status nearer the centre than the periphery of modern petrology'. The term lamprophyre was introduced by Gümbel (1874, p. 36) to describe a group of dark-coloured dyke rocks from the Fichtelgebirge of West Germany. It is derived from the Greek word *lampros* meaning bright or glistening. These are fitting terms for describing the mica-bearing rocks of the Fichtelgebirge.

Rocks of the lamprophyric clan usually contain more discrete phases than the common plutonic and volcanic rocks. They normally have high abundances of hydrous minerals, carbonates, sulphides, apatites and zeolites. In the comprehensive, hierarchical classification of igneous rocks recommended by Le Maitre *et al.* (1989, pp. 10–12) the lamprophyric rocks are regarded as special for the following reasons: (1) they are dark coloured and contain more than 35 per cent of the M modal parameter in the system QAPFM; (2) they are volatile-rich and contain high concentrations of most, if not all, of the

components fluorine, phosphorus, sulphur, chlorine, potassium, zinc, rubidium, strontium, zirconium, niobium, caesium, barium, the light lanthanoids (lanthanum to samarium), thorium, uranium, water and carbon dioxide, compared to normal plutonic and volcanic rocks; (3) they usually have a porphyritic, or pseudo-porphyritic (large xenocrysts), texture; (4) their phenocrysts are usually mafic (olivine, augite, hornblende, kaersutite and biotite/phlogopite/tetraferriphlogopite) and they are frequently euhedral in shape; (5) minerals of the feldspar, nepheline and sodalite groups may be present but are usually restricted to the groundmass; and (6) most of these rocks display deuteric alteration.

The lamprophyre clan is readily subdivided in the following way: (1) the kersantite-minette-spessartite-vogesite family; (2) the camptonite-monchiquite-sannaite family; (3) the alnöite-polzenite family; (4) the aillikite-damtjernite (damkjernite) family; and (5) the kimberlite-lamproite family (see Fig. 11.8). Rocks of the aillikite-damtjernite family can be readily separated from the other more common lamprophyres because they contain significant amounts of primary carbonate minerals and they

Table 11.14 Chemical compositions of selected primitive lavas from Roccamonfina, Roman Magmatic Province. After Giannetti and Ellam, 1994, pp. 24–25.

Rock type	Alkali basalt R-112: per cent	Potassic basalt R-96: per cent	Potassic trachybasalt R-53: per cent	Potassic trachybasalt R-62a: per cent	Potassic basanite R-35: per cent	Ultrapotassic tephrite R-45: per cent
SiO_2	45.84	48.57	47.02	47.55	45.3	46.06
TiO_2	0.91	0.73	0.93	0.91	1.04	1.06
Al_2O_3	14.12	15.12	16.67	17.85	16.75	16.03
Fe_2O_3	3.74	4.73	3.78	3.52	6.85	6.75
FeO	5.94	4.52	5.38	5.31	3.28	2.66
MnO	0.16	0.15	0.15	0.15	0.14	0.16
MgO	9.49	8.7	7.06	6.16	6.12	5.79
CaO	15.06	12.53	12.06	11.38	11.03	12.58
Na_2O	1.9	1.57	1.84	1.86	1.84	1.45
K_2O	1.32	1.83	3.36	3.91	5.24	5.56
P_2O_5	0.3	0.16	0.35	0.35	0.7	0.54
LOI	0.8	0.54	0.76	0.59	1.29	1.16
Total	99.58	99.15	99.36	99.54	99.58	99.8
Trace elements	ppm	ppm	ppm	ppm	ppm	ppm
Sc	44	39	37	32	30	27
V	275	177	241	235	271	237
Cr	383	211	143	107	78	100
Ni	29	83	60	44	51	48
Rb	215	89	175	179	502	405
Sr	1040	597	1289	1500	1606	1649
Zr	106	100	195	190	262	194
Nb	11	11	8	10	8	13
Ba	617	241	540	714	1022	859
Ce	66.8	57.1	97.1	102.9	202.6	167.2
Nd	37.9	32.2	49.6	50.5	91.4	77.2
Sm	7.71	6.26	9.48	9.69	15.72	13.78
Eu	1.9	1.5	2.31	2.43	3.43	3.07
Gd	6.41	5.52	7.59	7.83	11.73	10.52
Ho	0.92	0.95	1.06	1.04	1.36	1.41
Er	2.25	2.71	2.77	2.77	3.45	3.37
Yb	1.64	2.35	1.95	1.98	2.32	2.42
Lu	0.3	0.41	0.28	0.31	0.34	0.35

Table 11.15 CIPW normative composition of selected primitive lavas from Roccamonfina, Roman Magmatic Province. Major element chemistry obtained from Giannetti and Ellam, 1994, pp. 24–25.

Rock type	Alkali basalt R-112: per cent	Potassic basalt R-96: per cent	Potassic trachybasalt R-53: per cent	Potassic trachybasalt R-62a: per cent	Potassic basanite R-35: per cent	Ultrapotassic tephrite R-45: per cent
Orthoclase	7.91	11	20.17	23.38	13.02	5.66
Albite	2.64	13.51	3.24	3.37	0	0
Anorthite	26.47	29.3	27.75	29.16	22.46	21.2
Leucite	0	0	0	0	14.6	21.8
Nepheline	7.4	0	6.81	6.8	8.62	6.77
Diopside	38.16	26.54	24.95	20.91	23.53	31.65
Hypersthene	0	2.51	0	0	0	0
Olivine	11.9	12.47	11.58	11.02	10.95	6.69
Magnetite	3.05	2.88	2.88	2.77	3.12	2.88
Ilmenite	1.76	1.41	1.79	1.76	2.03	2.06
Apatite	0.71	0.38	0.83	0.83	1.67	1.29

feldspar	foid	biotite/ phlogopite augite ± olivine	hornblende augite ± olivine	amphibole Ti- augite olivine biotite/ phlogopite	melilite biotite/ ± Ti- augite ± olivine ± calcite
or >pl	-	minette	vogesite	-	-
pl >or	-	kersantite	spessartite	-	-
or >pl	fel >foid	-	-	sannaite	-
pl >or	fel >foid	-	-	camptonite	-
-	glass or foid	-	-	monchiquite	polzenite
-	-	-	-	-	alnöite

Fig. 11.8 Classification and nomenclature of the lamprophyres, After Le Maitre *et al.*, 1989, p. 11.

may grade into the carbonatites. Aillikites are defined as containing essential amounts of primary carbonate minerals but no feldspars or melilite. The damtjernites contain essential amounts of primary carbonate minerals with some feldspar.

Metais and Chayes (1963, pp. 156–157) investigated the chemical composition of the common lamprophyres and concluded that in terms of average major element composition minettes, vogesites, kersantites and spessartites seem almost indistinguishable. These rocks differ from basalts in that they are enriched in potassium (K_2O more than 2.5 per cent) and total water. The minettes have an average K_2O content of 5.5 per cent. Superficially the minettes appear to be chemically akin to the leucite tephrites and the leucite phonolitic tephrites. Rocks of the kersantite-minette-spessartite-vogesite family are usually found associated with epizonal granitic, syenitic and monzonitic rocks, but they are occasionally associated with alkaline complexes and carbonatites. It is the latter association that links all the various lamprophyres together as a natural group or clan.

11.9 Lamproites

Niggli (1923, p. 184) introduced the term lamproite to describe a group of lamprophyre-like subvolcanic and extrusive, igneous rocks enriched in both K and Mg. These lamproites may now be defined as a chemically and petrographically diverse group of mafic to ultramafic, peralkaline, potassium- and magnesium-rich lamprophyric rocks that characteristically contain some of the following phases; phlogopite, diopside,

potassic richterite, unusually iron-rich leucites and potassium feldspars, jeppeite $[(K,Ba)_2(Ti,Fe)_6O_{13}]$, priderite $[(K,Ba)(Ti,Fe)_8O_{16}]$, wadeite $[K_2CaZrSi_4O_{12}]$ and glass. Normally they do not contain primary plagioclase, kalsilite, melanite, melilite, monticellite, nepheline or sodalite group minerals (see Fig. 11.9). These rocks also tend to be somewhat enriched in the following chemical components: titanium, potassium, rubidium, strontium, zirconium, barium, the light lanthanoids and lead. They usually also have the following chemical characteristics: K_2O/Na_2O ratio greater than five, K_2O greater than Al_2O_3, MgO greater than five. The type areas for most of the common lamproites are either the West Kimberly region of Western Australia (Mt Cedric, Fitzroy River, Mamilu Hill and the Wolgidee Hills; Prider, 1960), or the Leucite Hills of southwestern Wyoming (Oreanda Butte and the state of Wyoming; Carmichael, 1967). The lamproites of Western Australia form a distinct group as they all have remarkably high titanium contents.

Diamond-bearing olivine lamproites have been described from several localities, including Prairie Creek, Arkansas and the east and west Kimberly areas of Western Australia (Atkinson *et al.*, 1984; Jaques *et al.*, 1986; Scott Smith and Skinner, 1984). The olivine lamproites grade via the leucite-bearing olivine-diopside lamproites to the well-known leucite lamproites. According to Atkinson *et al.* (1984, p. 213) there are 45 lamproite outcrops in the Ellendale Province of Western Australia, 27 are mainly composed of leucite lamproite, 14 are olivine lamproites and 4 are leucite-olivine lamproites. Most of the Ellendale lamproites contain diamonds, and the most diamondiferous of these rocks are the tuffaceous olivine lamproites. The uniquely diamondiferous

Dominant phase	Other essential phases	Rock name
clinopyroxene	leucite	**cedricite**
olivine (> 20%)	clinopyroxene & phlogopite	**olivine lamproite**
leucite	olivine, clinopyroxene & glass	**gaussbergite**
leucite	{clinopyroxene, olivine, potassium {feldspar & phlogopite	**jumillite**
potassic richterite	leucite	**mamilite**
phlogopite	leucite, clinopyroxene	**fitzroyite**
phlogopite	olivine, clinopyroxene & glass	**verite**
phlogopite	{clinopyroxene, potassic richterite, {olivine & leucite	**wolgidite**
phlogopite	leucite, clinopyroxene & glass	**wyomingite**

The term orendite has not been used because Sahama (1974, p. 96) used it as a general term for all the lamproites.

Fig. 11.9 A simple classification and nomenclature of the lamproites.

Argyle lamproite crops out in the Halls Creek Mobile Zone, northeast of the Ellendale Province.

At the present level of exposure the Argyle diatreme is composed of a variety of weakly bedded crystal-lithic tuffs that are cut by dykes of olivine lamproite. The tuffs usually consist of lamproite clasts set in a groundmass of olivine crystals, comminuted lamproitic material and rounded quartz grains derived from the disaggregation of earlier sediments. Olivine lamproites are essentially composed of phenocrysts of often altered olivine (10–25 per cent) and phlogopite (15–30 per cent), set in a fine grained potassium-rich groundmass.

11.10   Kimberlites and olivine lamproites

In his pioneering study of the southern African kimberlites Wagner (1914) divided them into two groups. They have variously been called: (1) the basaltic type, or Group One; and (2) the micaceous type or Group Two (cf. Berg and Allsopp, 1972; Smith, 1983; Smith *et al.*, 1985; Skinner, 1989). Basaltic is an inappropriate term to describe the Group One rocks as they have ultrabasic, or more exactly picritic or meimechitic, compositions. They typically contain more than 18 per cent MgO and no plagioclase. Before attempting to install the kimberlites in the magma/igneous rock system one has to figure out their chemical characteristics (Smith *et al.*, 1985, pp. 270–271), the relationship between the two groups of kimberlites and the relationship between the Group Two kimberlites and the olivine lamproites (see Tables 11.16 and 11.17).

Group Two kimberlites are regarded as unique to southern Africa where they crop out in a southwest to northeast trending belt that extends from Sutherland in the Cape Province to northern Swaziland. In this area they make up about 20 per cent of the kimberlites. According to Skinner (1989, p. 531) the Group Two kimberlites of the Kaapvaal Craton are dominated by primary phlogopite. It occurs as a phenocryst in the groundmass or as larger interstitial, poikilitic grains. Skinner also claims that several Group Two kimberlites contain serpentine and clay-mineral pseudomorphs that display the characteristic habit of melilite. This shows a possible link between the Group Two kimberlites and the alnöite-polzenite family of lamprophyres and the olivine melilitites. The alnöites appear to provide links between the carbonatites, lamprophyres, melilitites and possibly the kimberlites. McIver and Ferguson (1979) and Moore (1983) claim to have discovered definite links between kimberlites and the olivine melilitites. In 1928 Wagner suggested that the Group Two kimberlites might be called orangeites. Recently this term has been reintroduced by Mitchell.

The term olivine lamproite was introduced by Jaques *et al.* (1984, p. 230) to describe a group of lamproitic rocks from the Ellendale Province of Western Australia in which olivine was the dominant primary mineral. These rocks are strongly porphyritic and contain two generations of olivine. The olivine macrocrysts are typically anhedral to rounded, with undulose extinction and kink-banding. Common

Table 11.16 Mean major element composition of selected lamprophyric rocks.

	Group 1B Kimberlite Smith et al., 1985, p. 276	Aillikite-damkjernite Rock, 1986, p. 178	Group 1A Kimberlite Smith et al., 1985, p. 276	Alnöite-polzenite Rock, 1986, p. 178	Group 2 Kimberlite (orangeites) Smith et al., 1985, p. 276	Camptonite-monchiquite-sannaite Rock, 1987, p. 206	Olivine lamproite Jaques et al., 1986, p. 249	Kersantite-minette-spessartite-vogesite Rock, 1987, p. 206	Leucite lamproite (Noonkanbah) Jaques et al., 1986, p. 249
SiO_2	25.7	29	32.1	32.2	36.3	41.9	41.5	51.5	53.2
TiO_2	3	3.2	2	2.2	1	3	3.62	1.3	5.84
Al_2O_3	3.1	6.6	2.6	9.2	3.2	13.7	3.64	14	8.34
Fe_2O_3*	12.7	14.26	9.2	12.67	8.4	12.52	9	9.15	7.33
MnO	0.2	0.27	0.2	0.24	0.2	0.21	0.13	0.15	0.08
MgO	23.8	13.3	28.5	15.3	29.7	7.2	25	6.9	7.8
CaO	14.1	15.7	8.2	16	6	10.6	4.99	6.6	3.22
Na_2O	0.2	1	0.2	2.1	0.1	3.2	0.46	2.7	0.57
K_2O	0.6	2.1	1.1	2.1	3.2	2.3	4.12	3.8	9.89
P_2O_5	1.1	1.3	1.1	1.5	1.1	0.85	1.68	0.71	0.98
H_2O^+	7.2	4.5	8.6	3.9	5.3	3.6	6.36	2.7	2.98
CO_2	8.6	9.6	4.3	3.1	3.6	2.7	0.45	2.2	0.5
Total	100.3	100.83	98.1	100.51	98.1	101.78	100.95	101.71	100.73

Table 11.17 Trace element composition of selected lamprophyric rocks in parts per million.

	Group 1B Kimberlite Smith et al., 1985, p. 276	Aillikite-damkjernite-alnöite-polzenite Rock, 1987, p. 210	Group 1A Kimberlite Smith et al., 1985, p. 276	Group 2 Kimberlite (orangeites) Smith et al., 1985, p. 276	Camptonite-monchiquite-sannaite Rock, 1987, p. 210	Olivine lamproite (Ellendale) Jaques et al., 1986, p. 249	Kersantite-minette-spessartite-vogesite Rock, 1987, p. 210	Leucite lamproite (Noonkanbah) Jaques et al., 1986, p. 249
Sc	20	20	13	20	23	21	20	14
V	170	330	75	85	341	85	167	210
Cr	1000	487	1400	1800	162	1006	462	348
Co	79	46	83	85	44	70	35	33
Ni	800	485	1360	1400	88	1004	186	346
Cu	79	95	54	30	57	56	52	66
Zn	75	107	56	60	104	71	82	77
Ga	8	7	4	6	23	4	21	17
Rb	30	73	50	135	70	479	124	275
Sr	1020	1146	825	1140	1089	1312	896	1184
Y	30	31	13	16	34	16	25	20
Zr	385	351	200	290	344	1133	276	1144
Nb	210	116	165	120	102	184	18	123
Ba	850	1322	1000	3000	1064	10334	1900	9871
La	125	151	90	200	83	421	113	292
Ce	220	256	140	350	151	734	216	435
Pr		34			21		27	
Nd	100	131	90		64		122	
Sm		20			20		20	
Eu		4.7			5		4.5	
Gd		11			13		15.4	
Dy		6.9			5.8		4.7	
Ho		1.1			2.1		1.3	
Er		3			2.7		2.9	
Yb		2.5			2.3		1.9	
Lu		0.3			0.4		0.3	
Hf		6			7		10	
Pb	10		7	30	10	50	20	52
Th	27	9	18	30	13	60	20	23
U	6	8	4	5	5	2	5	3

accessory phases include chromite and perovskite. Groundmass phases include glass, phlogopite, diopside, chromite, apatite, perovskite, wadeite, barite, priderite and rare ilmenite.

Bergman (1987, p. 169), Dawson (1987, p. 100) and others have noted that the Group Two kimberlites and the olivine lamproites have many modal and chemical characteristics in common. Table 11.16 gives the mean major element compositions of the three main types of kimberlites (Groups 1A, 1B and 2) and the mean major element composition of the olivine lamproites from the West Kimberly area of Western Australia. This table shows a systematic increase in silica and potassium and a concomitant decrease in calcium and carbon dioxide as one progresses from Group One B, via Group One A and Group Two kimberlites (orangeites) to the olivine lamproites. An examination of the normative compositions of these groups reveals that the orangeites and the olivine lamproites are characteristically peralkaline and contain normative aegirine and K_2SiO_3. On average the orangeites contain about 1 per cent normative aegirine, whereas the olivine lamproites contain about 3 per cent of this normative mineral. The Group One kimberlites are not normally peralkaline, exceptions include a few Group One B kimberlites (Smith *et al.* 1985, p. 270).

There is probably a gradational modal and chemical transition between the various groups of kimberlites and the olivine lamproites; also gradational changes between the olivine lamproites and the other types of lamproites. This is the main reason for grouping the various kimberlites and lamproites into the kimberlite-lamproite family. Perhaps the orangeites should be called lamproitic kimberlites to emphasise the gradational nature of the kimberlite-lamproite family. The olivine lamproites might be called ellendaleites, after the Ellendale lamproite field in the West Kimberly area of Western Australia. The type rock could be the olivine lamproite from the Ellendale Four Intrusion (Jaques *et al.*, 1986, pp. 102–111, Table 35, Number 5).

Some kimberlitic and a few lamproitic rocks contain more than 50 per cent carbonate minerals (Dawson and Hawthorne, 1973, p. 77; Jaques *et al.*, 1986, p. 130). Such rocks are carbonatites with kimberlitic or lamproitic affinities. Perhaps one should call kimberlites that contain between 20 and 50 per cent carbonate minerals carbonatitic kimberlites. There is also a gradational relationship between the aillikites and the carbonatitic kimberlites.

11.11   Authentic kimberlites

Kimberlites are ultrabasic lamprophyric rocks and the following is a summary of their usual characteristics. Olivine, or altered olivine, is the most abundant phase. It normally occurs in at least two generations. The groundmass generation is characteristically forsteritic in composition, with the forsterite component being greater than 85. Some, or all, of the larger olivine crystals (megacrysts and macrocrysts) are xenocrysts. The xenocrysts are normally anhedral or rounded. Most of them exhibit evidence of deformation, such as kink-banding or wavy extinction. Kimberlites normally contain the minerals perovskite and monticellite. They do not contain glass or modal feldspars, nepheline, leucite, quartz, jeppeite, priderite or wadeite. They commonly contain non-groundmass ilmenites that are characteristically enriched in Mg (3–23 per cent MgO), Cr (greater than 1000 ppm), and Nb (greater than 100 ppm). Kimberlites (sensu stricto) do not contain normative aegirine. Most kimberlites (excluding orangeites) contain less than 35 per cent SiO_2 and they plot outside the normal TAS diagram. Kimberlites usually have MgO/Al_2O_3 ratios of more than six. This separates them from the other lamprophyres, except for the olivine lamproites.

Kimberlites are widely distributed throughout south, central and west Africa, where they are usually found emplaced into old, stable cratonic areas, such as the Kaapvaal Craton in the south, the Congo Craton in the centre and the West African Craton in the northwest. Another major kimberlite province is the Yakutian Province. It lies within the East Siberian Craton. Kimberlites are also known from Argentina, Australia, Brazil, Canada, China, Greenland, Finland, India and the USA. The ages of kimberlites range from Proterozoic to Cenozoic. In a few areas, such as the Kaapvaal Craton, kimberlites of three different ages are known. Most kimberlites are considered to lie on deep-seated fracture zones. In the Kaapvaal Craton a single fracture system, called the Lesotho trend, appears to control the emplacement of kimberlites in a zone that is about 600 km long and up to 280 km wide. The Kimberly area of South Africa contains a high concentration of kimberlites and it is on the intersection of the Lesotho trend with two other deep-seated fracture systems. Similar structural controls are believed to have been responsible for the location of most of the kimberlites of the Yakutian Province.

A consistent feature of kimberlite distribution is their tendency to concentrate in old continental cratonic areas. Beneath such areas one characteristically finds not only old crustal rocks but also thick sections of old subcrustal lithosphere. Such segments of lithosphere are usually composed of materials that can develop deep-seated fractures that tap magmas from greater depths than normal. Within both the Kaapvaal and East Siberian cratons the compositions of the various kimberlitic and lamprophyric rocks seem to conform to a systematic pattern. At the centres of the cratons the kimberlites contain significant quantities of pyrope garnet and diamond. They are surrounded by a zone in which the kimberlites are somewhat depleted in these minerals. Beyond this zone the kimberlitic rocks do not normally carry either of these minerals, but tend to be associated with other alkaline rocks such as olivine-leucite lamproites, alkalic picrites, meimechites, olivine melilitites, alnöites or carbonatites.

Prospecting has shown that the garnets, chromites and other indicator minerals found in diamond-rich pipes and in diamonds themselves are likely to have subtly different chemical compositions from those in barren kimberlites. The garnet from a diamondiferous pipe is generally lower in calcium (less than 4 per cent) and higher in chromium than other garnets. Chromite inclusions in diamonds usually contain remarkably high chromium contents (greater than 62.5 per cent). It is likely that these phases are usually only generated in the diamond stability field of the mantle. It is also claimed that the ilmenites from diamondiferous pipes are seldom highly oxidised. It is likely that when the ilmenites became oxidised the diamonds also become oxidised and are transformed into carbon dioxide. The investigation of the composition of garnets and chromites is thus regarded as an indication of the number of diamonds that are likely to be entrained in a proto-kimberlite magma. Also the study of the composition of ilmenites may provide a measure of how many diamonds will survive the journey to the surface.

Except Mwadui in Tanzania and Orapa in Botswana, few individual kimberlite diatremes are larger than 1 km² in outcrop area. Many occupy areas of less than one hectare. They often have rounded shapes, but some, such as the Hololo kimberlite in Lesotho, have elongate shapes. In three-dimensions the form of most diatremes is that of a narrow inverted cone that tapers gently with increasing depth (Hawthorne, 1973, p. 163). At depth these diatremes usually pass into dykes (see Fig. 11.10). Dia-

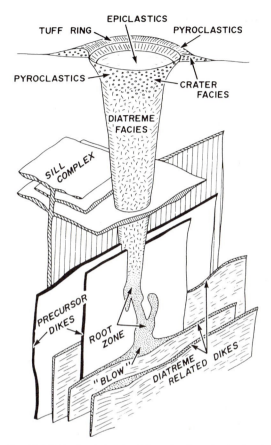

Fig. 11.10 Idealised sketch of the volcanic and subvolcanic features of a diatreme containing olivine lamproitic or kimberlitic rock. Adapted from Hawthorne, 1973, p. 163; and Mitchell, 1986.

tremes often reveal many different intrusive facies. For example, the large kimberlite body in the Kimberly Mine in South Africa contains 15 individual intrusive units (Wagner, 1914). Kimberlitic diatremes often crop out in clusters and kimberlitic dykes occur in swarms. Such dykes are usually long but narrow (about 2 m wide). A particularly long kimberlitic dyke crops out in the Winberg District of South Africa. It has been traced for over 65 km. Kimberlitic sills are much less common than kimberlitic dykes or diatremes.

Because kimberlitic magmas transport dense mantle-derived xenoliths and metastable diamond xenocrysts, it is likely that they move rapidly through the lithosphere. According to Anderson (1979) they probably moved at velocities of between 7 and 20 m/s. Experimental studies on the minerals commonly found in kimberlites show that they probably equilibrated with mantle phases at depths of at least 180 km. Haggerty (1994) has attempted to trace the source of some phases to even greater depths. He contends that some rare diamonds are plucked from the transition zone (400–670 km), whereas others nucleated at even greater depths. The evidence in support of his hypothesis is that some diamonds contain inclusions of very high-pressure minerals like majorite (Sautter *et al.*, 1991). Majorite, $Mg_3(Fe,Si,Al)_2(SiO_4)_3$, has a garnet structure, but a composition that is similar to hypersthene. Haggerty (1994, p. 66) has also speculated that other exotic inclusions found in diamonds, such as moissanite (SiC), may have crystallised in the D'' layer at the base of the mantle.

Kimberlites are a paradox as they have major element compositions that are similar to picrites or meimechites, yet they are also enriched in the incompatible elements. The origin of kimberlites has been tackled in two ways. One group of investigators has emphasised their links with the origin of the more common picrobasaltic rocks, while the other group has emphasised the importance of kimberlites in providing clues about how magmatic processes operate in the deep mantle. Wagner (1914) and some more recent authors have proposed that kimberlitic magmas were generated by a low degree of partial melting of mantle materials. This petrogenetic model has been called the incipient melting hypothesis. Experimental and geochemical data suggest that this petrogenetic model only works if the source materials, such as garnet-clinopyroxene harzburgites, had previously been enriched in the incompatible elements. This petrogenetic model seems acceptable for the generation of the unique orangeites and the olivine lamproites (Tainton and McKenzie, 1994, p. 789).

When pondering the origin of the other, globally distributed, kimberlitic rocks one has to keep in mind that each of these rocks contains a complex assemblage of phases that formed over an extraordinary range of depths and any one of these rocks might contain unique information about the nature of the deep mantle. It is thus not sufficient to suggest that they are products of the partial melting of previously enriched sub-continental lithosphere. One has to pursue their connections with the deep mantle because they appear to be the only rocks that can supply original data on this part of the planet. One should regard the Group One kimberlites as rocks that open the diamond window to the Earth's core, instead of

considering them to be rare alkaline rocks that can be generated by using the standard petrogenetic model devised to account for the origin of the potassic and ultrapotassic rocks. When considering the origin of kimberlites one also should be examining the challenging question of how the incompatible elements become concentrated in the old, sub-continental lithosphere.

Experimental studies have shown that melts of kimberlitic composition can be generated by the partial melting of suitable source materials at pressures greater than 5 GPa in the presence of carbon dioxide. Eggler and Wendlandt (1979, p. 330) have suggested that at pressures greater than this kimberlitic magma may be relatively abundant. They postulate that the rarity of kimberlites at the surface may be due to a lack of tectonic settings conducive to their ascent.

Another hypothesis that has been proposed for the origin of kimberlites is the residual liquid hypothesis (O'Hara and Yoder, 1967). It proposes that kimberlitic magma may be generated from a more normal basic or ultrabasic magma by high-pressure, fractional crystallisation. Harris and Middlemost (1970) proposed that kimberlites were generated by a two-stage process. In the first stage a tenuous magma enriched in volatiles, particularly species in the subsystem H-C-O, was generated by volatiles degassing from the deep mantle. As it moved upwards by zone melting from a depth of about 600 km incompatible elements were concentrated from a large volume of essentially solid mantle material into a small volume of magma. At higher levels (260 km) the hot, volatile and incompatible element-enriched tenuous magma induces partial melting in the mantle rocks producing a new picritic magma that is enriched in the incompatible elements. This magma is kimberlitic in composition and if it encounters a suitable deep-seated fracture system it rises to higher levels as a mechanical mixture of liquid magma, phenocrysts, xenocrysts, xenoliths and a separate low-viscosity fluid, or propellant, phase.

It now seems likely that the petrogenesis of most authentic kimberlites is a long and complex process. According to Haggerty (1994) the entrainment of the first of the many phases found in kimberlites begins at the core–mantle boundary. Outgassing from the outer liquid core triggers instability in the D'' layer resulting in the generation of a mantle plume. The plume or proto-magma begins its slow ascent through the lower mantle. As it rises the complementary processes of zone melting and wall rock reaction augment its content of incompatible and volatile elements. The proto-magma remains in equilibrium with the essential phases present in the materials through which it passes. When the initial batches of proto-magma reach a depth of about 260 km they are likely to induce a small amount of partial melting thus resulting in the production of a kimberlitic magma (Wyllie, 1980). On encountering the base of a thick block of continental lithosphere the early batches of magma probably will be forced to stop and their energy will be expended in producing metasomatism that locally enriches the sub-continental lithosphere in the incompatible and volatile elements. If batches of the kimberlitic magma evolve under an oceanic area they are likely to continue to rise to much shallower depths and in so doing their major element composition will gradually change as the magma equilibrates with the essential phases in the rocks it encounters. Eventually this magma is likely to become entrained in the major convective system that operates beneath the oceanic lithosphere. Such incompatible enriched pockets are likely to help in the production of alkaline rocks on intraplate oceanic island and seamounts.

Occasionally some kimberlitic magma impinges on the base of a thick block of continental lithosphere. This may cause structural adjustments in the overlying cooler and more rigid rocks. If this results in fracturing, the magma and any ambient fluid phase will move into the fractures. As the fracture system develops, successive batches of magma will be mobilised from deeper within the upper mantle. Eventually magmas of authentic kimberlitic composition will move into the vent system. This material is a quasi-magma that has a fluid propellant mixed with partly crystallised alkaline picritic silicate liquid and a range of xenocrysts and xenoliths. Movement through the lithosphere is likely to be rapid in those kimberlitic magmas that successfully carry diamonds to the surface. A long residence time in the lithosphere probably results in the resorption of many diamonds and other xenocrysts.

Source rocks for diamonds

According to Janse and Sheahan (1995) about 5000 kimberlitic and olivine lamproitic bodies have been discovered. Only 500 of them are diamondiferous, 50 have been mined for diamonds and 15 are currently being mined. Although these bodies are now the principal source of diamonds, they are not the primary source of the diamonds they contain (Helmstaedt and Gurney, 1995, p. 126). Diamonds are normally older than the kimberlitic or lamproitic rocks that host them. Most diamonds seem to have formed at an early stage in the development of the lithospheric mantle. They subsequently survived in cool lithospheric keels beneath Precambrian, particularly Archean, cratons. In southern Africa old diamonds from the lithospheric keel beneath the Kaapvaal Craton became entrained in proto-kimberlitic and proto-lamproitic magmas that erupted at various times between the Late Archean and the Cenozoic. A similar petrogenetic model has been proposed for the diamonds found in the kimberlitic rocks of the Siberian Platform (Spetsius, 1955). According to Helmstaedt and Gurney (1995, p. 141) the primary source rocks for most diamonds are low-calcium garnet harzburgites, or less frequently eclogites (see Section 3.4), but they do acknowledge that some diamonds found in a few unusual Group One kimberlites may be derived from deeper sources within the mantle. Bulanova (1995) has published a detailed review on the physical characteristics and origin of commercial-sized natural diamonds. He concluded that the diamonds found in kimberlitic rocks are likely to crystallise in a variety of mantle rock sources.

Diamonds, or graphitised diamonds, have also been reported from peridotites and pyroxenites in obduction complexes. The host rocks are usually either harzburgites, such as those from the Luobusa and Donqiao ophiolites of Tibet, or garnet pyroxenites, such as those found in the Beni Bousera and Ronda ophiolites of northern Morocco and southern Spain. Both of the Tibetan ophiolites are dismembered remnants of oceanic lithosphere. They occur within the suture zone marking the line of collision between Indian and Eurasian plates. In the Beni Bousera and Ronda ophiolites the diamonds are graphitised and they occur in garnet-pyroxene cumulate rocks that are probably derived from a magma that was generated within a subduction system (Nixon, 1995, p. 61).

Small quantities of diamond have been reported from various carbonatitic, lamprophyric and ultrapotassic rocks. It seems that the occurrence of diamonds in volcanic and subvolcanic rocks is essentially linked to the ability of their eruptive mechanisms to transport diamonds from where they occur naturally to the surface. Magmas of carbonatitic, lamprophyric and ultrapotassic composition are likely to fill this role, with magmas of olivine lamproitic and kimberlitic composition forming at the greatest depths. When rising to the surface they are the most likely to pass through rocks belonging to the diamond facies. Their eruptive mechanism is likely to promote the disaggregation of the diamond-bearing mantle rocks and provide a suitable method for transporting diamonds to the surface.

11.12 Petrogenesis of the alkaline rocks

This Chapter has shown that no single primary magma, or simple petrogenetic model, can produce all the alkaline rocks. This is because: (1) they crop out in most tectonic settings; (2) they belong to several petrological associations; (3) they have extremely diverse sodium/potassium ratios; and (4) rocks from each principal group of alkaline rocks are likely to contain different amounts of alkali metals. Many alkaline rocks, particularly those that are sodic and silica undersaturated, crop out on oceanic islands and seamounts. It is thus self-evident that at least some of these rocks evolved from magmas that were generated in source areas far removed from direct contact with the continental lithosphere. Excellent examples of oceanic intraplate alkaline magmatism are provided by the well documented, post-erosional basanites, nephelinites and nepheline melilitites from the Hawaiian Province (Clague, 1987, pp. 237–238).

It is usually suggested that the silica undersaturated alkaline rocks evolved from primary magmas generated by partial melting in the upper mantle. While the common sodic alkaline magmas are likely to be produced by partial melting of normal mantle materials, many potassic and ultrapotassic, silica undersaturated magmas are generated by the partial melting of subcontinental lithospheric materials that had been extensively modified by alkali metasomatism. According to Bailey (1987, p. 1) 'alkaline magmatism provides the best prima facie evidence of metasomatism and open-system conditions in the upper mantle'. This statement is particularly apposite when it is applied to the leucite- and kalsilite-bearing ultrapotassic rocks.

Arculus (1975) and others have shown that many primitive alkaline magmas can only be generated by partial melting in the upper mantle if carbon dioxide is present as the major volatile component. It is now generally assumed that alkali picrobasalts, basanites, melanephelinites, melilitites and various alkaline lamprophyric rocks, including the olivine lamproites and kimberlites, evolved from mantle-derived magmas. Most comendites, peralkaline trachytes and phonolites are produced by magmatic differentiation, possibly aided by processes such as wall-rock assimilation. Silica undersaturated magmas of the alkaline suite usually follow liquid lines of descent that converge on the phonolite field as predicted by experimental investigations that have targeted Petrogeny's residua system (Bowen, 1937; see Chapter 10). This system is too simple to simulate completely the differentiation trends of the diverse silica undersaturated alkaline magmas. Important reasons why this system is inadequate are: (1) peralkaline magmas characteristically precipitate sodic amphiboles and pyroxenes that are omitted from the simple residua system; and (2) the buoyancy of most feldspathoids in their parental magmas. The latter effect produces a special group of foid-cumulates that have high differentiation indices.

It is interesting to evaluate the origin of the primitive alkaline rocks that occur on oceanic islands and seamounts. Their chemical composition is usually more like that of the rocks of the small-volume, continental magmatic provinces than typical mid-oceanic ridge basalts. According to McKenzie and O'Nions (1995, p. 154) these primitive alkaline rocks frequently contain isotopic signatures that identify them as containing a component that was once part of the continental lithosphere. To explain how this happened, one might suggest that material from the base of the continental lithosphere was thermally eroded and entrained into a large-scale convective current operating in the upper mantle. Some entrained material is likely to have remained trapped for several convective cycles. If partially melted this fertilised mantle material could generate the magmas responsible for producing the oceanic island basalts and other less abundant alkaline rocks that erupt in oceanic intraplate settings.

Most Phanerozoic continental alkaline rocks are associated with large-scale doming and accompanying rifting. Some important characteristics of this type of magmatic activity include: (1) repeated alkaline magmatism in the same general area; (2) the presence of metasomatised mantle xenoliths in the more primitive rocks; and (3) the extrusion and intrusion of magmatic materials that are enriched in volatile components, particularly species of the subsystem H-C-O, the halogens and the incompatible elements. Some petrologists, like Bailey (1983), have suggested that on a global scale alkaline magmatism is promoted by the release of volatile-charged magmas from deep sources in the mantle. Deep-seated fractures that cut through the lithosphere act as channels for the movement of these magmas. Once in the lithosphere the composition of these magmas is likely to be controlled by wall-rock reactions and polybaric fractional crystallisation. Spera (1981, p. 62) has estimated that the amount of heat now being transported by low-viscosity fluids fluxing through the mantle is probably sufficient to trigger all current alkaline magmatism.

In the future this type of alkaline magmatism may become the dominant magmatic activity on Earth (see Chapter 13). This prediction is based on a simple model of a cooling planet. Currently most of the Earth's interior heat is dissipated in the upper mantle. Interior heat is continually being transferred to an incoming stream of cooler slabs of subducted lithosphere. Eventually mid-oceanic ridge magmatism will stop and subduction will grind to a halt. Heat from the deep mantle and core will then start to build up in the upper mantle. It may result in the development of extensive domes and rift systems, with the rifts becoming a focus for volcano-magmatic activity. Gradually as the Earth cools and the lithosphere thickens the primary seat of partial melting will be forced to move to successively deeper levels in the mantle. This will cause a shift in the composition of the primary magma being generated. During the initial stage the magma is likely to be picrobasaltic, alkali basaltic or basanitic in composition. Gradually the compositions of the primary magma will become more unusual, with olivine lamproitic and kimberlitic becoming more common.

This proposal is indirectly supported by the Earth's petrological record. It shows that the relative proportion of alkaline rocks has increased with time. Planetary petrologists are also interested in the future petrological development of the Earth. Currently they are eagerly awaiting more data on the magmatic rocks of Mars because they want to know more about magmatic processes and products on a cooling planet that has already passed through its crustal spreading and subduction stage. If a large volume of the younger magmatic rocks of Mars are alkaline this discovery will add strong support to the petrological predictions that have been presented. On the other hand, if a substantial proportion of these rocks are not alkaline this will mean that the quest for a model that explains the petrological evolution of the terrestrial planets still has a long way to go.

Any discussion of the origin and evolution of the alkaline rocks would be incomplete without a report on the carbonatites and their consanguineous alkaline magmatic and metasomatic rocks. The next chapter examines these rare silicate and non-silicate rocks. It is most important to try to understand these unusual and often bizarre rocks because they provide novel insights into: (1) the nature of the upper mantle; (2) several unusual magmatic differentiation processes, and (3) the often neglected petrological process of alkali metasomatism.

References

Allen, J. B. and Deans, T., 1965. Ultrabasic eruptive with alnöitic-kimberlitic affinities from Malaita, Solomon Islands. *Mineralogical Magazine*, Tilley Volume, 34, pp. 16–34.

Anderson, O.L., 1979. The role of fracture dynamics in kimberlite pipe formation. In Boyd, F.R. and Meyer, H.O.A. (editors), *Kimberlites, Diatremes and Diamonds: Their Geology, Petrology and Geochemistry*, pp. 344–353, Proceedings of the Second International Kimberlite Conference, American Geophysical Union, Washington, D.C., 400 pp.

Appleton, J.D., 1972. Petrogenesis of potassium rich lavas from the Roccamonfina Volcano, Roman Region, Italy. *Journal of Petrology*, 13, pp. 425–456.

Arculus, R.J., 1975. Melting behaviour of two basanites in the range 10–35 kbar and the effect of TiO_2 on the olivine-diopside reactions at high pressures. *Yearbook Carnegie Institution Washington*, 74, pp. 512–515.

Arndt, N., Lehnert, K. and Vasil'ev, Y., 1995. Meimechites: highly magnesian lithosphere-contaminated alkaline magmas from deep subcontinental mantle. *Lithos*, 34, pp. 41–59.

Atkinson, W.J., Hughes, F.E. and Smith, C.B., 1984. A review of the kimberlite rocks of Western Australia. In Kornprobst, J. (editor), *Kimberlites 1: Kimberlites and Related Rocks*, pp. 195–224, Elsevier, Amsterdam, 466 pp.

Bailey, D.K., 1974. Continental rifting and alkaline magmatism. In Sørenson, H. (editor), *The Alkaline Rocks*, pp. 148–159, John Wiley and Sons, London, 622 pp.

Bailey, D.K., 1983. The chemical and thermal evolution of rifts. *Tectonophysics*, 94, pp. 585–597.

Bailey, D.K., 1987. Mantle metasomatism – perspective and prospect. In Fitton, J.G. and Upton, B.G.J. (editors), *Alkaline Igneous Rocks*. pp. 1–13, Geological

Society of London Special Publication Number 30, 568 pp.

Baker, P.E., Gass, I.G., Harris, P.G. and Le Maitre, R.W., 1964. The volcanological report of the Royal Society expedition to Tristan da Cunha, 1962. *Philosophical Transactions of the Royal Society of London*, 256, pp. 439–578.

Barker, D.S. and Young, K.P., 1979. A marine Cretaceous nepheline basanite volcano at Austin, Texas. *Texas Journal of Science*, 31, pp. 5–24.

Bates, R.L. and Jackson, J.A. (editors) 1987. *Glossary of Geology*. Third Edition, American Geological Institute, Alexandria, Virginia, USA, 788 pp.

Bell, K. and Powell, J.L., 1969. Strontium isotope studies of alkaline rocks: the potassium rich lavas of Birunga and Toro-Ankole regions, east and central equatorial Africa. *Journal of Petrology*, 10, pp. 536–572.

Berg, G.W. and Allsopp, H.L., 1972. Low $^{87}Sr/^{86}Sr$ ratios in fresh South African kimberlites. *Earth and Planetary Science Letters*, 16, pp. 27–30.

Bergman, S.C. 1987. Lamproites and other potassium-rich igneous rocks: a review of their occurrence, mineralogy and geochemistry. In Fitton, J.G. and Upton, B.G.J. (editors), *Alkaline Igneous Rocks*, pp. 103–190, Geological Society of London Special Publication Number 30, 568 pp.

Blundell, D., Freeman, R. and Mueller, S. (editors), 1992. *A Continent Revealed: The European Geotraverse*. Cambridge University Press, Cambridge, 275 pp.

Borley, G.D., 1974. Oceanic islands. In Sørensen, H. (editor), *The Alkaline Rocks*, pp. 311–330, John Wiley and Sons, 622 pp.

Bowen, N.L., 1928. *The evolution of the igneous rocks*. Princeton University Press, Princeton, New Jersey, 334 pp.

Bowen, N.L., 1937. Recent high-temperature research on silicates and its significance in igneous geology. *American Journal of Science*, 33, pp. 1–21.

Brey, G., 1978. Origins of Olivine Melilitites – Chemical and Experimental Constraints. *Journal of Volcanology and Geothermal Research*, 3, pp. 61–88.

Bulanova, G.P., 1995. The formation of diamond. *Journal of Geochemical Exploration*, 53, pp. 1–23.

Carmichael, I.S.E., 1967. The mineralogy and petrology of the volcanic rocks from the Leucite Hills, Wyoming. *Contributions to Mineralogy and Petrology*, 15, pp. 24–66.

Cattermole, P., 1994. *Venus: The Geological Story*. UCL Press Ltd, London, 250 pp.

Clague, D.A., 1987. Hawaiian alkaline volcanism. In Fitton, J.G. and Upton, B.G.J. (editors), *Alkaline Igneous Rocks*, pp. 227–252, Geological Society of London Special Publication, Number 30, pp. 227–252.

Clague, D.A. and Frey, F.A., 1982. Petrology and trace element geochemistry of the Honolulu Volcanics, Oahu: implications of the oceanic mantle beneath Hawaii. *Journal of Petrology*, 23, pp. 447–504.

Cornelissen, A.K. and Verwoerd, W.M., 1975. The Bushmanland kimberlites and related rocks. *Physics and Chemistry of the Earth*, 9, pp. 71–80.

Cross, W., Iddings, J.P., Pirsson, L.V. and Washington, H.S., 1902. A quantitative chemico-mineralogical classification and nomenclature of igneous rocks. *Journal of Geology*, 10, pp. 555–690.

Cundari, A., 1973. Petrology of the leucite-bearing lavas in New South Wales. *Journal of the Geological Society of Australia*, 20 (4), pp. 465–492.

Dawson, J.B. 1987. The kimberlite clan: relationships with olivine and leucite lamproites, and inferences for upper-mantle metasomatism. In Fitton, J.G. and Upton, B.G.J. (editors), *Alkaline Igneous Rocks*, pp. 95–101, Geological Society of London Special Publication Number 30, 568 pp.

Dawson, J.B. and Hawthorne, J.B.1973. Magmatic sedimentation and carbonatitic differentiation in kimberlite sills at Benfontein, South Africa. *Journal of the Geological Society of London*, 129, pp. 61–85.

Dautria, J.M., Dupuy, C., Takherist, D. and Dostal, J., 1992. Carbonate metasomatism in the lithospheric mantle: peridotitic xenoliths from a melilititic district of the Sahara basin. *Contributions to Mineralogy and Petrology*, 111, pp. 37–52.

Downes, H., 1987. Tertiary and Quaternary volcanism in the Massif Central, France. In Fitton, J.G. and Upton, B.G.J. (editors), *Alkaline Igneous Rocks*, pp. 517–530, Geological Society of London, Special Publication, 30, 568 pp.

Edgar, A.D., 1991. Source regions for ultrapotassic mafic-ultramafic magmatism in southwestern Uganda region of the African Rift: implications from experimental studies. In Kampunzu, A.B. and Lubala, R.T. (editors) *Magmatism in Extensional Structural Settings: the Phanerozoic African Plate*, pp. 73–84. Springer-Verlag, Berlin, 637 pp.

Eggler, D.H. and Wendlandt, R.F., 1979. Experimental studies on the relationship between kimberlite magmas and partial melting of peridotite. In Boyd, F.R. and Meyer, H.O.A. (editors), *Kimberlites, Diatremes and Diamonds: Their Geology, Petrology and Geochemistry*. Proceedings of the Second International Kimberlite

Conference, Volume 1, pp. 330–338, American Geophysical Union, Washington D.C., 400 pp.

Erlank, A.J. and Finger, L.W., 1970. The occurrence of potassic richterite in mica nodule from the Wesselton kimberlite, South Africa. *Yearbook of the Carnegie Institution of Washington*, 68, pp. 442–443.

Fuster, J.M., Arana, V., Brandle, J.L., Navarro, M., Alonso, U. and Aparicio, A., 1968. *Geology and Volcanology of the Canary Islands: Tenerife*. Instituto 'Lucas Mallada', Madrid, Spain, 218 pp.

Giannetti, B. and Ellam, R., 1994. The primitive lavas of Roccamonfina volcano, Roman region, Italy: new constraints on melting processes and source mineralogy. *Contributions to Mineralogy and Petrology*, 116, pp. 21–31.

Goles, G.G., 1976. Some constraints on the origin of phonolites from the Gregory Rift, Kenya, and inferences concerning basaltic magmas in the Rift System. *Lithos*, 9, pp. 1–8.

Gregory, J.W., 1900. Contributions to the geology of British East Africa. Part 2, The geology of Mount Kenya. *Quarterly Journal of the Geological Society of London*, 56, pp. 205–222.

Gregory, J.W., 1921. *Rift Valleys and Geology of Africa*. Seeley, Service and Company Limited, London.

Gümbel, C.W. von, 1874. *Die paläolithischem Eruptivgesteine des Fichtelgebirges*. Gotha, München, Germany.

Gupta, A.K. and Yagi, K., 1980. *Petrology and Genesis of Leucite-Bearing Rocks*. Springer-Verlag, Berlin, 252 pp.

Haggerty, S.E., 1994. Superkimberlites: a geodynamic diamond window to the Earth's core. *Earth and Planetary Science Letters*, 122, pp. 57–69.

Harris, P.G. and Middlemost, E.A.K., 1970. The evolution of kimberlites. *Lithos*, 3, pp. 79–88.

Hawthorne, J.B., 1973. Model of a kimberlite pipe. In Ahrens, L.H., Dawson, J.B., Duncan, A.R. and Erlank, A.J. (editors), *International Conference on Kimberlites*, pp. 163–166, University of Cape Town, South Africa.

Hay, D.E., Wendlandt, R.F. and Wendlandt, E.D., 1995. The origin of Kenya rift plateau-type phonolites: evidence from geochemical studies for fusion of lower crust modified by alkali basaltic magmatism. *Journal of Geophysical Research*, 100 (B1), pp. 411–422.

Helmstaedt, H.H. and Gurney, J.J., 1995. Geotectonic controls of primary diamond deposits: implications for area selection. *Journal of Geochemical Exploration*, 53, pp. 125–144.

Hoernle, K. and Schmincke, H.U., 1993. The petrology of the tholeiites through melilite nephelinites on Gran Canaria, Canary Islands: crystal fractionation, accumulation and depths of melting. *Journal of Petrology*, 34, pp. 573–597.

Holmes, A., 1937. The petrology of Katungite. *Geological Magazine*, 74, pp. 200–219.

Holmes, A., 1942. A suite of volcanic rocks from south-west Uganda containing kalsilite (a polymorph of $KAlSiO_4$). *Mineralogical Magazine*, 26, pp. 197–217.

Holmes A. and Harwood, H.F., 1937. The petrology of the volcanic area of Bufumbira. *Memoirs of the Geological Survey of Uganda*, Entebbe, 3 (2), pp. 1–300.

Jackson, A.J.A. and Wright, T.L., 1970. Xenoliths in the Honolulu volcanic series, Hawaii. *Journal of Petrology*, 11, pp. 405–430.

Janse, A.J.A. and Sheahan, P.A., 1995. Catalogue of world wide diamond and kimberlite occurrences: a selective and annotative approach. *Journal of Geochemical Exploration*, 53, pp. 73–111.

Jaques, A.L., Lewis, J.D., Smith, C.B., Gregory, G.P., Ferguson, J., Chappell, B.W. and McCulloch, M.T. 1984. The diamond-bearing ultrapotassic (lamproitic) rocks of the West Kimberly Region, Western Australia. In Kornprobst, J. (editor) 1984, *Kimberlites: 1: Kimberlites and Related Rocks*, pp. 225–254, Proceedings of the Third International Kimberlite Conference, Claremont-Ferrand, Elsevier, Amsterdam, 466 pp.

Jaques, A.L., Lewis, J.D. and Smith, C.B. 1986. *The Kimberlites and Lamproites of Western Australia*, Geological Survey of Western Australia. Bulletin 132, Government Printing Office, Perth, 267 pp.

Johannsen, A., 1938. *A Descriptive Petrography of the Igneous Rocks*. Volume 4, Part 1, The feldspathoid rocks and Part 2 The Peridotites and Perknites, The University of Chicago Press, Chicago, 523 pp.

Johnson, R.W. (editor) 1989. *Intraplate Volcanism in Eastern Australia and New Zealand*. Cambridge University Press, Cambridge, 408 pp.

Kamenetsky, V.S., Sobolev, A.V., Joron, J.L. and Semet, M.P., 1995. Petrology and geochemistry of Cretaceous ultramafic volcanics from Eastern Kamchatka. *Journal of Petrology*, 36 (3), pp. 637–662.

Keller, G.R., Khan, M.A., Morgan, P., Wendlandt, R.F., Baldridge, W.S., Olsen, K.H., Prodehl, C. and Braile, L.W., 1991. A comparative study of the Rio Grande and Kenya rifts. *Tectonophysics*, 197, pp. 355–372.

Keller, G.R., Mechie, J., Braile, L.W., Mooney, W.D. and Prodehl, C., 1994. Seismic structure of the uppermost mantle beneath the Kenya Rift. *Tectonophysics*, 236, pp. 201–216.

Kjarsgaard, B.A., Hamilton, D.L. and Peterson, T.D., 1995. Peralkaline nepheline/carbonatite liquid immiscibility: comparison of phase compositions in experiments and natural lavas from Oldoinyo Lengai. In Bell, K. and Keller, J. (editors), *Carbonatite Volcanism: Oldoinyo Lengai and the Petrogenesis of Natrocarbonatites*, pp. 163–190, Springer-Verlag, Berlin, 210 pp.

Kogarko, L.N., Kononova, V.A., Orlova, M.P. and Woolley, A.R., 1995. *Alkaline Rocks and Carbonatites of the World. Part Two: Former USSR*. Chapman and Hall, London, 226 pp.

Kukharenko, A.A., Orlova, M.P., Bulakh, A.G., Bagdasarov, E.A., Rimskaya-Korsakova, O.M., Nephedov, E.I., Ilinskii, G.A., Ergeev, A.S. and Abakumova, N.B., 1965. *The Caledonian Complex of Ultrabasic Alkaline Rocks and Carbonatites of the Kola Peninsula and North Karelia*. Nedra, Moscow, 772 pp.

Kushiro, I., Satake, H. and Akimoto, S., 1975. Carbonate-silicate reactions at high pressure and possible presence of dolomite and magnesite in the upper mantle. *Earth and Planetary Science Letters*, 28, p. 116–120.

Kyle, P.R., 1981. Mineralogy and geochemistry of a basanite to phonolite sequence at Hut Point Peninsula, Antarctica, based on core from Dry Valley Drilling Project drill holes. *Journal of Petrology*, 22 (4), pp. 451–500.

Kyle, P.R. (editor), 1995. *Volcanological and Environmental Studies of Mount Erebus, Antarctica*. American Geophysical Union, Antarctic Research Series, 66, 162 pp.

Lacroix, A., 1917. Les laves a hauyne d'Auvergne et leurs enclaves homogenes: importance theorique de ces dernieres. *Compte Redu Hebdomadaire des Seances de l'Academie des Sciences*, Paris, 164, pp. 581–588.

Larouziere, F.D. de, 1989. *Dictionnaire des roches d'origine magmatique*. Manuels et Methodes, 20, Editions du BRGM, Orleans, France, 188 pp.

Le Bas, M.J., 1973. A norm for feldspathoidal and melilitic igneous rocks. *The Journal of Geology*, 81 (1), pp. 89–96.

Le Bas, M.J., 1977. *Carbonatite-Nephelinite Volcanism: An African Case History*. John Wiley and Sons, London, 347 pp.

Le Bas, M.J., 1980. Alkaline magmatism and uplift of continental crust. *Proceedings of the Geologists' Association* (London), 91(1 and 2), pp. 33–38.

Le Bas, M.J., 1989. Nephelinitic and basanitic rocks. *Journal of Petrology*, 30 (5), pp. 1299–1312.

Le Maitre, R.W., 1962. Petrology of volcanic rocks, Gough Island, South Atlantic. *Bulletin of the Geological Society of America*, 73, pp. 1309–1340.

Le Maitre, R.W., 1976. The chemical variability of some common igneous rocks. *Journal of Petrology*, 17 (4), pp. 589–637.

Le Maitre, R.W., 1984. A proposal by the IUGS Subcommission on the Systematics of Igneous Rocks for a chemical classification of volcanic rocks based on the total alkali silica (TAS) diagram. *Australian Journal of Earth Sciences*, 31 (2), pp. 243–255.

Le Maitre, R.W., Bateman, P., Dubek, A., Keller, J., Lameyre, J., Le Bas, M.J., Sabine, P.A., Schmid, R., Sørensen, H., Streckeisen, A., Woolley, A.R. and Zanettin, B. 1989. *A Classification of Igneous Rocks and Glossary of Terms: Recommendations of the IUGS Subcommission on the Systematics of Igneous Rocks*. Blackwell Scientific Publications, Oxford, 193 pp.

Lippard, S.J., 1973. The petrology of phonolites from the Kenya Rift. *Lithos*, 6, pp. 217–234.

Lloyd, F.E., Huntingdon, A.T., Davies, G.R. and Nixon, P.H., 1991. Phanerozoic volcanism of southwest Uganda: a case for regional and LILE enrichment of the lithosphere beneath a domed and rifted continental plate. In Kampunzu, A.B. and Lubala, R.T. (editors) *Magmatism in Extensional Structural Settings: The Phanerozoic African Plate*, pp. 23–72, Springer-Verlag, Berlin, 637 pp.

Maaloe, S., James, D., Smedley, P., Petersen, S. and Garmann, L.B., 1992. The Koloa volcanic suite of Kauai, Hawaii. *Journal of Petrology*, 33, pp. 761–784.

Macdonald, G.A., 1968. Composition and origin of Hawaiian lavas. In Coats, R.R., Hay, R.L. and Anderson, C.A. (editors), *Studies in Volcanology: A Memoir in Honour of Howel Williams*, pp. 477–522, The Geological Society of America, Memoir 116, 678 pp.

Maury, R.C. and Varet, J., 1980. Le volcanisme tertiaire et quaternaire en France. In Autran, A. and Dercourt, J. (editors), pp. 137–159, *Evolutions geologiques de la France*, memoire du BRGM number 107–1980, 354 pp.

McBirney A.R. and Aoki, K., 1968. Petrology of the Island of Tahiti. *Geological Society of America*, Memoir 116, pp. 523–556.

McIver, J.R. and Ferguson, J. 1979. Kimberlitic, melilitic, trachytic and carbonatite eruptives at Saltpetre Kop, Sutherland, South Africa. In Boyd, F.R. and Meyer, H.O.A. (editors) *Kimberlites, Diatremes, and Diamonds: Their Geology, Petrology, and Geochemistry*, pp. 111–128, Proceedings of the Second International Kimberlite Conference, American Geophysical Union, Washington, 400 pp.

McKenzie, D. and O'Nions, R.K., 1995. The source regions of ocean island basalts. *Journal of Petrology*, 36(1), pp. 133–159.

Mechie, J., Fuchs, K. and Altherr, R., 1994. The relationship between seismic velocity, mineral composition and temperature and pressure in the upper mantle – with an application to the Kenya Rift and its eastern flank. *Tectonophysics*, 236, pp. 453–464.

Metais, D. and Chayes, F., 1963. Varieties of lamprophyres. *Carnegie Institution of Washington Yearbook*, 63, pp. 196–199.

Mitchell, R.H. 1986. *Kimberlites: Mineralogy, Geochemistry and Petrology*. Plenum Press, New York, 442 pp.

Mitchell, R.H. and Bergman, S.C., 1991. *Petrology of Lamproites*. Plenum Press, New York, 447 pp.

Mitchell, R.H., Smith, C.B. and Vladykin, N.V., 1994. Isotopic composition of strontium and neodymium in potassic rocks of the Little Murun complex, Aldan Shield, Siberia. *Lithos*, 32, pp. 243–248.

Mitchell-Thome, R.C., 1970. *Geology of the South Atlantic Islands*. Gebrüder Borntraeger, Berlin, 367 pp.

Moore, A.E. 1983. A note on the occurrence of melilite in kimberlite and olivine melilitites. *Mineralogical Magazine and Journal of the Mineralogical Society*, London, 47, pp. 404–406.

Morris, E.M. and Pasteris, J.D. (editors), 1987. *Mantle Metasomatism and Alkaline Magmatism*. Special Paper of the Geological Society of America, 215, 383 pp.

Morris, E.M. and Pasteris, J.D. 1987. Prologue to the symposium on Alkalic Rocks and Kimberlites. In Morris, E.M. and Pasteris, J.D. (editors), *Mantle Metasomatism and Alkaline Magmatism*, pp. 1–4, Special Paper of the Geological Society of America, 215, 383 pp.

Niggli, P. 1923. *Gesteins und Mineralprovinzen*. Gebrüder Borntraeger, Berlin, Volume 1, 602 pp.

Nixon, P.H., 1995. The morphology and nature of primary diamondiferous occurrences. *Journal of Geochemical Exploration*, 53, pp. 41–71.

Nockolds, S.R., 1954. Average chemical composition of some igneous rocks. *Bulletin of the Geological Society of America*, 65, pp. 1007–1032.

O'Hara, M.J. and Yoder, H.S., 1967. Formation and fractionation of basic magmas at high pressure. *Scottish Journal of Geology*, 3, pp. 67–117.

Peccerillo, A., 1992. Potassic and ultrapotassic rocks: compositional characteristics, petrogenesis, and geological significance. *Episodes*, 15 (4), pp. 243–251.

Peccerillo, A. and Manetti, P., 1985. The potassic alkaline volcanism of central southern Italy: a review of the data relevant to petrogenesis and geodynamic significance. *Transaction of the Geological Society of South Africa*, 88, pp. 379–394.

Peterson, T.D. and Kjarsgaard, B.A., 1995. What are the parental magmas at Oldoinyo Lengai? In Bell, K. and Keller, J. (editors), *Carbonatite Volcanism; Oldoinyo Lengai and the Petrogenesis of Natrocarbonatites*, pp. 148–162, Springer-Verlag, Berlin, 210 pp.

Prider, R.T., 1960. The leucite lamproites of the Fitzroy Basin, Western Australia. *Journal of the Geological Society of Australia*, 6, pp. 71–118.

Rittmann, A., 1973. *Stable Mineral Assemblages of Igneous Rocks*. Springer-Verlag, Berlin, 262 pp.

Rock, N.M.S., 1986. The nature and origin of ultramafic lamprophyres: alnöites and allied rocks. *Journal of Petrology*, 27 (1), pp. 155–196.

Rock, N.M.S., 1987. The nature and origin of lamprophyres: an overview. pp. 191–226, in Fitton, J.G. and Upton, B.G.J. (editors), *Alkaline Igneous Rocks*. Geological Society of London Special Publication Number 30, 568 pp.

Rock, N.M.S., 1991. *Lamprophyres*. Blackie, Glasgow, 285 pp.

Rogers, N.W., De Mulder, M. and Hawkesworth, C.J., 1992. An enriched mantle source for potassium basanites: evidence from Karisimbi Volcano, Virunga Volcanic Province, Rwanda. Contributions to *Mineralogy and Petrology*, 11, pp.543–556.

Russell, H.D., Heimstra, S.A. and Groeneveld, D., 1955. The mineralogy and petrology of the carbonatite at Loolekop, Eastern Transvaal. *Transactions and Proceedings of the Geological Society of South Africa*, 57, pp. 197–208.

Sahama, T.G., 1974. Potassium-rich alkaline rocks. In Sørensen, H. (editor) *The Alkaline Rocks*, pp. 96–109, Wiley, London, 622 pp.

Sautter, V., Haggerty, S.E. and Field, S., 1991. Ultra-deep (> 300 km) ultramafic xenoliths: new petrologic evidence from the transition zone. *Science*, 252, pp. 827–830.

Scarfe, C.M., Luth, W.C. and Tuttle, O.F., 1966. An experimental study bearing on the absence of leucite in plutonic rocks. *American Mineralogist*, 51, pp. 726–735.

Scott Smith, B.H. and Skinner, E.M.W., 1984. A new look at Prairie Creek, Arkansas. In Kornprobst, J. (editor), *Kimberlites 1: Kimberlites and Related Rocks*, pp. 255–283, Elsevier, Amsterdam, 466 pp.

Shand, S.J., 1950. *Eruptive Rocks, Their Genesis, Composition, Classification, and Their Relation to Ore-*

Deposits with a Chapter on Meteorites. (Fourth Edition). Thomas Murby and Company, 488 pp.

Skinner, E.M.W., 1989. Contrasting Group 1 and Group 2 kimberlite petrology: towards a genetic model for kimberlites. In Ross, J. and others, *Kimberlites and Related Rocks*, pp. 528–544, Proceedings of the Fourth International Kimberlite Conference, Volume 1, Geological Society of Australia, Special Publication Number 14, 646 pp.

Smith, C.B., 1983. Pb, Sr and Nd isotopic evidence for sources of southern African Cretaceous kimberlites. *Nature*, 304, pp. 51–54.

Smith, C.B., Gurney, J.J., Skinner, E.M.W., Clement, C.R. and Ebrahim, N. 1985. Geochemical character of southern African kimberlites; a new approach based on isotopic constraints. *Transactions of the Geological Society of South Africa*, 88, pp. 267–280.

Sørensen, H. (editor), 1974a. *The Alkaline Rocks.* John Wiley and Sons, 622 pp.

Sørensen, H., 1974b. Introduction. In Sørensen, H. (editor), *The Alkaline Rocks*, pp. 3–11, John Wiley and Sons, 622 pp.

Sørensen, H., 1974c. Alkali syenites, feldspathoidal syenites and related lavas. In Sørensen, H. (editor), *The Alkaline Rocks*, pp. 22–52, John Wiley and Sons, 622 pp.

Spencer, A.B., 1969. Alkalic Igneous Rocks of the Balcones Province, Texas. *Journal of Petrology*, 10 (2), pp. 272–306.

Spera, F.J., 1981. Carbon dioxide in igneous petrogenesis: 2. Fluid dynamics of mantle metasomatism. *Contributions to Mineralogy and Petrology*, 77, pp. 56–65.

Spetsius, Z.V., 1955. Occurrence of diamond in the mantle: a case study from the Siberian Platform. *Journal of Geochemical Exploration*, 53, pp. 25–39.

Stoppa, F. and Lavecchia, G., 1992. Late Pleistocene ultra-alkaline magmatic activity in the Umbria-Latium region (Italy): An overview. *Journal of Volcanology and Geothermal Research*, 52, pp. 277–293.

Streckeisen, A.L., 1978. IUGS Subcommission on the Systematics of Igneous Rocks: Classification and Nomenclature of Volcanic Rocks, Lamprophyres, Carbonatites and Melilitic Rocks: Recommendations and Suggestions. *Neues Jahrbuch fur Mineralogie*, Stuttgart, Abhandlungen, 143, pp. 1–14.

Tainton, K.M. and McKenzie, D., 1994. The generation of kimberlites, lamproites, and their source rocks. *Journal of Petrology*, 35 (3), pp. 787–817.

Taljaard, M.S., 1936. South African melilite basalts and their relations. *Transactions and Proceedings of the Geological Society of Africa*, 39, pp. 281–316.

Villemant, B., Trigila, R. and De Vivo, B., 1993. Geochemistry of Vesuvius volcanics during 1631–1944 period. *Journal of Volcanology and Geothermal Research*, 58, pp. 291–313.

Wagner, P.A., 1914. The Diamond Fields of Southern Africa. *The Transvaal Leader*, Johannesburg, South Africa, 355 pp.

Wagner, P.A., 1928. The evidence of kimberlite pipes on the constitution of the outer parts of the Earth. *South Africa Journal of Science*, 25, pp. 27–48.

Washington, H.S., 1906. *The Roman Comagmatic Region.* Carnegie Institution of Washington, Publication Number 57, 140 pp.

Watt, W.S., 1966. Chemical analyses from the Gadar igneous province, South Greenland. *Rapport Grønlands Geologiske Undersøgels*, 6, 92 pp.

Weaver, B.L., Wood, D.A., Tarney, J. and Joron, J.L., 1987. Geochemistry of ocean island basalts from the South Atlantic: Ascension, Bouvet, St. Helena, Gough and Tristan da Cunha. In Fitton, J.G. and Upton, B.G.J. (editors), *Alkaline Igneous Rocks*, pp. 253–267. Geological Society of London Special Publication Number 30, 568 pp.

Weinberg, S., 1993. *Dreams of a Final Theory; the Search for the Fundamental Laws of Nature.* Vintage, London, 260 pp.

Wilson, M. and Downes, H., 1991. Tertiary-Quaternary extension-related alkaline magmatism in western and central Europe. *Journal of Petrology*, 32, pp. 811–849.

Wilson, M., Rosenbaum, J.M. and Dunworth, E.A., 1995. Melilitites: partial melts of the thermal boundary layer? *Contributions to Mineralogy and Petrology*, 119, pp. 181–196.

Woolley, A.R., 1987. *Alkaline Rocks and Carbonatites of the World: Part 1: North and South America.* British Museum (Natural History), London, 216 pp.

Woolley, A.R., 1989. The spatial and temporal distribution of carbonatites. In Bell, K. *Carbonatites: Genesis and Evolution*, pp. 15–35, Unwin Hyman, London, 210 pp.

Wyllie, P.J., 1980. The origin of kimberlite. *Journal of Geophysical Research*, 85, pp. 6902–6910.

Yoder, H.S., 1979. Melilite-bearing rocks and related lamprophyres, In Yoder, H.S. (editor) T*he Evolution of the Igneous Rocks: Fiftieth Anniversary Perspectives*, pp. 391–411, Princeton University Press, New Jersey, 588 pp.

CHAPTER 12

Carbonatites

...carbonatites may be effective metasomatising agents at both mantle and crustal levels.

Bell and Keller, 1995, p. vii.

12.1 Bizarre igneous rocks

When the Tanzanian volcano Oldoinyo Lengai erupted in 1960 the idea that carbonate lavas could be extruded from a volcano seemed bizarre (Dawson, 1962). However, as early as 1914 Reck had observed what he called soda mudflows on the volcano's flanks. In 1956 James reported carbonatites on the flanks and in the crater of Kerimasi Volcano to its immediate south (see Fig. 12.1). Petrological ideas have changed so fast in the last 35 years that it is now widely known that Kaiserstuhl Volcano on the Rhine River in the heart of the European Union is a Miocene carbonatite volcano (Wimmenauer, 1966). There is now an extensive literature on the geochemistry and petrology of carbonatites and comagmatic rocks (Pecora, 1956; Smith, 1956; Wyllie and Tuttle, 1960; Heinrich, 1966; Tuttle and Gittins, 1966; Le Bas, 1977; Woolley, 1987; Bell, 1989; Wyllie, 1989; Bailey, 1993; Bell and Keller, 1995; Kogarko et al., 1995).

A carbonatite is an igneous rock that contains more than 50 per cent primary carbonate minerals. The term 'karbonatite' was first used by Brögger (1921, p. 350) in his description of the magmatic carbonate-rich rocks of the Fen ring complex in southeastern Norway. Nowadays four main types of carbonatite are recognised (Le Maitre et al., 1989, p. 10). They are calcite-carbonatite (designated sövite when coarse-grained and alvikite when fine-grained), dolomite-carbonatite (designated beforsite when medium- to fine-grained), ferrocarbonatite ($MgO < (FeO + Fe_2O_3 + MnO)$ and $CaO / (CaO + MgO + FeO + Fe_2O_3 + MnO) < 0.8$ in weight per cent, after Le Bas, 1977, p. 37), and natrocarbonatite. The latter is essentially composed of carbonates of the alkali metals that include nyerereite and fairchildite. Natrocarbonatite is the type of carbonatite that erupts from Oldoinyo Lengai (Du Bois et al., 1963; Bell and Keller, 1995; Dawson et al., 1995). By weight this rock contains about 60 per cent sodium carbonate, 30 per cent calcium carbonate and 10 per cent potassium carbonate (see Table 12.1).

Hogarth (1989, p. 105) has shown that 280 minerals (including 44 secondary minerals) occur in carbonatite volcanic and subvolcanic complexes. This enormous number of minerals reflects both the wide range of chemical components found in these rocks and the wide spectrum of physical conditions in which they evolve. Excluding carbonates, other minerals that are often found in these rocks include apatite, barite, chalcopyrite, columbite, diopside, fluorapatite, fluorite, forsteritic olivine, gypsum, hematite, magnetite, melilite, monazite, monticellite, nepheline, perovskite, phlogopite, pyrite, pyrolusite, sodium-amphiboles, sodium-pyroxenes and vermiculite.

12.2 Abundance and distribution

According to Woolley (1989, p. 15) the literature of petrology contains about 330 descriptions of carbonatite-bearing volcanoes and intrusive complexes. Most are located on major lithospheric domes, or major lineaments in stable, intraplate areas. A few have developed on plate margins. Only four Archean carbonatites are known and Woolley (1989, p. 33) has shown that there is an increase in carbonatite activity through time. Half the known carbonatites occur in Africa and most of these rocks crop out in

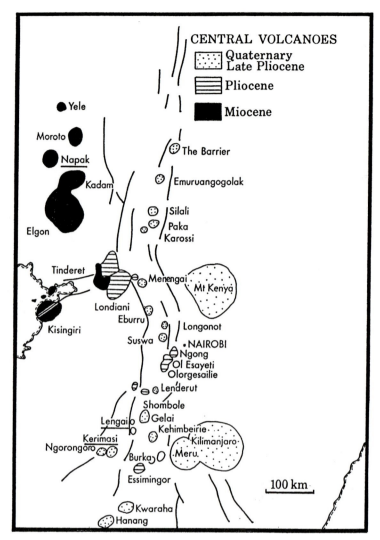

Fig. 12.1 Sketch map of part of the eastern branch of the African Rift System showing the location of the central volcanoes, including the carbonatite volcanoes Kerimasi, Lengai and Napak discussed in the text. After King and Chapman, 1972, p. 189.

the East African Rift System. There are also at least 20 alkaline complexes that contain carbonatites on the other side of the Atlantic Ocean in Brazil. A few carbonatites have been described from oceanic islands.

Notable oceanic carbonatites crop out on Fuerteventura, an eastern Canary island, and on seven of the islands of the Cape Verde Group (Republica de Cabo Verde). These islands are all located in the North Atlantic Ocean off the west coast of Africa. The Cape Verde islands are situated 620 km west of

Dakar in Senegal. They are young volcanic islands that developed on oceanic lithosphere. Fogo is one of these islands. It contains carbonatites and it is still volcanically active.

While most carbonatites crop out in continental settings, Le Bas (1984, p. 169) has made the interesting observation that carbonatites and associated syenitic and ijolitic rocks are relatively more abundant on oceanic islands than in continental settings. Few differences have been reported between oceanic and continental carbonatites in terms of their field setting

Table 12.1 Chemical composition of selected carbonatites.

	Calciocarbonatite Woolley & Kempe, 1989, pp. 8–9	Magnesiocarbonatite Woolley & Kempe, 1989, pp. 8–9	Ferrocarbonatite Woolley & Kempe, 1989, pp. 8–9	Natrocarbonatite Keller & Hoefs, 1995, p. 117
SiO_2	2.72	3.63	4.7	0.16
TiO_2	0.15	0.33	0.42	0.02
Al_2O_3	1.06	0.99	1.46	n.d.
Fe_2O_3	2.25	2.41	7.44	0.28
FeO	1.01	3.93	5.28	n.d.
MnO	0.52	0.96	1.65	0.38
MgO	1.8	15.06	6.05	0.38
CaO	49.12	30.12	32.77	14.02
Na_2O	0.29	0.29	0.39	32.22
K_2O	0.26	0.28	0.39	8.38
P_2O_5	2.1	1.9	1.97	0.85
H_2O^+	0.76	1.2	1.25	0.56
CO_2	36.64	36.81	30.74	31.55
BaO	0.34	0.64	0.8	1.66
SrO	0.86	0.69	0.88	1.42
F	0.29	0.31	0.45	2.5
Cl	0.08	0.07	0.02	3.4
S	0.41	0.35	0.96	n.d.
SO_3	0.88	1.08	1.08	3.72
Total	101.54	101.05	98.7	101.5
	ppm	ppm	ppm	
Li	0.1	n.d.	10	
Be	2.4	< 5	12	
Sc	7	14	10	
V	80	89	191	
Cr	13	55	62	
Co	11	17	26	
Ni	18	33	26	
Cu	24	27	16	
Zn	188	251	606	
Ga	< 5	5	12	
Rb	14	31	n.d.	
Y	199	61	204	
Zr	189	165	127	
Nb	1204	569	1292	
Mo	n.d.	12	71	
Ag	n.d.	3.2	3.4	
Cs	20	0.9	0.6	
La	608	764	2666	
Ce	1687	2183	5125	
Pr	219	560	550	
Nd	883	634	1618	
Sm	130	45	128	
Eu	39	12	34	
Gd	105	n.d.	130	
Tb	9	4.5	16	
Dy	34	n.d.	52	
Ho	6	n.d.	6	
Er	4	n.d.	17	
Tm	1	n.d.	1.8	
Yb	5	9.5	15.5	
Lu	0.7	0.08	n.d.	
Hf	n.d.	3.2	n.d.	
Ta	5	21	0.9	
W	n.d.	10	20	
Au	n.d.	n.d.	12	
Pb	56	89	217	
Th	52	93	276	
U	8.7	13	7.2	

and petrography. Kogarko *et al.* (1992, p. 67) claim that there may be some geochemical differences because the Cape Verde carbonatites are enriched in lithium, scandium, titanium, vanadium, cobalt, yttrium, zirconium and the lanthanoids, and depleted in strontium, niobium, barium and lead, as compared to normal continental carbonatites.

Intrusive carbonatites, like those from the Fen Complex, usually form the cores of small ring complexes that are about $25 \, km^2$ in area. Field relationships usually reveal that they developed in close association with silica-undersaturated alkaline rocks, such as nepheline syenites, ijolites, urtites, melteigites, jacupirangites and occasionally damtjernites (damkjernites) and related lamprophyric rocks. The melteigites, ijolites and urtites are all nepheline-bearing foidolites. Damtjernites are melanocratic nepheline-bearing lamprophyres that usually contain phenocrysts of phlogopite, clinopyroxene, olivine and nepheline set in a fine-grained groundmass of clinopyroxene, phlogopite, calcite, nepheline, orthoclase, chlorite, magnetite, apatite, picotite, pyrite, sphene and melanite. Intense alkali metasomatism, or fenitisation, is characteristic of most carbonatite-ijolite ring complexes. These fenites may be either potassium-rich or sodium-rich. Some fenite aureoles, such as those surrounding the principal carbonatite complexes of Malawi, contain both sodium-rich and potassium-rich varieties (Woolley, 1982, p. 14).

Extrusive carbonatites erupt either as flows or as tephra. They are normally interbedded with co-magmatic olivine-poor nephelinites (nepheline greater than mafic phases) and melanephelinites (nepheline less than mafic phases). Carbonatite volcanoes are found in Uganda, Western Kenya, Northern Tanzania, Catanda in Angola, Rufunsa in Zambia, Kaiserstuhl in Germany, Polino in the Umbria-Latium Province of Italy, the United Arab Emirates and Khanneshin Volcano in southern Afghanistan. As was noted in Section 11.6 on the Western Rift of the East African Rift System there are about 50 Quaternary monogenetic carbonatitic tephra cones in the Fort Portal area of western Uganda.

12.3 Oldoinyo Lengai

Oldoinyo Lengai, or the Mountain of God as it is known to the local Masai people, is the only volcano that has been observed to erupt carbonatitic lavas and tephra (Dawson, 1989; Bell and Keller, 1995).

It is the youngest of the Neogene-Quaternary volcanoes of northern Tanzania. Some of its oldest pyroclastic rocks crop out in Olduvai Gorge where they have been dated at between 400 000 and 150 000 years old. Oldoinyo Lengai is situated south of Lake Natron in the Arusha area where it forms an almost perfectly symmetrical, steep-sided, cone-shaped edifice that rises about 2 km above the floor of the Eastern Rift Valley (see Fig. 12.1). It has a basal diameter of about 12 km and a volume of approximately $60 \, km^3$. Ninety per cent of the volcano is composed of yellow tuffs, agglomerates and lesser volumes of lavas of phonolitic and nephelinitic composition. This basal unit is overlain by mica and pyroxene crystal tuffs, black nephelinitic tuffs, minor melanephelinitic flows, variegated carbonatitic tuffs and modern natrocarbonatitic lavas and carbonate-silicate tuffs.

According to Donaldson *et al.* (1987) Oldoinyo Lengai is constructed from volcanic materials derived from many small batches of magma. Each batch had a slightly different composition because it evolved in a somewhat similar, but not identical, way. The primary magma that evolved to produce the natrocarbonatite lavas and pyroclastic rocks of the 'Lengai trend' was probably a carbonated, sodium-rich, olivine melilitite of moderate peralkalinity (Kjarsgaard *et al.*, 1995, p. 187). While in transit to the surface, fractional crystallisation increased its peralkalinity and carbonate content. This resulted in the production of a 'highly peralkaline carbonated wollastonite nephelinite magma' (Kjarsgaard *et al.*, 1995, p. 188). The normative composition of this hypothetical parental material (Kjarsgaard *et al.*, 1995, p. 166) is rich in nepheline, leucite, aegirine, $Na_2O.SiO_2$ and larnite. At low pressures (about 100 MPa) the magma is presumed to have separated into two immiscible liquids: a natrocarbonatite magma and a conjugate silicate magma of peralkaline nephelinitic composition. This petrogenetic model does not solve the general problem of carbonatite genesis, because most carbonatites are calcite-carbonatites and many of them appear to be comagmatic with melanephelinites, nephelinites, or ijolites.

At the 1990 International Volcanological Congress in Mainz in Germany a special session was dedicated to Oldoinyo Lengai and its remarkable natrocarbonatite lavas and tephra (see Table 12.1). Bell and Keller (1995, p. 3) reviewed the discussions that took place during the meeting. According to them the meeting concluded that the carbon, oxygen, strontium, neodymium and lead isotopic ratios obtained

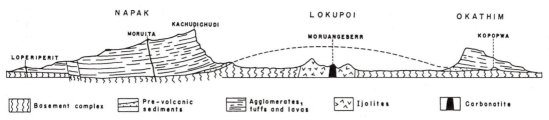

Fig. 12.2 Generalised cross section through the Napak Volcanic Complex of Karamoja, northeastern Uganda. This section spans a distance of about 30 km. After King, 1949, p. 3.

from these rocks showed that they were probably derived from a magma that originated in the mantle; also that the natrocarbonatites were likely to be produced by extensive magmatic differentiation (Keller and Hoefs, 1995).

During eruption the natrocarbonatite lavas are black in daylight but have a weak incandescent glow at night. Extrusion temperature is probably about 585°C. This is close to the lower limit of incandescence. The natrocarbonatite lavas are hydroscopic and rapidly absorb atmospheric moisture. This reaction produces a white outer crust. The lavas are typically porphyritic and contain large clear plates of nyerereite, $Na_2Ca(CO_3)_2$, rounded brown grains of another alkali carbonate (gregoryite?), set in a fine-grained, dark matrix. Plutonic xenoliths are included in some volcanic breccias and they show that there is probably a peralkaline igneous-metasomatic complex beneath Oldoinyo Lengai.

12.4   Additional carbonatitic lavas

Vesicular calcite-carbonatite lava flows have been described from the Fort Portal area of Western Uganda. These extrusive rocks are thought-provoking because they lack associated silicate lavas or tephras and they carry dense xenocrysts and xenoliths of deep crustal and upper mantle origin. It seems inappropriate to link the origin of these rocks directly with those of the natrocarbonatites. Bailey (1993, p. 637) claims that most effusive carbonatitic magmas were generated in the upper mantle. He also links their origin with the alnöite-polzenite, the aillikite-damtjernite (damkjernite) families of lamprophyres and the carbonate-rich kimberlites. This idea is not only currently, but also historically, important because Brögger proposed such a petrogenetic connection

between the carbonatites and damtjernites of the Fen complex of southeastern Norway in 1921.

The Miocene (18–13 million years) Kaiserstuhl Volcano near Freiburg in the Rhine graben of Germany contains agglutinated calcite-carbonatite lapilli tuffs. These lapilli have a teardrop-shape that they acquired when explosively erupted into the air (Keller, 1989, p. 70). Extrusive dolomite-carbonatites (beforsites) have been reported from the Rufunsa volcanoes of southeastern Zambia (Bailey, 1966). Quench droplets found in some of these volcanic vents contain iron-free, high-manganese, high-strontium dolomite. These rocks also contain xenocrysts of spinel that probably originated in the upper mantle (Bailey, 1989). There are no silicate rocks associated with these carbonatites. According to Bailey (1993, p. 649) all the characteristics of these volcanic dolomite carbonatites are consistent with a primary origin.

The subvolcanic carbonatitic rocks of the Rufunsa Province exhibit the following intrusive sequence calcite carbonatite → dolomite carbonatite → ferrocarbonatite. According to Le Bas (1977, p. 285) this is the normal intrusive sequence in the carbonatites of East Africa. Napak Volcanic Complex of Karamoja in northeastern Uganda provides irrefutable evidence of the link between extrusive and intrusive carbonatites (King, 1949). Extensive erosion has removed the central part of this volcano. Its remnants now consist of a central sub-volcanic carbonatite-ijolite ring complex surrounded by a sequence of nephelinitic pyroclastic rocks and lavas. These volcanic rocks had once formed the flanks of a lofty volcano (see Fig. 12.2).

12.5   Geochemistry of carbonatites

As one would expect, the chemical compositions of the main groups of carbonatites differ significantly from those of the common silicate magmatic rocks (see Table 12.1). Their compositions also differ from those of the sedimentary limestones as the magmatic carbonatites usually contain significantly more lithium, strontium, yttrium, barium, lanthanum, cerium, lead, thorium and uranium. When the compositions of carbonatites are compared with the estimated composition of the primitive mantle (Taylor and McLennan, 1985, p. 264) the carbonatites are found to be very strongly enriched in carbon, phosphorus, niobium, tantalum, thorium and uranium, strongly enriched in strontium, barium, lanthanum, cerium, neodymium, samarium, europium and gold, and enriched in sulphur, potassium, calcium, rubidium, yttrium, zirconium, caesium, gadolinium, terbium, holmium, ytterbium and lutetium. Because carbonatites normally have this distinctive composition one might regard the elements carbon, fluorine, phosphorus, sulphur, potassium, rubidium, strontium, yttrium, zirconium, niobium, caesium, barium, lanthanum, cerium, neodymium, samarium, europium, gadolinium, terbium, holmium, ytterbium, lutetium, tantalum, lead, thorium and uranium, as the carbonatophile elements. Carbonatites are the major economic source of niobium, and the rare earth elements, and an important source of phosphorus.

The Araxa carbonatite of Minas Gerais Province, Brazil, is the principal producer of niobium. It produces about 42 000 tonnes of pyrochlore concentrates annually (Woolley, 1987, pp. 181–182). The circular central part of the complex is about 4.5 km in diameter and it is surrounded by a fenitisation aureole that is up to 2.5 km wide. Within the central unit the principal rock is dolomite carbonatite or beforsite. This unit also contains smaller bodies of calcite carbonatite or sövite. These rocks cut and grade into an outer collar of glimmerite. The latter is a rock essentially composed of phlogopite. Within the carbonatites there are bands and veins of phoscorite. They are essentially composed of apatite, but they also may contain a variety of other phases such as magnetite, phlogopite, pyrochlore and carbonates. Most of the phoscorites from Araxa are enriched in bariopyrochlore. This mineral may contain about 63 per cent Nb_2O_5, 17 per cent BaO, 3.3 per cent $(REE)_2O_3$ and 2.3 per cent ThO_2.

The largest mine in a carbonatite complex and one of the largest open-cast mining operations in the world is in the Phalaborwa Carbonatite Complex in northern South Africa. It is pipe-like in form and covers an area of about 16 km^2. Within the complex

Table 12.2 Estimated major element composition and CIPW norm of the parental magma of the carbonatites of western Kenya. After Le Bas, 1977, p. 289.

SiO_2	42.01	Orthoclase	16.14
TiO_2	2.41	Albite	3.67
Al_2O_3	12.41	Anorthite	0.56
Fe_2O_3	6.77	Nepheline	23.87
FeO	4.96	Diopside	31.21
MnO	0.31	Wollastonite	2.67
MgO	5.75	Magnetite	9.73
CaO	12.21	Ilmenite	4.55
Na_2O	5.64	Apatite	1.64
K_2O	2.69	Calcite	3.2
P_2O_5	0.67	Water	2.76
CO_2	1.41		
H_2O^+	2.76		

there are large bodies of copper-bearing carbonatite and phoscorite set in clinopyroxenites and syenites. Economic quantities of phosphorus, copper, zirconium, silver, platinum, gold, uranium and vermiculite, $(Mg, Fe^{2+},Al)_3(Al,Si)_4O_{10}(OH)_2.4H_2O$, are all won from the complex. It has been proposed that the large quantities of copper sulphides found in the complex developed because the potassium-rich, peralkaline, ultrabasic parental magma, scavenged the copper from the basement rocks that were locally enriched in this element (Eriksson, 1989, p. 248).

12.6   Carbonatite petrogenesis

Any account of the petrogenesis of carbonatites must examine the relative abundance and composition of any consanguineous magmatic and metasomatic silicate rocks. For example, from published descriptions of nephelinite-carbonatite volcanoes and ijolite-carbonatite intrusive complexes, one can construct a composite model of a typical carbonatite-nephelinite-ijolite eruptive complex. By using this model one can estimate the relative volumes and possibly the relationships between the various rock units in the carbonatite-nephelinite-ijolite rock association. Le Bas (1977, pp. 288–289) has estimated the composition of the parental magma of the carbonatitic and associated rocks of western Kenya. He added 64 per cent melanephelinite, 19 per cent nephelinite, 6 per cent phonolite, 4 per cent melilitite, 4 per cent ijolite and 3 per cent natrocarbonatite (Le Bas, 1977, pp. 288–289). Such a parental magma would be a melanephelinite as is revealed in Table 12.2.

The primary magmas responsible for most carbonatite complexes are probably sodium-rich melanephelinites, or potassium-rich aillikites, but there are a wide range of other magmas that belong to the alnöite-polzenite, aillikite-damtjernite and kimberlite-lamproite families that also may evolve to produce carbonatites. Most of these magmas are likely to segregate from metasomatically enriched source rocks in the upper mantle. According to Haggerty (1989, p. 546) the lithospheric mantle probably has two broad zones of metasomatically enriched rocks. At depths greater than about 75 km potassium enrichment is dominant, whereas at lesser depths the rocks are enriched in sodium. Aillikitic magmas probably segregate from source rocks that have been metasomatically enriched in potassium, whereas melanephelinitic magmas probably segregate from source rocks enriched in sodium. Suitable source rocks are likely to be depleted in garnet and clinopyroxene, but somewhat enriched in a carbonate phase and minerals like phlogopite, apatite, zircon and monazite.

Partial melting is likely to be triggered by deep-seated fracturing that promotes decompression melting and the release and focussing of previously trapped volatiles. The segregation and ascent of these extremely reactive magmas is likely to be rapid. On moving up through the lithosphere, the drop in confining pressure probably induces volatile-rich carbonatitic fluid to separate from the silicate primary magmas. Wall-rock reaction, fracturing and fractional crystallisation would occur during ascent of both silicate and carbonatite magmas. At approximately 2.5 km the reduced lithostatic pressure would enable the carbonatitic magma to expand explosively and brecciate the country rocks. It seems apposite to suggest that many rocks from carbonatite complexes were produced in a fractured reaction chamber, rather than in a normal, more passive, magma chamber.

It is suggested that: (1) the fenites and some syenitic rocks are products of alkali-metasomatism; (2) the other syenitic rocks are either mobilised fenites, or products of reaction between silica-oversaturated crustal rocks and a phonolitic magma; (3) the damkjernites, aillikites, melanephelinites, melilitites and possibly the ijolites are all magmatic rocks that evolved at depth and may represent primary magmas; (4) the nephelinites and phonolites are products of higher level fractional crystallisation; (5) the melteigites and jacupirangites are essentially pyroxene cumulates; and (6) the nepheline-rich urtites are essentially products of the cumulus processes and evolved at high-levels in the crust. It is thus proposed that many carbonatites are derived by liquid immiscibility from a range of foiditic or lamprophyric alkaline magmas. It is difficult to fault Bailey's (1993, p. 649) declaration that carbonatites are of multiple origins. They reflect different aspects of carbon activity in the mantle and that any attempt to explain all phenomena associated with carbonatites in a single hypothesis is likely to be unsuccessful.

References

Bailey, D.K., 1966. Carbonatite volcanoes and shallow intrusions in Zambia, In Tuttle, O.F. and Gittins, J., (editors) *Carbonatites*, pp. 127–154, Wiley, New York, 591 pp.

Bailey, D.K., 1989. Carbonatite melt from the mantle in the volcanoes of south-east Zambia. *Nature*, 388, pp. 415–418.

Bailey, D.K., 1993. Carbonatite magmas. *Journal of the Geological Society*, London, 150, pp. 637–651.

Bell, K. (editor), 1989. *Carbonatites: genesis and evolution.* Unwin Hyman, London, 618 pp.

Bell, K. and Keller, J. (editors), 1995. *Carbonatite Volcanism: Oldoinyo Lengai and the Petrogenesis of Natrocarbonatites.* Springer-Verlag, Berlin, 210 pp.

Brögger, W.C., 1921. Die Eruptivgesteine des Kristiania gebietes, IV, Das Fengebiet in Telemark, Norwegen. *Norske Videnskaps-Akademi i Oslo, Matematisk-naturvidenskapelig*, Klasse 9, pp. 1–408.

Dawson, J.B., 1962. The geology of Oldoinyo Lengai. *Bulletin Volcanologique*, 24, pp. 349–387.

Dawson, J.B., 1989. Sodium carbonatite extrusions from Oldoinyo Lengai, Tanzania: implications for carbonatite complex genesis. In Bell, K. (editor), *Carbonatites: Genesis and Evolution*, pp. 255–277, Unwin Hyman, London, 618 pp..

Dawson J.B., Keller, J. and Nyamweru, C., 1995. Historic and recent eruptive activity of Oldoinyo Lengai. In Bell, K. and Keller, J. (editors), *Carbonatite Volcanism: Oldoinyo Lengai and the Petrogenesis of Natrocarbonatites*, pp. 4–22, Springer-Verlag, Berlin, 210 pp.

Donaldson, C.H., Dawson, J.B., Kanaris-Sotiriou, R., Batchelor, R.A. and Walsh, J.N., 1987. The silicate lavas of Oldoinyo Lengai. *Neues Jahrbuch für Mineralogie, Abhandlungen*, Stuttgart, 156, pp. 247–279.

Du Bois, C.G.B., Furst, J., Guest, N.J. and Jennings, D.J., 1963. Fresh natrocarbonatite lava from Oldoinyo L'Engai. *Nature*, 197, pp. 445–446.

Eriksson, S.C., 1989. Phalaborwa: A saga of magmatism, metasomatism and miscibility. In Bell, K. (editor), *Carbonatites: Genesis and Evolution*, pp. 221–254, Unwin Hyman, London, 618 pp.

Haggerty, S.E., 1989. Mantle metasomes and the kinship between carbonatites and kimberlites. In Bell, K. (editor), *Carbonatites: Genesis and Evolution*, pp. 546–560, Unwin Hyman, London, 618 pp.

Heinrich, E.W., 1966. *The Geology of Carbonatites*. Rand McNally and Company, Chicago, 607 pp.

Hogarth, D.D., 1989. Pyrochlore, apatite and amphibole: distinctive minerals in carbonatite. In Bell, K. (editor), *Carbonatites: Genesis and Evolution*, pp. 105–148, Unwin Hyman, London, 618 pp.

James, T.C., 1956. Carbonatites and rift valleys in East Africa. *Tanganyika Geological Survey*, Unpublished Report TCJ/34.

Keller, J., 1989. Extrusive carbonatites and their significance. In Bell, K. (editor), *Carbonatites: Genesis and Evolution*, pp. 70–88, Unwin Hyman, London, 618 pp.

Keller, J. and Hoefs, J., 1995. Stable isotopic characteristics of Recent natrocarbonatites from Oldoinyo Lengai. In Bell, K. and Keller, J. (editors), *Carbonatite Volcanism: Oldoinyo Lengai and the Petrogenesis of Natrocarbonatites*, pp. 113–123, Springer-Verlag, Berlin, 210 pp.

King, B.C., 1949. *The Napak Area of Karamoja, Uganda*. Uganda Geological Survey, Memoir Number 5, 57 pp.

King, B.C. and Chapman, G. R., 1972. Volcanism of the Kenya rift valley. *Philosophical Transactions of the Royal Society of London*, A, 271, pp. 185–208.

Kjarsgaard, B.A., Hamilton, D.L. and Peterson, T.D., 1995. Peralkaline nephelinite/carbonatite liquid immiscibility: comparison of phase compositions from Oldoinyo Lengai. In Bell, K. and Keller, J. (editors), *Carbonatite Volcanism: Oldoinyo Lengai and the Petrogenesis of Natrocarbonatites*, pp. 163–190, Springer-Verlag, Berlin, 210 pp.

Kogarko, L.N., Ryabukhim, V.A. and Volynets, M.P., 1992. Cape Verde Island Carbonatite Geochemistry. *Geochemistry International*, 29 (12), pp. 62–74.

Kogarko, L.N., Kononova, V.A., Orlova, M.P. and Woolley, A.R., 1995. *Alkaline Rocks and Carbonatites of the World. Part Two: Former USSR*. Chapman and Hall, London, 226 pp.

Le Bas, M.J., 1977. *Carbonatite-Nephelinite Volcanism: An African Case History*. John Wiley and Sons, London, 347 pp.

Le Bas, M.J., 1984. Oceanic carbonatites. In Kornprobst, J. (editor), *Kimberlites 1: Kimberlites and Related Rocks*, pp. 169–178, Proceedings of the Third International Kimberlite Conference, Volume 1, Elsevier Science Publishers, Amsterdam, 466 pp.

Le Maitre, R.W., Bateman, P., Dubek, A., Keller, J., Lameyre, J., Le Bas, M.J., Sabine, P.A., Schmid, R., Sørensen, H., Streckeisen, A., Woolley, A.R. and Zanettin, B. (editors), 1989. *A Classification of Igneous Rocks and Glossary of Terms: Recommendations of the IUGS Subcommission on the Systematics of Igneous Rocks*. Blackwell Scientific Publications Limited, Oxford, 193 pp.

Pecora, W.T., 1956. Carbonatites: a review. *Bulletin of the Geological Society of America*, 67, pp. 1537–1556.

Reck, H. 1914. *Oldoinyo L'Engai, ein tatiger Vulkan im Gebiete der Deutsch-Ostafricanischen Bruchstufe*. Branca-Festschrift, Borntraeger, Leipzig, pp. 373–409.

Smith (or Campbell-Smith) W.C., 1956. A review of some problems of African carbonatites. *Quarterly Journal of the Geological Society of London*, 112, pp. 189–220.

Taylor, S.R. and McLennan, S.M., 1985. *The Continental Crust: Its Composition and Evolution*. Blackwell Scientific Publications, Oxford, 312 pp.

Tuttle, O.F. and Gittins, J., (editors) 1966. *Carbonatites*. Wiley, New York, 591 pp.

Wimmenauer, W., 1966. The eruptive rocks and carbonatites of the Kaiserstuhl, Germany. In Tuttle, O.F. and Gittins, J. (editors), *Carbonatites*, pp. 183–204, Wiley, New York, 591 pp.

Woolley, A.R., 1982. A discussion of carbonatite evolution and nomenclature, and the generation of sodic and potassic fenites. *Mineralogical Magazine*, 46, pp. 13–17.

Woolley, A.R., 1987. *Alkaline Rocks and Carbonatites of the World. Part One, North and South America*, British Museum (Natural History), London, 216 pp.

Woolley, A.R., 1989. The spatial and temporal distribution of carbonatites. In Bell, K. (editor), *Carbonatites: Genesis and Evolution*, pp. 15–37, Unwin Hyman, London, 618 pp.

Woolley, A.R. and Kempe, D.R.C., 1989. Carbonatites: nomenclature, average chemical compositions, and element distribution. In Bell, K. (editor), *Carbonatites: Genesis and Evolution*, pp. 1–14, Unwin Hyman, London, 618 pp.

Wyllie, P.J., 1989. Origin of carbonatites: evidence from phase equilibrium studies. In Bell, K. (editor), *Carbonatites: Genesis and Evolution*, pp. 500–545, Unwin Hyman, London, 618 pp.

Wyllie, P.J. and Tuttle, O.F., 1960. The system $CaO-CO_2-H_2O$ and the origin of carbonatites. *Journal of Petrology*, 1, pp. 1–46.

CHAPTER 13

Origins – the role of magma in planetary evolution

The kind of beauty that we find in physical theories...the beauty of simplicity and inevitability – the beauty of perfect structure, the beauty of everything fitting together, of nothing being changeable, of logical rigidity. It is a beauty that is spare and classic....

Weinberg, 1993, p. 119.

13.1 Summing up

Grand ideas in all fields of research change our perception of the world and perhaps of our place in the cosmos. The new images of the Solar System acquired during the last three decades have profoundly and permanently changed our perception of the Earth and its place in the Solar System. It is anticipated that our new perspective will free us from our ingrained Earth chauvinism. If we are to understand the planets we have to discover how the magma/igneous rock system works when confronted with a range of novel physical and chemical conditions.

To track down the origin and cosmic abundance of the chemical elements one has to delve into the field of cosmology. Currently most cosmologists support the big-bang theory. Evidence in support of this theory includes: (1) the Hubble-Sandage observation that the galaxies recede at speeds proportional to their distance from the observer; (2) the predictions the theory makes about the abundance of elements in the early cosmos; (3) the predictions it makes about the number of different types of elementary particles in the cosmos; (4) the predictions it makes about the present background radiation in the cosmos; and (5) the predictions it makes about large-scale fluctuations in the cosmic background radiation (Parker, 1993).

In 1931 Georges Lemaitre published his short but innovative letter to the journal *Nature* in which he presented an outline of the big-bang theory. His letter was written in response to an earlier paper by Sir Arthur Eddington on the end of the cosmos. Lemaitre (1931, p. 706) stated that he could 'conceive the beginning of the universe in the form of a unique atom, the atomic weight of which is the total mass of the universe'. This primordial atom was considered unstable and Lemaitre suggested that it would 'divide in smaller and smaller atoms by a kind of super-radio-active process'.

A more refined theory was published in 1948 by George Gamow, Ralph Alpher and Hans Bethe (the alpha-beta-gamma authors; Calder, 1977, p. 168). This paper is important because it examines the way some light elements are created as a consequence of the big-bang. For example, nine-tenths of the helium presently found in the cosmos was produced immediately after the big-bang. The big-bang created time, space and matter. This event occurred at a finite time in the past when all energy was condensed into a single, super-dense and super-hot body. Since then the cosmos has continued to expand. If one extrapolates backward to when the whole cosmos was compressed into a singularity, one discovers that the age of the cosmos is circa 14 giga-years old.

In 1965 it was discovered that the cosmos is filled with a cold background glow of microwave radiation. This radiation is equivalent to the black body radiation of an object at 2.7 K. This discovery implies that the cosmos contains about 400 million photons/m^3 or 10^8 photons for every atom. There is no simple way of accounting for this quantity of radiation unless it is regarded as a relic of when the entire cosmos was hot and dense. Since the big-bang the wavelengths of this radiation have been stretched and red shifted. All bodies including our own radiate energy and the peak intensity of this radiation differs with the temperature of the body. The peak intensity for humans is in the infrared. Some cool stars appear red, hotter stars appear blue, and very hot objects, such as supernovae, radiate strongly in the X-ray band.

Over the past few years NASA's Cosmic Background Explorer satellite (COBE) has gradually built up a whole-sky chart of the intensity of the background radiation. This chart shows the structure of the cosmos as it appeared some 14 giga-years ago, or more precisely some 300 000 years after the big-bang. The image of the 'primordial fireball' resembles a sphere covered by an intricate pattern. This pattern has been retrieved by studying minor fluctuations in temperature (30 millionths of a degree) above and below the mean cosmic background temperature of 2.73 K. The pattern has been poetically described as tiny ripples in the fabric of space-time. Ripples that later developed into superclusters of galaxies and the other major structures in the present cosmos.

The image recorded by COBE captures the instant when the cosmos suddenly ceased to be opaque to radiation. This occurred when the temperature of the cosmos had decreased enough for neutral atoms to form and the cosmos ceased to be composed of plasma that acted as a radiation trap. Recent studies of the large-scale structure of the cosmos show that it has a huge bubble-like structures that may be 150 million light-years in diameter and it is draped with chains of superclusters of galaxies. These structures are interspersed with 'great walls' of galaxies.

The current tale of creation pieced together by cosmologists like Weinberg (1983) in his thought-provoking book *The First Three Minutes* suggest that at first there was a vacuum. In the vacuum there was a ferment of quantum activity and it teemed with virtual particles that popped in and out of existence in a way that seems reminiscent of *Alice in Wonderland*. It was also full of complex inter-actions (see Parker, 1993, p. 265). Immediately after the big-bang to 10^{-43} s all four forces of nature (the strong nuclear, the electromagnetic, the weak nuclear and gravity) were unified. During this initial era the temperature was about 10^{32} K. At 10^{-43} s there was a transition and gravity broke away. The cosmos then entered the GUT era (GUT = Grand Unified Theory: strong nuclear + electromagnetic + weak nuclear) that lasted until 10^{-34} s when the strong nuclear force broke free. Finally at 10^{-12} s the electro-magnetic and weak nuclear forces parted company (Pagels, 1985).

In 1979 Guth was engaged in theoretical research on magnetic monopoles. His research predicted that at the end of the GUT era huge numbers of dense monopoles would be produced and their density would rapidly lead to the collapse of the cosmos. To overcome this problem he suggested a way to retarding monopole production. This would happen

if the cosmmos was supercooled. He then had to examine the consequences of supercooling and found that during this process huge amounts of heat would be stored in a false vacuum. This would produce a negative pressure resulting in astoundingly rapid expansion of the cosmos. The proposed rate of expansion was much faster than previously proposed and it has been called inflation (Guth, 1989).

After a very short period of inflation the cosmos re-established its more normal rate of expansion. Inflation probably operated for only 10^{-32} s, and dur-ing this time the diameter of the cosmos increased by a factor of at least 10^{50}. Guth has called the cosmos 'the ultimate free lunch' implying that it emerged from a vacuum with the matter energy being equal and opposite to the gravitational energy and the sum of all this energy was zero. This connotes that the present density of the cosmos is likely to be equal to the critical density that separates its collapse from indefinite expansion. The ratio of the actual average density of the cosmos to this critical density is called omega. Omega must be very close to one, otherwise the cosmos would have already collapsed if it was somewhat greater than one, or it would cur-rently appear to be almost empty, if omega was somewhat less than one.

13.2   Age and origin of the chemical elements

The big-bang was the first stage in element synthesis, or nucleogenesis. During this stage the main nuclear species to form were hydrogen 1, deuterium 2, helium 3, helium 4, and lithium 7. The initial abundance of these nuclides is often difficult to estimate because much of the present interstellar medium has been astrated, or processed by one or more star. This results in the destruction of 'fragile' nuclides, such as deuterium. Weinberg (1983) has described how the initially extremely hot cosmos cooled via a 'quark soup' to a temperature of one thousand mil-lion degrees at the end of the first three minutes. This temperature was cool enough for protons and neu-trons to begin to form complex nuclei such as deu-terium 2. Only 2 per cent of the elements in our galaxy are heavier than helium 4. These heavier nuclides can and probably were created by nuclear reactions that are known to occur in our galaxy.

Nuclei must touch for fusion to occur. If this is to happen they must collide at velocities that overcome the electrical repulsion exerted by the protons in the nuclei. Broecker (1985, p. 42) has likened this process to throwing ping-pong balls at a fan. The more atoms are heated the higher their velocity. Hydrogen fusion occurs when gravitational collapse heats up a proto-star to a temperature of about $2 \times 10^7°C$. The fusion of hydrogen to helium provides a long stable era in the life of a star like the Sun. Our Sun probably will remain in this stable state for another five giga-years.

When about 12 per cent of the mass of a main sequence star has been converted to helium the star becomes unstable. The central layers contract, whereas the outer layers expand to produce a red giant star. In the contracting central zone temperatures rise until helium fusion occurs. Three helium 4 nuclei merge to form one carbon 12 nucleus at a temperature of about $2 \times 10^8°C$. In stars of large mass there is essentially a cycle of fuel depletion, collapse, heating, followed by the ignition of another nuclear fuel. Carbon nuclei merge to become magnesium 24. At even higher temperatures silicon 28, sulphur 32, argon 36 and calcium 40 form by fusion. These processes are commonly known as hydrogen burning, helium burning, carbon burning, oxygen burning and neon burning. At a temperature of about $4 \times 10^9°C$ the equilibrium process operates and creates the nuclides of the iron peak, such as iron 56 and nickel 62.

S- and R-Processes

To generate nuclides of higher atomic weight energy has to be added to promote fusion. In the atmospheres of several classes of giant stars fusion, and related processes, release a flux of neutrons that creates new nuclides by successive neutron capture. In the S-Process neutron capture takes place somewhat slowly. It is akin to the processes that operate in a nuclear reactor. Adequate time is available between neutron capture for most radioisotopes to undergo radioactive decay. The S-Process normally starts with the iron-group of nuclides and operates with the successive capture by individual nuclei of large numbers of neutrons and the emission of beta-particles. This process ends with the creation of lead and bismuth. After these nuclides have formed further slow neutron capture results in the emission

of alpha-particles. The nuclides lead 206, lead 207, lead 208 and lead 209 act as a 'sink' that stops the S-Process.

The S-Process accounts for the production of about 150 nuclides. It operates in the atmospheres of giant stars for approximately 100 000 years when they are at their most luminous and there is a steady supply of neutrons. Currently some elements that are produced in significant amounts by the S-Process such as zinc, barium and rubidium can be detected in the atmospheres of several classes of giant stars.

In the rapid, or R-Process, neutron capture is so swift that unstable nuclei can act as part of a chain reaction that creates nuclei of higher mass number. To create uranium and thorium the chain reaction has to pass through polonium 212. It has a half-life of seven millionths of a second. This means that if polonium is to be used as a stepping stone the R-Process has to operate at explosive speed. The R-Process operates effectively for only about ten seconds during the explosion evolution of a supernovae. An explosion of this type occurred immediately before the formation of the Solar System and created the various heavy isotopes like those of uranium. The spectral lines of technetium, an element that lacks stable isotopes, have been detected in light derived from the remnants of recent supernovae. Some nuclides, particularly proton-rich nuclides, are not created by either the S- or R-processes. They are created by other minor processes such as the addition of protons to existing nuclides (P-process), or cosmic-ray spallation.

Some light elements such as deuterium, lithium, beryllium and boron are readily destroyed in the interiors of stars, but are also easily created by the action of cosmic rays on carbon, nitrogen and oxygen nuclides in interstellar space. In our galactic halo above or below the galactic disk there are stars that are chemically anomalous, with extremely low abundances of elements heavier than helium. They are interpreted as old stars that are about ten giga-years old. They did not inherit any significant quantities of heavy elements.

All the hydrogen and most of the helium found in the present cosmos were created cosmologically immediately after the big-bang. The heavier elements and some helium was produced by later astrophysical processes. Individual processes of nucleogenesis have created abundance peaks that appear to match the abundance peaks observed in the Sun and other Solar System materials.

Table 13.1 Solar System abundance of the elements in cosmic abundance units (CAU), or number atoms per million silicon atoms. After Anders and Grevesse, 1989.

Proton number	Element	CAU	Proton number	Element	CAU
1	H	27 900 000 000	47	Ag	0.486
2	He	2 720 000 000	48	Cd	1.61
3	Li	57.1	49	In	0.184
4	Be	0.73	50	Sn	3.82
5	B	21.2	51	Sb	0.309
6	C	10 100 000	52	Te	4.81
7	N	3 130 000	53	I	0.9
8	O	23 800 000	54	Xe	4.7
9	F	843	55	Cs	0.372
10	Ne	3 440 000	56	Ba	4.49
11	Na	57 400	57	La	0.446
12	Mg	1 074 000	58	Ce	1.136
13	Al	84 900	59	Pr	0.1669
14	Si	1 000 000	60	Nd	0.8279
15	P	10 400	61	Pm	n.d.
16	S	515 000	62	Sm	0.2582
17	Cl	5240	63	Eu	0.0973
18	Ar	101 000	64	Gd	0.33
19	K	3770	65	Tb	0.0603
20	Ca	61 100	66	Dy	0.3942
21	Sc	34.2	67	Ho	0.0889
22	Ti	2400	68	Er	0.2508
23	V	293	69	Tm	0.0378
24	Cr	13 500	70	Yb	0.2479
25	Mn	9550	71	Lu	0.0367
26	Fe	900 000	72	Hf	0.154
27	Co	2250	73	Ta	0.0207
28	Ni	49 300	74	W	0.133
29	Cu	522	75	Re	0.0517
30	Zn	1260	76	Os	0.675
31	Ga	37.8	77	Ir	0.661
32	Ge	119	78	Pt	1.34
33	As	6.56	79	Au	0.187
34	Se	62.1	80	Hg	0.34
35	Br	11.8	81	Tl	0.184
36	Kr	45	82	Pb	3.15
37	Rb	7.09	83	Bi	0.144
38	Sr	23.5	84	Po	n.d.
39	Y	4.64	85	At	n.d.
40	Zr	11.4	86	Rn	n.d.
41	Nb	0.698	87	Fr	n.d.
42	Mo	2.55	88	Ra	n.d.
43	Tc	n.d.	89	Ac	n.d.
44	Ru	1.86	90	Th	0.0335
45	Rh	0.344	91	Pa	n.d.
46	Pd	1.39	92	U	0.009

13.3 Cosmic abundance data

Many eminent geochemists (Goldschmidt, 1937; Suess and Urey, 1956; Goles, 1969; Anders and Ebihara, 1982; Anders and Grevesse, 1989; Newsom, 1995) have produced cosmic abundance tables (see Table 13.1). When producing such tables it is usually

assumed that the chondritic meteorites can provide data on the non-volatile part of the Solar System (see Section 2.11), whereas the Solar photosphere can supply abundance data on the volatile elements (Newsom, 1995, pp. 167–168). To bring these sets of abundance data together as a single table, it is usual to select and fix the value of one non-volatile element that has been precisely determined in both

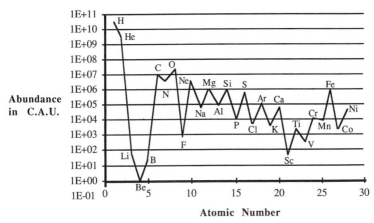

Abundance in C.A.U.

Atomic Number

Fig. 13.1 Cosmic abundance diagram for the first 28 chemical elements (atomic number versus relative abundance of the elements) in cosmic abundance units (number of atoms per million silicon atoms).

sets of data. The element selected is usually silicon and traditionally cosmic abundance data are quoted in cosmic abundance units or the number of atoms per million silicon atoms.

The range in the cosmic abundances of the chemical elements is enormous, with hydrogen being a thousand million times more abundant than lithium, the element below it in the periodic table. If one constructs a cosmic abundance diagram (atomic number, Z, versus relative abundance), one discovers a rapid exponential decrease in abundance as one proceeds from hydrogen to approximately molybdenum (atomic number 42), after that the curve flattens out for the heavier elements (see Fig. 13.1). The graph has a very jagged appearance because, aside from four exceptions, elements of even atomic number are more abundant than their neighbours of odd atomic number. This attribute of elemental abundance is known as the Oddo-Harkins Rule.

Elements that display particularly high abundances are hydrogen, helium, carbon, nitrogen, oxygen, neon, magnesium, silicon, sulphur, argon, calcium, chromium, iron and nickel. Lithium, beryllium and boron are all strongly depleted when compared with the other light elements. The general decline in abundance with increasing atomic number is clearly interrupted by a sizable peak centred around atomic number 26, or iron. It is important to remember that both the regularities and irregularities displayed by the cosmic abundance diagram developed because of the element's nuclear properties, not their chemical properties, and this pattern of abundance has had an enormous influence on the composition of the planets and their constituent rocks.

13.4   Planet-forming processes

In astronomy statistical analysis is a particularly valuable tool for investigating the origin and evolution of the stars. Until very recently this approach was denied to investigators trying to figure out how planets form. This was because they only had unequivocal information about one planetary system. We now know of at least seven planets that orbit stars other than the Sun. The first of these planets was discovered in orbit around 51 Pegasi, in the constellation Pegasus. Its mass is about half that of Jupiter, yet it is only 0.05 astronomical units from its host star. One astronomical unit is approximately equal to the semimajor axis of the Earth's orbit, or $1.495\,978\,70 \times 10^{11}$m. The pattern of giant planets in tight orbits around their host stars seems to be unexceptional in our neighbourhood of the cosmos.

Theoretical calculations reveal that planets of the size of Jupiter can survive at distance of 0.05 astronomical units from their host stars. The foremost problem posed by the new data from alien planetary systems is, do these giant planets form in orbits close to their host stars or are they generated out beyond about three astronomical units and they move later into tighter orbits around their host stars. At present the latter idea is favoured. It seems easier to form a giant planet in a manner similar to that proposed for

Jupiter and to speculate that at this stage in the evolution of an alien planetary system there may be still enough mass left over from the embryonic molecular cloud to produce gravitational interactions capable of drawing the giant planet towards its host star.

Now to return to the Solar System where the planetary bodies lie in a plane and rotate in the same direction. This arrangement underpins the idea that the Sun and its planets have a common origin. The flatness of the system probably means that all its component parts evolved from a rotating disk of dust and gas.

Refractory inclusions (CAIs) from the Allende carbonaceous chondrite give model lead 207/lead 206 ages of 4.559 ± 0.004 giga-years. This is the oldest solid material known in the Solar System. Some meteorites contain evidence of even earlier events, such as the production of the short-lived radioactive nuclides manganese 53, iodine 129, samarium 146 and plutonium 244 in a supernoval explosion circa 24 million years before the formation of the refractory inclusions. The Solar Nebula probably separated from a larger molecular cloud that was between two and five parsec (pc = 3.0857×10^{13} km) in diameter and contained a thousand to ten thousand atoms per cubic centimetre (Taylor, 1992, p. 50). Molecular clouds of this type provide the principal sites for the genesis of stars of the size of the Sun. Young T Tauri stars and Herbig-Haro objects are currently developing from such clouds.

Once a separate rotating disk has developed the particulate materials, including metals, sulphides and silicate phases, begin to settle towards its midplane. During transient heating episodes droplets of silicate melt or chondrules form. Instabilities are likely to develop in the rotating nebular resulting in the clumping together of assorted phases that become aggregations, planetesimals and even planetoids. A planetesimal is a body of rock or ice that forms in molecular cloud and is under 10 km in diameter (Mitton, 1991, p. 294). Planetoids are larger bodies that form when planetesimals coalesce.

The Sun contains 99.9 per cent of the mass of the Solar System, yet it only contains 2 per cent of the system's angular momentum. One would not expect a naturally formed disk to be axisymmetric. In such a disk differential rotation would produce gravitational torque and transfer angular momentum outwards and mass inwards. Another suggestion is that some dispersed particles and gas from the disk became ionised and trapped in the magnetic field of the proto-Sun. Such ionised materials would co-rotate

with the proto-Sun so transferring angular momentum hydromagnetically from the centre to the surrounding plasma (Alfven and Arrhenius, 1976). At various times in the early history of the Solar System the velocity of the outermost rim of the disk would become high enough for centrifugal force to compensate for the gravitational attraction of the proto-Sun. This would result in the discarding of aggregations and planetesimals in relatively stable orbit. In this scenario the central mass of the Solar Nebula contracts to form the Sun, and part of the remaining materials in orbit around the Sun accrete to form planetary bodies. This model was first suggested by the Marquis de Laplace in 1796 (Taylor, 1992; Wetherill, 1991; Wood, 1979). In the Solar System the planetary accretionary process is estimated to have been completed in about a 100 million years. This means that the construction of the terrestrial planets was essentially complete by circa 4.45 giga-years ago.

T Tauri stage

Deuterium fusion would begin when the mass of the proto-Sun exceeded 0.3 solar masses and this process would soon be followed by hydrogen fusion. At this stage the proto-Sun would be violently active and unstable. It probably would pass through a Herbig-Haro stage with the ejection of strong bipolar jets of material along its rotational axis. Such outflows of material are likely to provide another mechanism for transferring angular momentum from the proto-star to the surrounding disk. Before the Sun became a stable member of the main sequence it probably passed through a T Tauri stage. This stage may have lasted for about a million years. During this epoch a strong solar wind would have stopped the inward flow of material to the Sun. The strong solar wind also would clear the inner Solar Nebular of gasses and dust. Heating from the frequent outbursts of solar flares would aid in driving volatile components out of the inner nebula. Water ice would begin to accumulate at, and beyond, the 'snow line'. The location of this line would fluctuate but eventually stabilise somewhere between the Belt Asteroids and Jupiter. It is probable that the formation of the Sun predated the accretion of the terrestrial planets because they accreted in an environment that had already been mainly cleared of elements like hydrogen and helium.

Giant gaseous planets

Rapid accretion occurred immediately beyond the snow line, as this is where the volatiles driven from the inner nebula condensed. The mass of proto-Jupiter soon became so large that it attracted all the materials in its vicinity, including light gases. This rapid accretion prevented the growth of large planets in the positions now occupied by Mars and the Belt Asteroids. The other giant gaseous planets, Saturn, Uranus and Neptune, formed, more slowly in the outer regions of the nebula. During the evolution of the Jovian planets there were many huge collisions that produced planetary 'sub-nebulae' from which the major satellites of these giant planets formed. The capture of icy planetoids completed the Jovian planets inventory of satellites. Within the snow line the smaller, rocky terrestrial planets accreted the planetesimals, and planetoids, that remained in their somewhat restricted feeding-zones. The feeding-zones of both Mars and Mercury were strongly depleted by their massive neighbours Jupiter and the Sun.

It is now usually proposed that the terrestrial planets accreted from a variety of massive planetoids, instead of small particles and planetesimals. Taylor (1992, p. 25) estimated that between 50 and 75 per cent of the present mass of the Earth accreted from massive planetoids. These planetoids had already gone through a melting stage that had separated silicate, sulphide and metal phases at relatively low pressures. This means that there may have been little high-pressure equilibration between the materials of the Earth's core and the mantle (Taylor and Norman 1990). If this is correct sulphur is likely to be the main light element in the core. Core formation in the terrestrial planets probably started as soon as the planet was large enough for melting to occur.

It is likely that the Moon formed from a catastrophic collision between the proto-Earth that had already developed a core and a planetoid that was about 15 per cent of the Earth's mass. During this event most of the material that now makes up the Moon melted and the surface became a magma/lava ocean. Venus also experienced huge impacts and one such collision reversed its rotation.

The Belt Asteroids belong to two zones, an inner zone of differentiated planetesimals and planetoids and an outer zone of primitive planetesimals and planetoids. This zoning probably shows that melting and differentiation in the Solar Nebula was, at least in part, controlled by distance from the Sun. This observation has been interpreted as showing that as the terrestrial planets occur closer to the Sun than the Belt Asteroids they probably accreted from differentiated planetesimals. The type-1 carbonaceous chondrites are interesting because they preserve isotopic anomalies. For example the relative abundances of oxygen 16 relative to oxygen 17 and 18 shows that the Solar Nebula was initially cool, because at higher temperatures such anomalies would be homogenised. The primitive nebula was probably oxidised, but in the inner nebula a reduction process seems to have operated producing native metals. This process was possible set in train by the heat produced by the proto-Sun.

Planetesimals in collision

Wetherill (1991, pp. 45–58) used a computer to generate a model of how the Solar System formed. During his first stage the planetesimals collide and grow into planetary embryos or planetoids. This stage takes place rapidly in about ten thousand to a hundred thousand years. There is very little mixing of materials from different regions of the Solar System. Eventually, when planetoids have incorporated most of the material in their neighbourhood, their growth slows down. In the second, longer lasting stage (10^8 years), the planetoids collide with one another to form the terrestrial planets and the cores of the giant planets. During this stage there is widespread mixing of materials between the various planetoids and planetesimals and large impacts occurred such as the one responsible for the ejection from the proto-Earth of much of the material now incorporated into the Moon.

The Wetherill (1991) model predicts that much of the Earth was molten before it completed its growth. A corollary of this is that the differentiation of the Earth into a metallic core and a less dense silicate mantle took place while the Earth was still growing. At depth within the molten planet pressure would increase the melting points of the solid phases, and one would expect the temperature of a substantial part of the planet to be raised above 2200°C. At such temperatures the distribution of trace elements between phases would be governed by high temperature partition coefficients unlike those normally measured in the laboratory. If core-mantle segregation took place at very high temperatures (2500°–4500°C) this would explain the paradox of the supposed over-abundance of siderophile elements in the mantle. According to Murthy (1991, p. 3) this spurious problem disappears if one uses high temperature partition coefficients.

13.5 Future of the Sun and its system of planets

When studying the materials that form the planets of the Solar System one must keep reminding oneself that the amount of unused energy and material in the Solar System is prodigious. It is hoped that in the next few hundred years humankind will progress from being a society that attempts to control the resources of a single planet to a society that intelligently manages the resources of a planetary system.

To conclude we will indulge in some geopoetry (Hess, 1959) and speculate about the future of the Sun, the Earth and the Solar System. Our knowledge of the Solar System predicts that during the next billion years the life-forms of Earth will be stressed by periodic major meteoroid and cometary impacts, and countless catastrophic volcanic eruptions. The Earth is now a complex, self regulating heat pump. It uses fluids to carry heat from the core and lower mantle into the upper mantle, lithosphere, hydrosphere and atmosphere. Currently, most cooling takes place in the upper mantle that is continually receiving cool slabs of subducted lithosphere. Once the present style of mid-oceanic ridge volcanism ends, subduction will cease, plate tectonics will halt, and magmas will be forced to carry much of their heat directly through to the surface. Hotspot, or volcano-magmatic activity, through a stationary lithosphere will dominate. This is likely to result in the growth of crustal domes and large volcanic edifices, perhaps comparable in size to those found on Mars. As the Earth cools and the lithosphere thickens the primary magmas will equilibrate with the ambient phases at successively deeper levels in the mantle and their compositions will shift. The dominant primary magmas will move across from the subalkaline into the midalkaline and finally the alkaline fields.

The Sun is slowly growing more luminous and releasing more energy. It is now about 30 per cent brighter than when it stabilised about 4.5 giga-years ago. It probably will brighten by some 10 per cent in the next billion years. This warming will bring the Earth to the threshold of a moist greenhouse cata-strophe. The seas, or vestiges of the seas, probably will remain for another 3.5 giga-years. Ultimately, when the water vapour from the seas is dissociated in the upper atmosphere, the Earth will resemble modern Venus. In about six billion years time, the Sun will start to expand and transform into a huge red giant star that will engulf Mercury. Jupiter will then be released from its ice covering revealing its rocky nucleus for the first time. As the Sun releases vast amounts of energy its mass will decrease. The Earth's orbit will expand until it approaches the orbit currently occupied by Mars. Even at this great distance the surface temperature of Earth will exceed 1300°C. This will trigger the melting and mixing of the crustal rocks and the generation of a global magma-lava ocean.

Later the Sun will cool and contract to form a dwarf star. The Earth's magma-lava ocean will cool and the surface will be covered by a vast global sheet of inhomogeneous lava with the bulk composition of a basaltic icelandite enriched in incompatible elements. Many of the latter elements will have been brought to the surface by alkaline magmas after the demise of plate tectonics. With further cooling and contraction the vast lava plains will be disrupted by extensive faulting and the generation of prominent fault scarps like those currently observed on Mercury. In the words of a Hindu prophesy the Earth will become '. . . a ruined world, a world burnt out, a corpse upon the road of night'.

References

Alfven, H. and Arrhenius, G., 1976. *Evolution of the Solar System*. NASA, Washington, DC 599 pp.

Anders, E. and Ebihara, M., 1982. Solar-system abundances of the elements. *Geochimica et Cosmochimica Acta*, 46, pp. 2363–2380.

Anders, E. and Grevesse, N., 1989. Abundance of the elements: meteoritic and solar. *Geochimica et Cosmochimica Acta*, 53, pp. 197–214.

Broecker, W.S., 1985. *How to Build a Habitable Planet*. Eldigio Press, Palisades, N.Y. 291 pp.

Calder, N., 1977. *The Key to the Universe: a Report on the New Physics*. British Broadcasting Corporation, London, 196 pp.

Goldschmidt, V.M., 1937. *Geochemische Verteilungsgesetze der Element, 9, Die Mengenverhaltnisse der Elemente und der Atom-Arten*. Norsk Videnskaps-Akademi i

Oslo, Matematisk-Naturvidenskapelig Klasse, Number 4.

Goles, G., 1969. Cosmic abundance. In Wedepohl, K.H., *Handbook of Geochemistry*, pp. 116–133, Volume 1, Springer-Verlag, Berlin, 442 pp.

Guth, A. H., 1989. Starting the Universe: the Big Bang and cosmic inflation. In Cornell, J. (editor), *Bubbles, Voids, and Bumps in Time: The New Cosmology*, pp. 105–146, Cambridge University Press, Cambridge, 190 pp.

Hess, H.H., 1959. Nature of the great ocean ridges. International Oceanographic Congress, Abstracts. *American Association for the Advancement of Science*, Washington, DC, pp. 33–34.

Laplace, P.S. de, 1796 is now available in a facsimile reprint: Laplace, P.S. de, 1984. Exposition du systeme du mode. In *Corpus des Oeuvres de Philosophie en Langue Français*, Fayard, Paris.

Lemaitre, G. 1931. The beginning of the World from the point of view of quantum theory. *Nature*, 127, Number 3210, p. 706.

Mitton, J., 1991. *A Concise Dictionary of Astronomy.* Oxford University Press, Oxford, 423 pp.

Murthy, V.R., 1991. Early differentiation of the Earth and the problem of mantle siderophile elements: a new approach, *Science*, 253, pp. 303–306.

Newsom, H.E., 1995. Composition of the Solar System, Planets, Meteorites, and Major Terrestrial Reservoirs. In Ahrens, T.J. (editor), *Global Earth Physics: A Handbook of Physical Constants*, pp. 159–189, American Geophysical Union, Reference Shelf 1, Washington, 376 pp.

Pagels, H.R., 1985. *Perfect Symmetry: The Search for the Beginning of Time*. Michael Joseph, London, 390 pp.

Parker, B., 1993. *The Vindication of the Big Bang: Breakthroughs and Barriers*, Plenum Press, New York, 361 pp.

Suess, H.E. and Urey, H.C., 1956. Abundance of the elements. *Reviews of Modern Physics*, 28, pp. 53–74.

Taylor, S.R., 1992. Solar System Evolution: *A New Perspective*. Cambridge University Press, Cambridge, 307 pp.

Taylor, S.R. and Norman, M.D., 1990. Accretion of differentiated planetesimals to the Earth. In Newsom, H.E. and Jones, J.H. (editors), *Origin of the Earth*. Oxford University Press, New York, pp. 29–43.

Weinberg, S., 1983. *The First Three Minutes: A Modern View of the Universe*. Flamingo edition, Fontana Paperbacks, London, 204 pp.

Weinberg, S., 1993. *Dreams of a Final Theory*. Vintage, London, 260 pp.

Wetherill, G.W., 1991. From Planetesimals to Planetary Embryos, Planets, and Meteorites. *Carnegie Institution of Washington Yearbook*, 90, pp. 45–58.

Wood, J.A., 1979. *The Solar System*. Prentice-Hall Inc., Englewood Cliffs, New Jersey, 196 pp.

INDEX